Ecophysiology and Ecology of
Grassland

Ecophysiology and Ecology of Grassland

Editors

Bingcheng Xu
Zhongming Wen

MDPI • Basel • Beijing • Wuhan • Barcelona • Belgrade • Manchester • Tokyo • Cluj • Tianjin

Editors
Bingcheng Xu
Institute of Soil and Water
Conservation
Northwest A&F University
Yangling
China

Zhongming Wen
College of Grassland Agriculture
Northwest A&F University
Yangling
China

Editorial Office
MDPI
St. Alban-Anlage 66
4052 Basel, Switzerland

This is a reprint of articles from the Special Issue published online in the open access journal *Plants* (ISSN 2223-7747) (available at: www.mdpi.com/journal/plants/special_issues/grassland_ecology).

For citation purposes, cite each article independently as indicated on the article page online and as indicated below:

LastName, A.A.; LastName, B.B.; LastName, C.C. Article Title. *Journal Name* **Year**, *Volume Number*, Page Range.

ISBN 978-3-0365-7985-6 (Hbk)
ISBN 978-3-0365-7984-9 (PDF)

Contents

plants

MDPI

Editorial

Periodical Progress in Ecophysiology and Ecology of Grassland

Bingcheng Xu [1,*] and Zhongming Wen [2,*]

[1] State Key Laboratory of Soil Erosion and Dryland Farming on the Loess Platea, Northwest A&F University, Yangling 712100, China
[2] College of Grassland Agriculture, Northwest A&F University, Yangling 712100, China
* Correspondence: bcxu@ms.iswc.ac.cn (B.X.); zmwen@ms.iswc.ac.cn (Z.W.)

Citation: Xu, B.; Wen, Z. Periodical Progress in Ecophysiology and Ecology of Grassland. *Plants* **2023**, *12*, 2244. https://doi.org/10.3390/plants12122244

Received: 5 June 2023
Accepted: 6 June 2023
Published: 8 June 2023

As one of the most important ecosystems on the planet, grasslands serve a variety of purposes in ecology, economy, culture and entertainment. Due to overuse, insufficient input, poor maintenance and climate change, natural grasslands have become one of the most degraded ecosystems and are experiencing numerous challenges. To better understand the dynamics of biomass elaboration and the renewal of natural grassland, it is necessary to conduct a systematical analysis of ecophysiological and ecological traits of grassland communities and their key species. With this framework, "Ecophysiology and Ecology of Grassland" focuses on recent advancements concerning integrated research from species to ecosystem levels in terms of natural grasslands as well as artificial grasslands, in response to human disturbances, abiotic stresses and climate change. The following is a brief summary of the topics and applications that comprise this Special Issue.

Fang et al., (2021) [1] studied the spatial patterns and potential drivers of leaf stoichiometry and herb biomass from 15 sites spreading from south to north along a 500 km latitudinal gradient on the Loess Plateau. The findings demonstrated a strong relationship between herb biomass and leaf N and P contents and environmental driving factors, including slope, soil P content and latitude, altitude, mean annual rainfall and mean annual temperature, which can be used to inform future ecological restoration efforts and policy adjustments in the region as well as to offer basic regional data for global-scale research.

Zhang and Wang et al., (2022) [2] investigated the responses of plants, soil bacteria, and fungal diversity to climate change (especially rainfall patterns and air temperature) in the desert grassland of the Ningxia Hui Autonomous Region of China. The results indicated that increased precipitation promoted root biomass growth more than aboveground living biomass, and changing precipitation and increasing temperature, as well as their interaction, primarily altered the plant diversity, soil bacteria and fungal diversity but had no significant impact on plant biomass production, organic carbon, total nitrogen and total phosphorus contents of plants. The study sheds light on the diversity and variability of desert grassland plants and soil microorganisms in relation to climate warming and precipitation change.

Zhang and Xie et al., (2022) [3] studied the dynamics of grassland vegetation and compared the effect of drought in the Mongolian Plateau (MP) from 2000 to 2013 using a multi-index method that included coverage (Fv), surface bareness (Fb), and net primary production (NPP). Fv and NPP exhibited an increasing trend (0.18 vs. 0.43), while Fb showed a decreasing trend, with a value of −0.16. Generally, the grassland in the MP showed a tendency to recover, and the response is region-specific (positive reaction mainly distributed in the middle of MP). The results provided a scientific foundation for guiding ecological, environmental improvement and drought prevention in typical farming and pastoral areas.

Song et al., (2022) [4] investigated the effect of the water deficit on the eco-adaptation of common grassland species and conducted a water control experiment to unveil the ecophysiology of *Glycyrrhiza uralensis* in response to different water stress gradients. The species is the dominant species in the natural restoration of the desert steppe. *G. uralensis* was shown to maintain water content and turgor pressure under water stress, promote root biomass accumulation, and improve water use efficiency, indicating that it possesses a water-conservation strategy to avoid dehydration and tolerate drought. The study also investigated the biomass allocation, water use efficiency, and physiological and morphological characteristics of *G. uralensis*, which provided scientific support for adopting the species to restore the grassland.

Guo et al., (2022) [5] investigated the leaf stoichiometry of the widely distributed species *Stellera chamaejasme* from the Inner Mongolian Plateau (IM) and Qinghai–Tibet Plateau (QT) in China and evaluated its relationships with environmental variables. There was no discernible difference between leaf C, N, and P content in *S. chamaejasme* from the QT and IM, but the leaf K concentration was significantly higher in QT than that in IM, and there was no significant correlation between leaf ecological stoichiometry of *S. chamaejasme* and soil physicochemical properties. According to the findings, *S. chamaejasme* can adapt to changing environments by adjusting its relationships with climatic or soil factors to improve its survival in degraded grasslands.

Jin et al., (2022) [6] investigated the photosynthetic characteristics and leaf economic traits of three dominant species (two grass species: *Bothriochloa ischaemum* and *Stipa bungeana*; one leguminous subshrub: *Lespedeza davurica*) in a semiarid grassland community on the Loess Plateau of China. The community was continuously treated with nitrogen (N) and phosphorus (P) inputs for three years, and the study suggests that N and P addition shifted leaf economic traits towards a greater light harvesting ability and elevated photosynthesis in the three dominant species due to evident N and P synergetic effects, and this was achieved by species-specific responses in leaf functional traits. The results provide insights into grassland restoration and the assessment of community development in the context of atmospheric N deposition and intensive agricultural fertilization.

In the same experiment as Jin et al., (2022) [6], Yang et al., (2022) [7] used trait-based approaches to analyze and compare the relative contributions of plant functional traits to grassland productivity under N and/or P addition on the semiarid Loess Plateau. The results showed that the linkages between plant functional traits and the relative biomass of species differed under different N and P addition levels, and dominant species traits could predict ecosystem functioning (productivity) on the semiarid grassland. The study validated the mass ratio hypothesis and highlighted the close linkages between community-level functional traits and grassland productivity. The study advances our understanding of the mechanisms underpinning biodiversity–ecosystem functional relationships and has significant implications for semiarid grassland management.

Zuffo et al., (2022) [8] compared the tolerance of nine cultivars belonging to five species of perennial tropical forage grasses in Brazil growing in pots with controlled soil water conditions. After comparing and evaluating of twelve tolerance indices, the mean production (MP), drought resistance index (DI), stress tolerance index (STI), geometric mean production (GMP), yield index (YI), modified stress tolerance (k_2STI) and harmonic mean (HM) were selected as the most suitable parameters for identifying forage grass cultivars with greater water stress tolerance and a high potential for shoot biomass production under severe water stress condition. The study is significant in terms of selecting metrics for assessing and identifying water-stress-tolerant genotypes.

Duan et al., (2022) [9] evaluated plant functional traits (PFTs) of 171 herb plots in 57 sites (from varied topography and herb types) using 29 variables categorized into four types from a typical Soil and Water Conservation Demonstration Park (SWDP) on the Loess Plateau of China. The study attempted to quantify the effects of topographic conditions, soil factors and vegetation structure on PFTs. The results showed that the topographic conditions and soil properties had direct effects on plant functional traits, with slope having the most weight in topographic conditions and maximum water capacity (MWC) having the most the highest weight in soil properties, followed by soil water content (SWC). The study adds to our understanding of the mechanisms of resource utilization, competition and adaptation of plants in heterogeneous habitats through quantifying the association between distinct factors.

Mugloo et al., (2023) [10] reported the yield and nutritional contents of grass and legume species in Kashmir Valley's rangelands. The study area included grazed, protected, and seed-sown sites. The results showed that aboveground biomass (AGB) and total biomass yield were highest in the protected sites of the central Kashmir region, whereas belowground biomass (BGB) was highest in the protected sites of the southern Kashmir region. The study indicated that moderate grazing had moderate effects on biomass production, and species in different regions responded differently to disturbance, implying site-specific and species-specific traits in response to disturbance and climate change.

In a one-year plot experiment, Ma et al., (2023) [11] simulated combinations of different species compositions (1, 2, 4, 6 species and 8 species), and monitored the phenology of alfalfa and determined the related functional traits such as light acquisition traits (plant height and relative height, leaf mass and area, leaf length and width, and specific leaf area) and nutrient acquisition traits (leaf carbon content (LCC), leaf nitrogen content (LNC), leaf C/N ratio (LCC/LNC), biomass and abundance, relative biomass and abundance) after N addition. The results showed that the effect of N addition and plant diversity on flowering phenology was driven by the intraspecific variation in functional traits, and such alteration would influence the flower numbers and plant reproductive strategy.

He et al., (2023) [12] reviewed the studies on the structure and stability of grassland ecosystems, summarized the progress and future directions, and suggested ideas for improving the ecosystem service capacity of grasslands for Karst Desertification Control (KDC). Based on the background of KDC and its geographical characteristics, they proposed three insights to optimize the spatial allocation, enhance the grassland stability for rocky desertification control and coordinate the relation between grassland structure and stability. This study provided support for grassland managers and relevant policymakers to improve the structure, stability, and service capacity of grassland ecosystems in the karst region.

As Guest Editors, we sincerely thank all the authors who contributed to our Special Issue. We appreciate the efforts of all the reviewers and academic editors in evaluating the selected manuscripts and upholding the high standards of peer review. We anticipate that the papers in this Special Issue will serve as a springboard for future work on the ecophysiology and ecology of grasslands in the face of anthropogenic activities and climate change.

Conflicts of Interest: The authors declare no conflict of interest.

References

1. Fang, Z.; Han, X.; Xie, M.; Jiao, F. Spatial distribution patterns and driving factors of plant biomass and leaf n, p stoichiometry on the Loess Plateau of China. *Plants* **2021**, *10*, 2420. [CrossRef] [PubMed]
2. Zhang, Y.; Wang, Z.; Wang, Q.; Yang, Y.; Bo, Y.; Xu, W.; Li, J. Comparative assessment of grassland dynamic and its response to drought based on multi-index in the Mongolian Plateau. *Plants* **2022**, *11*, 310. [CrossRef] [PubMed]
3. Zhang, Y.; Xie, Y.; Ma, H.; Zhang, J.; Jing, L.; Wang, Y.; Li, J. The influence of climate warming and humidity on plant diversity and soil bacteria and fungi diversity in desert grassland. *Plants* **2021**, *10*, 2580. [CrossRef] [PubMed]
4. Song, K.; Hu, H.; Xie, Y.; Fu, L. The effect of soil water deficiency on water use strategies and response mechanisms of *Glycyrrhiza uralensis* Fisch. *Plants* **2022**, *11*, 1464. [CrossRef] [PubMed]
5. Guo, L.; Liu, L.; Meng, H.; Zhang, L.; Silva, V.J.; Zhao, H.; Wang, K.; He, W.; Huang, D. Biogeographic Patterns of Leaf Element Stoichiometry of *Stellera chamaejasme* L. in Degraded grasslands on Inner Mongolia Plateau and Qinghai-Tibetan Plateau. *Plants* **2022**, *11*, 1943. [CrossRef] [PubMed]
6. Jin, Y.; Lai, S.; Chen, Z.; Jian, C.; Zhou, J.; Niu, F.; Xu, B. Leaf photosynthetic and functional traits of grassland dominant species in response to nutrient addition on the Chinese Loess Plateau. *Plants* **2022**, *11*, 2921. [CrossRef] [PubMed]
7. Yang, Y.; Chen, Z.; Xu, B.; Wei, J.; Zhu, X.; Yao, H.; Wen, Z. Using trait-based methods to study the response of grassland to fertilization in the grassland in semiarid areas in the Loess Plateau of China. *Plants* **2022**, *11*, 2045. [CrossRef] [PubMed]
8. Zuffo, A.M.; Steiner, F.; Aguilera, J.G.; Ratke, R.F.; Barrozo, L.M.; Mezzomo, R.; Santos, A.S.D.; Gonzales, H.H.S.; Cubillas, P.A.; Ancca, S.M. Selected indices to identify water-stress-tolerant tropical forage grasses. *Plants* **2022**, *11*, 2444. [CrossRef] [PubMed]
9. Duan, G.; Wen, Z.; Xue, W.; Bu, Y.; Lu, J.; Wen, B.; Wang, B.; Chen, S. Agents Affecting the plant functional traits in national soil and water conservation demonstration park (China). *Plants* **2022**, *11*, 2891. [CrossRef] [PubMed]

10. Mugloo, J.A.; Khanday, M.U.D.; Dar, M.U.D.; Saleem, I.; Alharby, H.F.; Bamagoos, A.A.; Alghamdi, S.A.; Abdulmajeed, A.M.; Kumar, P.; Abou Fayssal, S. Assessment of selected grass and legume biomass species in high altitude rangelands of Kashmir Himalaya Valley (J&K), India: Evaluation of yield and nutritional profile. *Plants* **2023**, *12*, 1448. [PubMed]
11. Ma, Y.; Zhao, X.; Li, X.; Hu, X.; Wang, C. Intraspecific variation in functional traits of *Medicago sativa* determine the effect of plant diversity and nitrogen addition on flowering phenology in a one-year common garden experiment. *Plants* **2023**, *12*, 1994. [CrossRef]
12. He, S.; Xiong, K.; Song, S.; Chi, Y.; Fang, J.; He, C. Research progress of grassland ecosystem structure and stability and inspiration for improving its service capacity in the Karst desertification control. *Plants* **2023**, *12*, 770. [CrossRef] [PubMed]

Article

Spatial Distribution Patterns and Driving Factors of Plant Biomass and Leaf N, P Stoichiometry on the Loess Plateau of China

Zhao Fang [1], **Xiaoyu Han** [1], **Mingyang Xie** [2] and **Feng Jiao** [1,2,*]

[1]   Institute of Soil and Water Conservation, Northwest A&F University, Xi'an 712100, China; fangzhao16@mails.ucas.ac.cn (Z.F.); 2020060524@nwafu.edu.cn (X.H.)
[2]   Institute of Soil and Water Conservation, Chinese Academy of Sciences and Ministry of Water Resource, Xi'an 712100, China; Xiemingyang19@mails.ucas.ac.cn
*   Correspondence: jiaof@ms.iswc.ac.cn

Citation: Fang, Z.; Han, X.; Xie, M.; Jiao, F. Spatial Distribution Patterns and Driving Factors of Plant Biomass and Leaf N, P Stoichiometry on the Loess Plateau of China. *Plants* **2021**, *10*, 2420. https://doi.org/10.3390/plants10112420

Academic Editor: Jess K. Zimmerman

Received: 2 October 2021
Accepted: 4 November 2021
Published: 9 November 2021

Publisher's Note: MDPI stays neutral with regard to jurisdictional claims in published maps and institutional affiliations.

Abstract: Understanding the geographic patterns and potential drivers of leaf stoichiometry and plant biomass is critical for modeling the biogeochemical cycling of ecosystems and to forecast the responses of ecosystems to global changes. Therefore, we studied the spatial patterns and potential drivers of leaf stoichiometry and herb biomass from 15 sites spanning from south to north along a 500 km latitudinal gradient of the Loess Plateau. We found that leaf N and P stoichiometry and the biomass of herb plants varied greatly on the Loess Plateau, showing spatial patterns, and there were significant differences among the four vegetation zones. With increasing latitude (decreasing mean annual temperature and decreasing mean precipitation), aboveground and belowground biomass displayed an opening downward parabolic trend, while the root–shoot ratio gradually decreased. Furthermore, there were significant linear relationships between the leaf nitrogen (N) and phosphorus (P) contents and latitude and climate (mean annual rainfall and mean annual temperature). However, the leaf N/P ratio showed no significant latitudinal or climatic trends. Redundancy analysis and stepwise regression analysis revealed herb biomass and leaf N and P contents were strongly related to environmental driving factors (slope, soil P content and latitude, altitude, mean annual rainfall and mean annual temperature). Compared with global scale results, herb plants on the Loess Plateau are characterized by relatively lower biomass, higher N content, lower P content and a higher N/P ratio, and vegetative growth may be more susceptible to P limitation. These findings indicated that the remarkable spatial distribution patterns of leaf N and P stoichiometry and herb biomass were jointly regulated by the climate, soil properties and topographic properties, providing new insights into potential vegetation restoration strategies.

Keywords: leaf nitrogen (N) and phosphorus (P) contents; N:P ratio; biomass; herb community; driving factor; Loess Plateau

1. Introduction

Aboveground biomass (AGB), belowground biomass (BGB) and the ratio of roots to shoots (R/S) are regarded as important parameter of vegetation biomass, playing critical roles in estimating terrestrial ecosystem productivity and in global climate models [1–4]. Because of the influences of intensive anthropogenic activities, the atmospheric CO_2 concentration has increased over the past 60 years [5]. Studies have shown that grassland ecosystems account for 1/4 of the Earth's land surface and 1/10 of global carbon stocks, which fixed the majority of atmospheric CO_2 [6,7]. Thus, a better understanding of plant biomass is essential for understanding vegetation dynamics, terrestrial ecosystem carbon (C) stocks and their response to environmental changes [8–11]. Previous studies have demonstrated that plant biomass has important implications for community structure and ecosystem function and is affected by environmental factors [12–14]. The varying responses of plant biomass on these environmental factors are complex, and less is known about the interactive effects of potential environmental driving factors (climate, soil properties, topographic properties, etc.) on the herb plant biomass of the Loess Plateau.

Nitrogen (N) and phosphorus (P) are the basic nutrients that limit plant growth [15], and they have close interactive relationships in terrestrial ecosystems [16]. N and P cycles are generally regarded as the main limiting factors for the productivity of an ecosystem, while the ratio of these two elements in leaf tissue may indicate whether a system is limited by N, P, or both [17]. Accordingly, the N/P ratio in plant leaves has become a focal indicator of plant nutrient limitations, adaptation strategies and ecosystem function at the community level and has been studied intensely. In general, an N/P value < 14 indicates that plant growth is mainly limited by N; an N/P value > 16 indicates that plant growth is mainly limited by P, and an N/P value between 14 and 16 indicates that plant growth is limited by both P and N or is not deficient in either nutrient [18]. So far, the increasing efforts are devoted to identifying the geographical pattern of leaf N and P stoichiometry and its relationship with environmental factors at local, regional or global scales [15,19–23]. There are, however, only few studies in Loess Plateau about the geographical pattern of leaf N and P stoichiometry cited in international scientific literatures [24].

The Loess Plateau in China, with an area of 6.2×10^5 km^2, is a unique area due to its distinctive landscape and deep loess deposits; this area also experiences the most intense soil erosion in the world [25]. In the last century, due to population increases and the intensification of human activities, the Loess Plateau has experienced the most serious vegetation damage and ecological degradation [26]. To improve this dilemma, the Chinese government launched a series of nationwide conservation projects, such as the 'Grain for Green' project. The Loess Plateau, with diverse vegetation types and the varying hydrothermal conditions extending from south to north, offers a unique opportunity for determining the critical factors affecting leaf N, P and plant biomass across vegetation transect. However, the knowledge gap regarding the relationships among leaf stoichiometry, plant biomass and vegetation growth in different types of vegetation has not been filled. Changes in leaf stoichiometry and biomass are affected and restricted by various environmental factors, and differences along latitudinal gradients often lead to corresponding changes in environmental factors, such as temperature, light and soil moisture, etc. It is still necessary to study the effects of potential drivers on leaf N and P stoichiometry and herb biomass in different vegetation recovery zones.

In this study, we tested the responses of leaf N and P stoichiometry and plant biomass to latitude and climate in different vegetation zones at fifteen sampling sites spanning from south to north along a 500 km-long latitudinal gradient on the Loess Plateau. Our objectives were to understand the effects of potential drivers (climate, soil properties, topographic properties) on leaf N and P stoichiometry and herb biomass to reveal the limiting conditions of nutrient restrictions, to provide the basis for future ecological restoration and polices on the Loess Plateau and to offer basic data for global scale research.

2. Materials and Methods

2.1. Study Site

From south to north, the following six representative areas in different vegetation zones on the Loess Plateau were selected as the study areas: Fuxian, Ganquan and Ansai counties in Yan'an city, Jingbian and Hengshan counties and Yuyang district in Yulin city in Shaanxi Province. According to previous studies [27], we established fifteen sampling sites along a 500-km-long latitudinal gradient in the Loess Plateau, which were distributed in forest zone (FZ), forest–steppe zone (FS), steppe zone (SZ) and steppe–desert zone (SD) (Figure 1). The study region was in the midlatitude temperate zone (107°97′–109°87′ E, 35°95′–38°36′ N), with an altitudinal range of about 1015–1600 m above sea level (m.a.s.l), a mean annual temperature (MAT) of 8.8 °C, a mean annual precipitation (MAP) of 505.3 mm, an annual sunshine time of 2395.6 h (the sunshine percentage is 54%) and an annual frost-free period of 157 days. Regional climate conditions, in recent years, have exhibited a warming and wetting trend, with distinct wet and dry seasons for precipitation. The main soil types were loess, loess sandy, and aeolian sandy soils according to the Genetic Soil Classification of China [28]. The land uses include forestland, grassland, and farmland. In this area, the main herb species are *Bothriochloa ischaemum*, *Stipa bungeana*, *Cleistogenes caespitosa*, *Lespedeza davurica*, *Astragalus melilotoides*, *Artemisia sacrorum*, *Heteropappus altaicus* and *Potentilla tanacetifolia*; other concrete description of vegetation and soil conditions can be found in the appendix materials (Table S1).

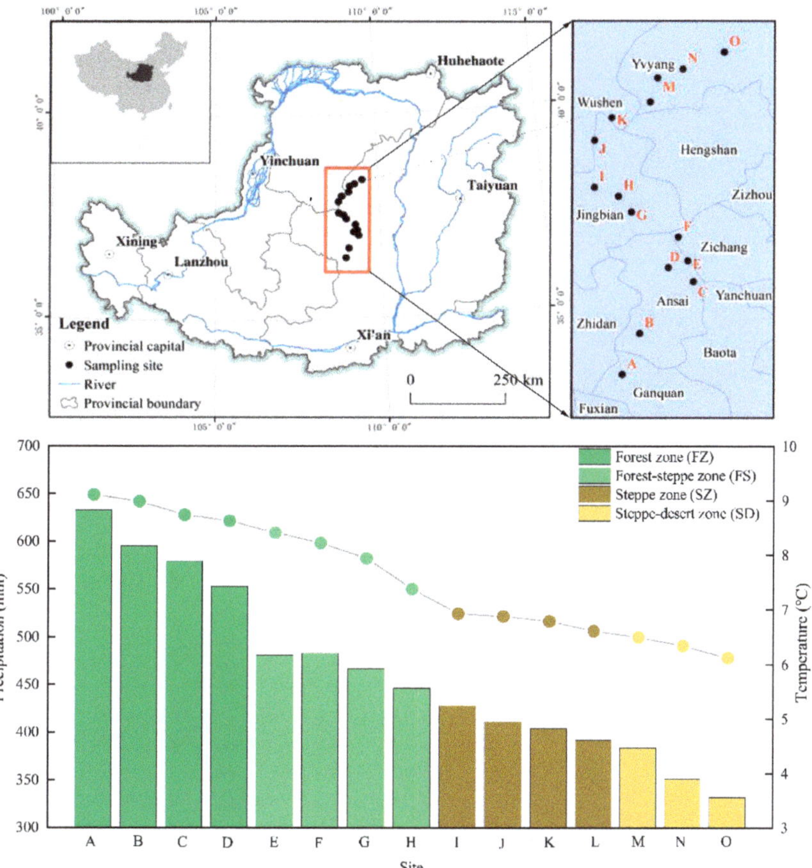

Figure 1. Location of the study area and illustration of sampling sites of mean monthly temperature and precipitation (1990–2010) along a 500-km-long latitudinal gradient on the Loess Plateau. All climate data (MAT and MAP) was obtained from China Meteorological Data Sharing Service System (http://data.cma.cn/ (accessed on 10 May 2021)).

2.2. Experimental Design and Soil Sampling

The vegetation measurements and sampling were conducted in late August 2012, when the vegetation reached maximum biomass and cover. At each site, three sampling plots (1 m × 1 m) with homogeneous vegetation and landform conditions were established to identify all species and investigate the vegetation coverage, height and density. We calculated the community diversity index (included Shannon–Wiener index, Margalef richness index and Pielou evenness index) according to the method of [29]. The AGB of the plants was collected by clipping the plants at ground level. Moreover, all target dominant plant leaves in every quadrat were collected individually to determine the leaf N and

P contents at the community level. The BGB (roots < 2 mm in diameter) was collected randomly from the upper 30 cm soil layer with a soil drilling sampling corer (9 cm in diameter). The litter was cleared before sampling. The roots were rinsed with a large amount of deionized water, dried at 70 °C for 48 h to a constant weight and weighed on an analytical balance. The dried leaves were ground to pass through a 0.15 mm sieve for elemental analysis.

Within each quadrat, three soil samples for 0–30 cm were randomly sampled by a soil auger (diameter of 5 cm) and then thoroughly mixed to form one composite sample, and a total of 45 soil samples (3 quadrats × 15 sites) were collected. We patiently removed debris and fine roots by hand when all soil samples were naturally air-dried in the lab, and then they were sieved through 2-mm and 0.15-mm mesh for different element analysis with a ball mill. The total N content of the leaves was measured with a CHNS/O Elemental Analyzer (PekinElmer, Boston, MA, USA), and the total P content was analyzed colorimetrically after H_2SO_4-H_2O_2-HF digestion using the molybdate/stannous chloride method [30]. Soil organic C and total N and P contents were determined using standard testing methods as described by Jiao et al. [31]. The contents of nutrient contents were expressed as $mg \cdot g^{-1}$ on a dry mass basis.

2.3. Data Analysis

Data, including all tables and figures, are presented as mean ± standard deviation (SD). A linear mixed-effect model (LMM) with Tukey's multiple comparisons test ($p < 0.05$) was conducted to assess the differences of leaf N, P stoichiometry and biomass among four vegetation zones in SPSS 22.0. The "vegetation zone" was used as the fixed factor and "sampling site" was used as a random factor. Linear or curvilinear (quadratic) regressions were adopted to explore the relationships of independent variables (MAT, MAP and latitude) with the dependent variables (leaf N content, leaf P content, leaf N:P ratio, AGB, BGB, and R/S ratio). All data were checked for normality and homogeneity of variance before conducting the parametric tests. Redundancy analysis (RDA) and stepwise regression analysis (SRA) were performed to identify the critical factors of herb biomass and leaf N and P stoichiometry. RDA was conducted in Canoco 5.0, and all figures were created in Origin 2018.

3. Results

3.1. Spatial Variation of Plant Biomass and Leaf N, P Stoichiometry along the Latitude Gradient

Plant biomass of the herb communities on the Loess Plateau across of the sampling sites ranged from 54.60 to 204.32 g/m^2 (CV = 27.8%) for AGB, 78.88 to 829.64 g/m^2 (CV = 48.8%) for BGB and 0.93 to 4.50 (CV = 13.5%) for R/S. As shown in Figure 2, with increasing latitude (decreasing mean annual temperature, MAT and decreasing mean precipitation, MAP), AGB and BGB displayed an opening downward parabolic trend (Figure 2A–F). However, the root–shoot ratio (R/S) decreased linearly with increasing latitude (decreasing MAT and decreasing MAP) (Figure 2G–I). By contrast, leaf N and P stoichiometry of herbs on the Loess Plateau across the sampling sites ranged from 20.45 to 31.96 mg/g (CV = 17.1%) for leaf N content, 1.22 to 1.62 mg/g (CV = 13.9%) for leaf P content and 16.90 to 19.94 (CV = 9.94%) for the leaf N/P ratio. The mean leaf N, P and N/P values were 25.79 mg/g, 1.37 mg/g and 18.71, respectively. Leaf N and P contents exhibited significant relations to latitude, MAT and MAP, but not N/P ratio. Besides, linear regression showed that the leaf N and P contents were substantially correlated with the latitude and increased with increasing latitude (decreasing MAT and decreasing MAP). However, the leaf N/P ratio was not positively correlated with environmental variables (latitude, MAT and MAP) (Figure 3).

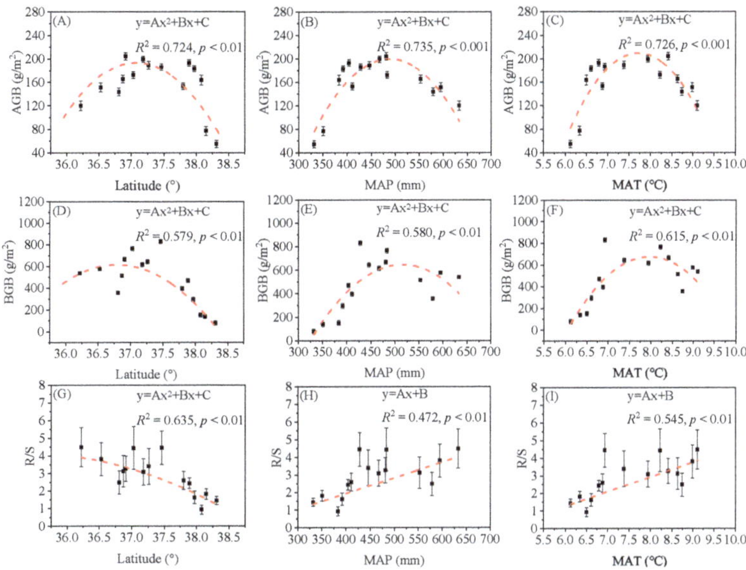

Figure 2. Relationships of AGB, BGB and R/S with MAT, MAP and absolute latitude. Note: (**A–C**), AGB: above-ground biomass; (**D–F**), BGB: below-ground biomass; (**G–I**), R/S: root-to-shoot ratio. The red lines indicate the fits of the linear model of AGB, BGB and R/S and environmental gradient (latitude, MAT and MAP). All climate data (MAT and MAP) was obtained from China Meteorological Data Sharing Service System (http://data.cma.cn/ (accessed on 10 May 2021)).

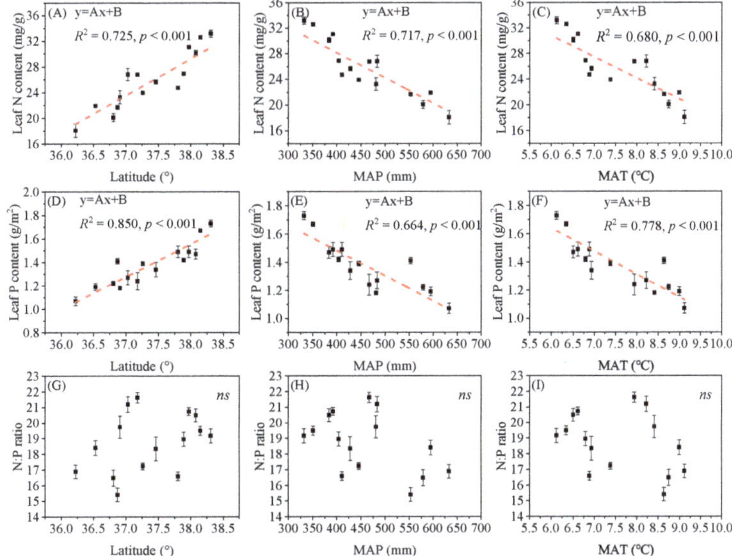

Figure 3. Linear regression relationships of Leaf N, P and the leaf N/P ratio with MAT, MAP and absolute latitude. Note: (**A–C**), Leaf N: leaf nitrogen content; (**D–F**), Leaf P: Leaf phosphorus content; (**G–I**), N:P ratio: Leaf N/P ratio. The red lines indicate the fits of the linear model of Leaf N, P and the leaf N/P ratio and environmental gradient (latitude, MAT and MAP). All climate data (MAT and MAP) was obtained from China Meteorological Data Sharing Service System (http://data.cma.cn/ (accessed on 10 May 2021)).

3.2. Differences in Herb Biomass and Leaf N, P Stoichiometry among Vegetation Types

To reveal the differences in herb biomass between different vegetation types in the Loess Plateau, the results of LMM showed that AGB, BGB and R/S of the steppe-desert zone were significantly lower than those in the other three vegetation zones (Figure 4A–C and Table S4) ($p < 0.05$). The biomass order of the herb communities (AGB, BGB) was FS > SZ > FZ > SD, while the root–shoot ratio (R/S) ranked as FS > FZ > SZ > SD. The biomass allocation of FZ, FS and SZ was predominantly concentrated in the belowground part (2 to 4 times), except for SD. Additionally, we compared the differences in the leaf N and P contents and N/P ratio among the different vegetation zones and found that the plant leaf N and P contents in the steppe–desert were significantly higher than those in the other three vegetation zones (Figure 4D,E and Table S4), and the lowest leaf N/P ratio was found in the forest zone ($p < 0.05$) (Figure 4F and Table S4).

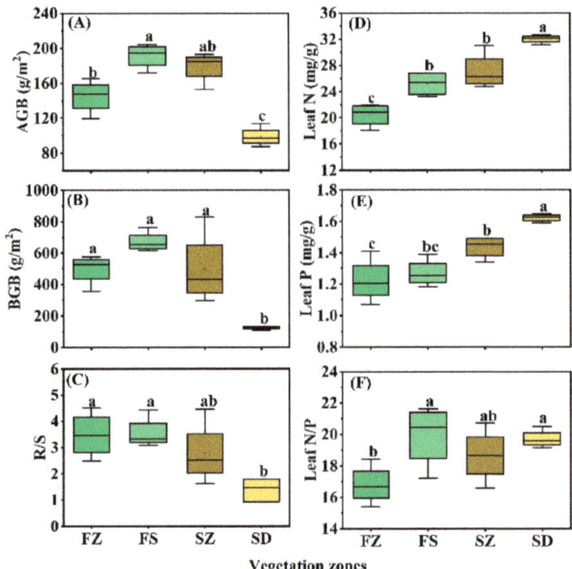

Figure 4. Boxplots of leaf N, P stoichiometry and plant biomass in each of vegetation zones along latitudinal gradients. Note: (**A**), AGB: above-ground biomass; (**B**), BGB: below-ground biomass; (**C**), R/S: root-to-shoot ratio; (**C,F**), FZ: forest zone; FS: forest–steppe zone; SZ: steppe zone; SD: steppe–desert zone. (**D**), Leaf N: leaf N content; (**E**), Leaf P: leaf P content. The boundaries of the box represent the lower (25%) and upper (75%) quartiles of data, respectively, and the median is represented by the line inside the box. Different lowercase letters above the bars indicate significant differences (ANOVA, $p < 0.05$) among different vegetation zones.

3.3. Factors Driving Herb Biomass and Leaf N, P Stoichiometry

RDA analysis showed that plant biomass and leaf N and P stoichiometry of herbs were determined by topographic properties, soil C:N:P stoichiometry and climatic conditions. A total of 85.67% and 5.93% were accounted for by the first two axes, respectively (Figure 5). Topographic properties and species diversity were strongly correlated with AGB, BGB and R/S. The BGB, R/S, and leaf N and P contents were mainly correlated to the soil C:N:P stoichiometry and climatic conditions, while the N/P ratio of leaves was less affected by the above factors. Moreover, we performed a stepwise regression to detect the critical factors of topographic properties, soil C:N:P stoichiometry and climatic conditions that affected plant biomass and leaf N and P stoichiometry. Taken together, these results demonstrate that the slope; soil P content; latitude, altitude and MAT; MAP and latitude were the key driving factors of AGB, BGB, R/S, leaf N and P, respectively. However, no critical factor had a strong effect on leaf N/P (Table 1).

Table 1. Stepwise regression analysis (SRA) used to identify the key factors of plant biomass and leaf N and P stoichiometry.

Variables	Equations	R^2	Sig.
AGB	AGB = −4.929SLO + 243.061	0.764	0.001 **
BGB	BGB = 722.128SOP − 107.608	0.740	0.002 **
R/S	R/S = −4.025LAT + 0.02ALT − 1.518MAT	0.927	0.000 **
Leaf N	Leaf N = − 0.045MAP + 46.653	0.915	0.000 **
Leaf P	Leaf P = 0.258LAT − 8.279	0.905	0.000 **
Leaf N/P	—	—	—

AGB, above-ground biomass; BGB, below-ground biomass; R/S, root-to-shoot ratio; SLO, slope; SOP, Soil P content; LAT, latitude; ALT, altitude; MAT, mean annual temperature; MAP, mean annual precipitation (** $p < 0.01$;). All climate data (MAT and MAP) was obtained from China Meteorological Data Sharing Service System (http://data.cma.cn/ (accessed on 10 May 2021)).

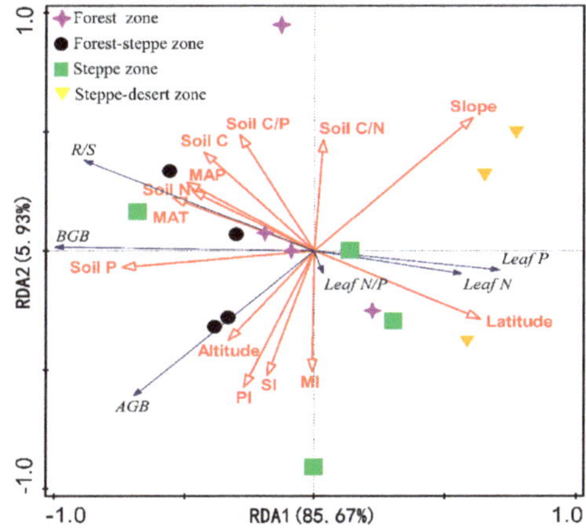

Figure 5. Redundancy analysis (RDA) of the relationship between plant biomass, leaf N and P stoichiometry and environmental driving factors in different vegetation zones. Note: AGB, above-ground biomass; BGB, below-ground biomass; R/S: root-to-shoot ratio; Leaf N: leaf nitrogen content; Leaf P: leaf phosphorus content; Leaf N/P: leaf nitrogen/ phosphorus ratios; Soil C: soil organic carbon; Soil N: soil total nitrogen; Soil P: soil total phosphorus; Soil C/N: soil carbon/nitrogen ratios; Soil C/P: soil carbon/ phosphorus ratios; Soil N/P: soil nitrogen/ phosphorus; MAP: mean annual precipitation; MAT: mean annual temperature; SI: Shannon–Wiener diversity index; MI: Margalef richness index; PI: Pielou evenness index. All climate data (MAT and MAP) was obtained from China Meteorological Data Sharing Service System (http://data.cma.cn/ (accessed on 10 May 2021)).

4. Discussion

4.1. The Spatial Patterns of Herb Biomass and Response to Environmental Factors

On the Loess Plateau, precipitation and temperature decrease toward the northwest with increasing latitude (Figure 1), and therefore, the vegetation zones change from forest to desert with increasing latitude from southeast to northwest [27]. Differences in environmental conditions (temperature, precipitation, etc.) are often linked to varying latitudes and, in turn, could be used to explain the distribution of vegetation types across the Loess Plateau. The results from our study suggest that the biomasses of the herbaceous communities in the different vegetation zones in the loess hilly region were in the order of forest–steppe zone > grassland zone > forest zone > steppe–desert zone, and the variations in AGB and BGB with increasing latitude (decreasing mean annual temperature, MAT and decreasing mean precipitation, MAP) showed an opening downward parabolic trend

(Figure 2). Besides, compared with other temperate grasslands in the world [3,32], the AGB, BGB and R/S ratio of the Loess Plateau were relatively lower but only slightly higher than China [33] and the Inner Mongolia grassland [34] (Table S2). Hydrothermal factors are important factors that restrict the biomass of herbaceous communities in semiarid areas [35,36], which contributed to the lower biomass of herbaceous vegetation on the Loess Plateau. The most common environmental stressor affecting plant growth in arid and semiarid regions is an insufficient water supply [37]. To better adapt to arid environments, the plant might allocate more resources to root growth to obtain more nutrients and water [12,38]. Changes in precipitation may shape the various physiological traits of plants by affecting the moisture regimes in arid and semiarid regions [39,40]. Thus, decreasing precipitation with increasing latitude will change the soil water availability [41] and indirectly affect the herbaceous biomass. Besides, temperature may influence various metabolic processes in plants by affecting the activities of enzymes, such as water and mineral absorption, material synthesis, transformation, transportation, and distribution, and further affecting the function of cells, thereby affecting the normal metabolic activities of plants. On the other hand, aggravated local water scarcity after afforestation and soil evaporation caused by increasing temperatures may be unfavorable for the growth of herbaceous vegetation in forest zones [42]. The steppe–desert zone and steppe zone have relatively less rainfall and lower temperatures; herbaceous plant growth is affected by hydrothermal factors, thus restraining plant growth, resulting in the lower AGB and BGB of the herbaceous community. However, with increasing latitude, precipitation increases improve soil water availability, and large amounts of litter accumulate in the forest zone and forest–steppe zone; these two factors jointly regulate soil nutrient dynamics via decomposition and element release from surface litter. Compared with SZ and SD, soil nutrients and hydrothermal conditions in FS are more conducive to the growth of herbaceous vegetation, thus showing the highest AGB and BGB.

RDA and SRA revealed herb biomass were strongly related to environmental driving factors, among them the slope, soil P content and latitude, altitude, MAT, which were key factors affecting AGB, BGB and R/S in the Loess Plateau, respectively. It is well known that the productivity of grasslands is mainly affected by soil water availability rather than directly by rainfall [37]. The slope is believed to play an important role in the distribution of soil moisture, which will significantly affect the productivity and vegetation patterns of grasslands [43,44]. Soil P content, as a key factor affecting belowground biomass, is related to the obvious P limitation of herbaceous vegetation on the Loess Plateau [45]. The difference in environmental driving factors of AGB and BGB causes the R/S to be affected by multiple environmental factors (latitude, altitude, MAT).

4.2. The Spatial Patterns of Leaf N and P Stoichiometry and Response to Environmental Factors

Confirming the results of previous studies [20,22], the leaf N and P contents increased with increasing latitude (decreasing MAT and decreasing MAP), which is a phenomenon that can be explained well by the temperature-plant physiological hypothesis [21]. Besides, our research also reveals that MAP and latitude were the key driving factors of leaf N and P, respectively. This is mainly because herb plants usually absorb highly mobile available nitrogen (such as nitrate nitrogen and ammonia nitrogen), but higher rainfall easily causes these highly mobile nitrogens to be leached, making the available nitrogen that can be absorbed and utilized decreases and resulting in a decrease in the N content of leaves [46].

With increasing latitude, the hydrothermal conditions change, and low temperatures affect the RNA efficiency of N- and P-rich enzymes, which reduces the plant biochemical reaction rate; however, increases in leaf N and P contents can compensate for the loss caused by the lower biochemical reaction rate [21]. Therefore, plants need to maintain high leaf N and P levels to offset the low temperature-induced inhibition of metabolic reactions, thus representing an adaptation of plant tissues to low temperature and environmental changes [21]. However, the correlation between N:P and latitude was not significant. On the one hand, our study area represented a relatively small latitude range; on the other hand, the focus of our study was limited to herbal communities. Both factors contributed to the leaf N:P characteristics, and the results were not consistent with those of previous studies [21,47].

To determine the leaf nutrient levels and leaf N/P characteristics of the herbaceous vegetation on the Loess Plateau in depth, we compared the results of this study with other scholars' findings; we found that the leaf N concentrations of herbaceous plants on the Loess Plateau were significantly higher than the level at the global scale [20,22,48] but were also slightly higher than that found on the Loess Plateau [24]. Unlike previous research [20,22,48], our research object was only limited to herbs and the sample size was relatively small, which may account for the discrepancy in our study.

Furthermore, the leaf P content on the Loess Plateau was significantly lower than that at the global scale [20,22,48] but was also slightly lower than on the Loess Plateau [24] and throughout China [22]. The N/P threshold is often used as an indicator of relative N and P restriction [17–19,49]. Compared with the results obtained in studies at the global scale [21], the higher N/P and lower leaf P contents of plants on the Loess Plateau [24] and throughout China [22,23] further indicate that the growth of plants in China is more restricted by P (Table S3).

5. Conclusions

In summary, we examined the spatial patterns of herb biomass and leaf N and P stoichiometry and their influencing factors on the regional scale, providing important information about better vegetation management and restoration on the Loess Plateau, and supplementing basic data for global scale research. Our results suggested that the leaf N and P stoichiometry and herb biomass on the Loess Plateau showed remarkable spatial distribution patterns and were regulated by the climate, soil properties and topographic properties. Compared with global scale results, the vegetative growth of the herb community on the Loess Plateau is more susceptible to P limitation. Further, more future research should focus on exploring the driving mechanisms of vegetation dynamics under increasing warming and wetting trend of the Loess Plateau.

Supplementary Materials: The following are available online at https://www.mdpi.com/article/10.3390/plants10112420/s1, Table S1. Basic information of the experimental plots; Table S2. Comparisons of AGB, BGB and R/S of Loess Plateau with other temperate grasslands in the world. Table S3. Comparisons of leaf N and P stoichiometry around the world. Table S4. Linear mixed-effect model showing the differences in herb biomass and leaf N and P stoichiometry on the Loess Plateau.

Author Contributions: Conceptualization and methodology, Z.F., X.H., M.X. and F.J.; data curation, Z.F. and F.J.; investigation, Z.F., X.H. and M.X.; writing—original draft, Z.F.; writing—review and editing, Z.F. and F.J. All authors have read and agreed to the published version of the manuscript.

Funding: This study was supported by the National Key Research and Development Program of China (2016YFA0600801) and the National Special Program on Basic Works for Science and Technology of China (2014FY210130).

Institutional Review Board Statement: Not applicable.

Informed Consent Statement: Not applicable.

Data Availability Statement: All available data can be obtained by contacting the corresponding author.

Acknowledgments: We thank the two anonymous reviewers that provided us with valuable comments that clearly improved our manuscript.

Conflicts of Interest: The authors declare no conflict of interest.

References

1. Müller, I.; Schmid, B.; Weiner, J. The effect of nutrient availability on biomass allocation patterns in 27 species of herbaceous plants. *Perspect. Plant Ecol.* **2000**, *3*, 115–127. [CrossRef]
2. Enquist, B.J.; Niklas, K.J. Global Allocation Rules for Patterns of Biomass Partitioning in Seed Plants. *Science* **2002**, *295*, 1517–1520. [CrossRef]
3. Mokany, K.; Raison, R.J.; Prokushkin, A.S. Critical analysis of root: Shoot ratios in terrestrial biomes. *Glob. Chang. Biol.* **2006**, *12*, 84–96. [CrossRef]
4. Sims, L.; Pastor, J.; Lee, T.; Dewey, B. Nitrogen, phosphorus and light effects on growth and allocation of biomass and nutrients in wild rice. *Oecologia* **2012**, *170*, 65–76. [CrossRef]
5. Revelle, R.; Suess, H.E. Carbon Dioxide Exchange Between Atmosphere and Ocean and the Question of an Increase of Atmospheric CO_2 during the Past Decades. *Tellus* **1957**, *9*, 18–27. [CrossRef]
6. Scurlock, J.M.O.; Johnson, K.; Olson, R.J. Estimating net primary productivity from grassland biomass dynamics measurements. *Glob. Chang. Biol.* **2002**, *8*, 736–753. [CrossRef]
7. Hui, D.; Jackson, R.B. Geographical and interannual variability in biomass partitioning in grassland ecosystems: A synthesis of field data. *New Phytol.* **2006**, *169*, 85–93. [CrossRef]
8. Melillo, J.M.; McGuire, A.D.; Kicklighter, D.W.; Moore, B.; Vorosmarty, C.J.; Schloss, A.L. Global climate change and terrestrial net primary production. *Nature* **1993**, *363*, 234–240. [CrossRef]
9. Scurlock, J.M.O.; Cramer, W.; Olson, R.J.; Parton, W.J.; Prince, S.D. Terrestrial NPP: Toward a consistent data set forglobal model evaluation. *Ecol. Appl.* **1999**, *9*, 913–919. [CrossRef]
10. Churkina, G.; Running, S.W.; Schloss, A.L.; Participants Potsdam, N.P.P.M.I. Comparing global models of terrestrial net primary productivity (NPP): The importance of water availability. *Glob. Chang. Biol.* **1999**, *5*, 46–55. [CrossRef]
11. Gao, X.X.; Dong, S.K.; Xu, Y.D.; Fry, E.L.; Li, Y.; Li, S.; Shen, H.; Xiao, J.N.; Wu, S.N.; Yang, M.Y.; et al. Plant biomass allocation and driving factors of grassland revegetation in a Qinghai-Tibetan Plateau chronosequence. *Land Degrad. Dev.* **2021**, *32*, 1732–1741. [CrossRef]
12. MCCARTHY, M.C.; ENQUIST, B.J. Consistency between an allometric approach and optimal partitioning theory in global patterns of plant biomass allocation. *Funct. Ecol.* **2007**, *21*, 713–720. [CrossRef]
13. Van der Putten, W.H.; Bardgett, R.D.; de Ruiter, P.C.; Hol, W.H.; Meyer, K.M.; Bezemer, T.M.; Bradford, M.A.; Christensen, S.; Eppinga, M.B.; Fukami, T.; et al. Empirical and theoretical challenges in aboveground-belowground ecology. *Oecologia* **2009**, *161*, 1–14. [CrossRef] [PubMed]
14. Anderson, L.J. Aboveground-belowground linkages: Biotic interactions, ecosystem processes, and global change. *Eos Trans. AGU* **2011**, *92*, 222. [CrossRef]
15. Güsewell, S. N: P ratios in terrestrial plants: Variation and functional significance. *New Phytol.* **2004**, *164*, 243–266. [CrossRef] [PubMed]
16. Vitousek, P. Nutrient cycling and nutrient use efficiency. *Am. Nat.* **1982**, *119*, 553–572. [CrossRef]
17. Koerselman, W.; Arthur, F.M.M. The Vegetation N:P Ratio: A New Tool to Detect the Nature of Nutrient Limitation. *J. Appl. Ecol.* **1996**, *33*, 1441–1450. [CrossRef]
18. Tessier, J.T.; Raynal, D.J. Use of nitrogen to phosphorus ratios in plant tissue as an indicator of nutrient limitation and nitrogen saturation. *J. Appl. Ecol.* **2003**, *40*, 523–534. [CrossRef]
19. Thompson, K.E.N.; Parkinson, J.A.; Band, S.R.; Spencer, R.E. A comparative study of leaf nutrient concentrations in a regional herbaceous flora. *New Phytol.* **1997**, *136*, 679–689. [CrossRef]
20. Reich, P.B.; Ellsworth, D.S.; Walters, M.B.; Vose, J.M.; Gresham, C.; Volin, J.C.; Bowman, W.D. Generality of Leaf Trait Relationsgips: A Test Across Six Biomes. *Ecology* **1999**, *80*, 1955–1969. [CrossRef]
21. Reich, P.B.; Oleksyn, J. Global patterns of plant leaf N and P in relation to temperature and latitude. *Proc. Natl. Acad. Sci. USA* **2004**, *101*, 11001–11006. [CrossRef]
22. Han, W.; Fang, J.; Guo, D.; Zhang, Y. Leaf nitrogen and phosphorus stoichiometry across 753 terrestrial plant species in China. *New Phytol.* **2005**, *168*, 377–385. [CrossRef] [PubMed]
23. Ren, S.-J.; Yu, G.-R.; Tao, B.; Wang, S.-Q. Leaf nitrogen and phosphorus stoichiometry across 654 terrestrial plant species in NSTEC. *Huan Jing Ke Xue* **2007**, *28*, 2665–2673. (In Chinese)
24. Zheng, S.; Shangguan, Z. Spatial patterns of leaf nutrient traits of the plants in the Loess Plateau of China. *Trees* **2007**, *21*, 357–370. [CrossRef]
25. Fu, B.-J.; Wang, Y.-F.; Lu, Y.-H.; He, C.-S.; Chen, L.-D.; Song, C.-J. The effects of land-use combinations on soil erosion: A case study in the Loess Plateau of China. *Prog. Phys. Geog.* **2009**, *33*, 793–804. [CrossRef]
26. Jiao, F.; Wen, Z.-M.; An, S.-S. Changes in soil properties across a chronosequence of vegetation restoration on the Loess Plateau of China. *Catena* **2011**, *86*, 110–116. [CrossRef]
27. Yamanaka, N.; Hou, Q.-C.; Du, S. Vegetation of the Loess Plateau. In *Restoration and Development of the Degraded Loess Plateau, China*; Tsunekawa, A., Liu, G., Yamanaka, N., Du, S., Eds.; Springer: Tokyo, Japan, 2014; pp. 49–60.
28. Shi, X.Z.; Yu, D.S.; Xu, S.X.; Warner, E.D.; Wang, H.J.; Sun, W.X.; Zhao, Y.C.; Gong, Z.T. Cross-reference for relating Genetic Soil Classification of China with WRB at different scales. *Geoderma* **2010**, *155*, 344–350. [CrossRef]

29. Liu, Y.; Zhu, G.; Hai, X.; Li, J.; Shangguan, Z.; Peng, C.; Deng, L. Long-term forest succession improves plant diversity and soil quality but not significantly increase soil microbial diversity: Evidence from the Loess Plateau. *Ecol. Eng.* **2020**, *142*, 105631. [CrossRef]
30. Sparks, D.L.; Page, A.L.; Helmke, P.A.; Loeppert, R.H.; Soltanpour, P.N.; Tabatabai, M.A.; Johnston, C.T.; Sumner, M.E. *Methods of soil Analysis. Part. 3—Chemical Methods*; Soil Science Society of America: Madison, WI, USA, 1996; pp. 869–919.
31. Jiao, F.; Wen, Z.-M.; An, S.-S.; Yuan, Z. Successional changes in soil stoichiometry after land abandonment in Loess Plateau, China. *Ecol. Eng.* **2013**, *58*, 249–254. [CrossRef]
32. Coupland, R.T. *Grassland Ecosystems of the World: Analysis of Grasslands and Their Uses*, 1st ed.; Cambridge University Press: New York, NY, USA, 1979; pp. 235–236.
33. Yang, Y.; Fang, J.; Ma, W.; Guo, D.; Mohammat, A. Large-scale pattern of biomass partitioning across China's grasslands. *Glob. Ecol. Biogeogr.* **2010**, *19*, 268–277. [CrossRef]
34. Ma, W.; Yang, Y.; He, J.S.; Zeng, H.; Fang, J.Y. Above- and belowground biomass in relation to environmental factors in temperate grasslands, Inner Mongolia. *Sci. China Ser. C Life Sci. Chin. Acad. Sci.* **2008**, *38*, 84–92. [CrossRef] [PubMed]
35. Xin, Z.; Xu, J.; Zheng, W. Spatiotemporal variations of vegetation cover on the Chinese Loess Plateau (1981–2006): Impacts of climate changes and human activities. *Sci. China Ser. D* **2008**, *51*, 67–78. [CrossRef]
36. Kardol, P.; Campany, C.E.; Souza, L.; Norby, R.J.; Weltzin, J.F.; Classen, A.T. Climate change effects on plant biomass alter dominance patterns and community evenness in an experimental old-field ecosystem. *Glob. Chang. Biol.* **2010**, *16*, 2676–2687. [CrossRef]
37. Zhang, X.; Zhao, W.W.; Liu, Y.X.; Fang, X.N.; Feng, Q. The relationships between grasslands and soil moisture on the Loess Plateau of China: A review. *Catena* **2016**, *145*, 56–67. [CrossRef]
38. Fan, L.; Ding, J.; Ma, X.; Li, Y. Ecological biomass allocation strategies in plant species with different life forms in a cold desert, China. *J. Arid Land* **2019**, *11*, 729–739. [CrossRef]
39. Cheng, X.; An, S.; Li, B.; Chen, J.; Lin, G.; Liu, Y.; Luo, Y.; Liu, S. Summer rain pulse size and rainwater uptake by three dominant desert plants in a desertified grassland ecosystem in northwestern China. *Plant. Ecol.* **2006**, *184*, 1–12. [CrossRef]
40. Xu, H.; Li, Y. Water-use strategy of three central Asian desert shrubs and their responses to rain pulse events. *Plant. Soil* **2006**, *285*, 5–17. [CrossRef]
41. West, A.G.; Hultine, K.R.; Burtch, K.G.; Ehleringer, J.R. Seasonal variations in moisture use in a pinon-juniper woodland. *Oecologia* **2007**, *153*, 787–798. [CrossRef]
42. Zhao, A.; Zhang, A.; Liu, J.; Feng, L.; Zhao, Y. Assessing the effects of drought and "Grain for Green" Program on vegetation dynamics in China's Loess Plateau from 2000 to 2014. *Catena* **2019**, *175*, 446–455. [CrossRef]
43. Thompson, S.E.; Katul, G.G.; Porporato, A. Role of microtopography in rainfall-runoff partitioning: An analysis using idealized geometry. *Water Resour. Res.* **2010**, *46*. [CrossRef]
44. Wei, P.; Pan, X.; Xu, L.; Hu, Q.; Zhang, X.; Guo, Y.; Shao, C.; Wang, C.; Li, Q.; Yin, Z. The effects of topography on aboveground biomass and soil moisture at local scale in dryland grassland ecosystem, China. *Ecol. Indic.* **2019**, *105*, 107–115. [CrossRef]
45. Fang, Z.; Li, D.-D.; Jiao, F.; Yao, J.; Du, H.-T. The Latitudinal Patterns of Leaf and Soil C:N:P Stoichiometry in the Loess Plateau of China. *Front. Plant. Sci.* **2019**, *10*, 85. [CrossRef] [PubMed]
46. Cregger, M.A.; McDowell, N.G.; Pangle, R.E.; Pockman, W.T.; Classen, A.T. The impact of precipitation change on nitrogen cycling in a semi-arid ecosystem. *Funct. Ecol.* **2014**, *28*, 1534–1544. [CrossRef]
47. Kerkhoff, A.J.; Enquist, B.J.; Elser, J.J.; Fagan, W.F. Plant allometry, stoichiometry and the temperature-dependence of primary productivity. *Glob. Ecol. Biogeogr.* **2005**, *14*, 585–598. [CrossRef]
48. Elser, J.J.; Fagan, W.F.; Denno, R.F.; Dobberfuhl, D.R.; Folarin, A.; Huberty, A.; Interlandi, S.; Kilham, S.S.; McCauley, E.; Schulz, K.L.; et al. Nutritional constraints in terrestrial and freshwater food webs. *Nature* **2000**, *408*, 578–580. [CrossRef]
49. Niklas, K.J.; Owens, T.; Reich, P.B.; Cobb, E.D. Nitrogen/phosphorus leaf stoichiometry and the scaling of plant growth. *Ecol. Lett.* **2005**, *8*, 636–642. [CrossRef]

Article

Comparative Assessment of Grassland Dynamic and Its Response to Drought Based on Multi-Index in the Mongolian Plateau

Yanzhen Zhang [1,2], Zhaoqi Wang [3], Qian Wang [4], Yue Yang [5], Yaojun Bo [1], Weizhou Xu [1,*] and Jianlong Li [2,*]

[1] College of Life Sciences, Yulin University, Yulin 719000, China; zhangyanzhen@yulinu.edu.cn (Y.Z.); byj212@123.com (Y.B.)
[2] Department of Ecology, School of Life Science, Nanjing University, Nanjing 210023, China
[3] State Key Laboratory of Plateau Ecology and Agriculture, Qinghai University, Xining 810016, China; wangzhaoqi_818@163.com
[4] School of Environment and Planning, Liaocheng University, Liaocheng 252059, China; qianwang@lcu.edu.cn
[5] Nanjing Institutes of Environmental Sciences, Ministry of Environmental Protection of the People's Republic of China, Nanjing 210042, China; yangyue@nies.org
* Correspondence: xuweizhou@yulinu.edu.cn (W.X.); lijianlongnju@163.com (J.L.)

Citation: Zhang, Y.; Wang, Z.; Wang, Q.; Yang, Y.; Bo, Y.; Xu, W.; Li, J. Comparative Assessment of Grassland Dynamic and Its Response to Drought Based on Multi-Index in the Mongolian Plateau. *Plants* **2022**, *11*, 310. https://doi.org/10.3390/plants11030310

Academic Editors: Bingcheng Xu and Zhongming Wen

Received: 20 October 2021
Accepted: 21 December 2021
Published: 25 January 2022

Publisher's Note: MDPI stays neutral with regard to jurisdictional claims in published maps and institutional affiliations.

Abstract: This study applied grassland related multi-index and assessed the effects of climate change by investigating grassland responses to drought. This process was performed to study grassland vegetation dynamic accurately and evaluate the effect of drought in the Mongolian Plateau (MP). The spatial–temporal characteristics of grassland dynamic in terms of coverage (F_v), surface bareness (F_b), and net primary production (NPP) from 2000 to 2013 were explored. We implemented the maximum Pearson correlation to analyze the grassland vegetation in response to drought by using self-calibrating Palmer Drought Severity Index (scPDSI). Results show that F_v and NPP present an increasing trend (0.18 vs. 0.43). F_b showed a decreasing trend with a value of −0.16. The grassland F_v and NPP positively correlated with scPDSI, with a value of 0.12 and 0.85, respectively, and F_b was −0.08. The positive correlation between F_v and NPP accounted for 84.08%, and the positive correlation between F_v and scPDSI accounted for 93.88%. On the contrary, the area with a negative correlation between F_b and scPDSI was 57.43%. The grassland in the MP showed a recovery tendency. The increase in grassland caused by positive reaction was mainly distributed in the middle of Mongolia (MG), whereas that caused by counter response was mainly distributed in the east and west MG and northeast Inner Mongolia autonomous region of China (IM). The relevant results may provide useful information for policymakers about mitigation strategies against the inverse effects of drought on grassland and help to ease the losses caused by drought.

Keywords: the Mongolia Plateau; grassland degradation; vegetation coverage; surface bareness degree; self-calibrating Palmer Drought Severity Index (scPDSI)

1. Introduction

As the earth's largest terrestrial ecosystem, grassland plays an important role in ecosystem cycles [1–3]. Evaluating the dynamic change in grassland ecosystem quantitatively is urgent because grassland provides many economic products and ecological services [4,5]. Previous research investigated the impact of climate change on the grassland vegetation dynamic by using different indicators. The indicators to evaluate grassland vegetation dynamic by remote sensing technology mainly include the normalized difference vegetation index (NDVI), enhanced vegetation index (EVI), vegetation coverage (F_v), and net primary productivity (NPP) [6–8]. Some recent studies have proposed ground bareness (F_b) as another important parameter of global land cover change [9,10]. As an opposite concept of F_v, F_b contains the attribute of surface reflectivity and temperature information of grassland vegetation rather than a complementary set of coverage. In recent years, research has focused on drought events by using the combination of identified NPP and NDVI [11]. Compared with single index analysis, the vegetation dynamic inversion based on multi-index can help to improve the reliability of results due to the diversities of analysis.

Drought is a natural phenomenon where the availability of water is significantly lower than normal for a long period and the supply cannot meet the existing demand [12]. With global warming, drought is quickly becoming a devastating environment incident [13]. The International Disaster Database estimated that droughts attributed approximately 5% of the natural disasters over the globe, and the losses caused by drought disasters accounted for more than 30% of those of the natural hazards [14]. Drought has been a crucial scientific issue in the domain of climate research due to its negative effects on water resources, livestock husbandry development, and local economy [15–17]. The influence of drought on terrestrial ecosystems is becoming increasingly acute [18]. The grassland dynamics and its response to driving factors are always investigated by researchers because grassland is more susceptible to droughts than other ecosystems [19]. Previous studies explored the impact of drought on grassland vegetation dynamic at multiple regions. Some researchers evaluated the NPP distribution and response to drought in Europe [20]. Their results suggest that rainfall deficit and extreme summer heat reduce the vegetation productivity in Eastern and Western Europe, respectively. Another study strengthened the conclusion of drought-induced reduction in NPP over the past decade in central Asia [21]. Therefore, a better understanding in grassland vegetation dynamic and its feedback on climate change will improve the local economic development, especially for the typical farming and pastoral areas.

The Mongolian Plateau (MP) is a typical arid and semiarid area, with natural grassland as the dominant vegetation type. It often suffered from different conditions of drought due to the decreasing water resource supply and climate change [22,23]. Droughts over the last century induced a heap of negative effects, such as water resource shortages, threat of food shortages, and vegetation degradation [24–27]. Therefore, quantitative assessment of grassland vegetation dynamic and the effect of droughts is urgent. In accordance with the recent analysis, the summer drought has contributed to the increasing extreme droughts since the 1990s [28]. Some researchers have proven that the self-calibrating Palmer Drought Severity Index (scPDSI) is suitable than other drought indexes when considering the impact of precipitation and temperature on the soil moisture in Inner Asia [29]. Another research from Wang revealed that the global grassland scPDSI value has a slightly increasing trend with a rate of 0.0119 per year [30]. However, there is still a lack of research on grassland vegetation dynamic and its response to droughts of the MP.

There are many related previous studies focused on single or two vegetation indexes to evaluate the grassland dynamic and its response to climate factors [31–33]. To enrich vegetation related research indicators, we selected F_v, F_b, and NPP to reflect the grassland vegetation dynamic for improving the reliability of conclusions. We evaluated the grassland response to droughts during the study period. A combined analysis of the three indexes in different drought severity areas was quantitatively assessed to enhance the credibility of the results. The results may provide a scientific basis for guiding ecological environment improvement and drought prevention for typical farming and pastoral areas in the world.

2. Results

2.1. Spatial and Temporal Distribution of F_v, F_b, and NPP

The spatial distribution of long-term mean grass F_v, F_b, and NPP in the Mongolia Plateau is shown in Figure 1. The grass F_v value is relatively higher in northern and northeastern MP, while lower in southwestern and western MP (Figure 1A). On the contrary, F_b greater than 60% distributed over the southwestern and western MP, while F_b less than 40% mainly distributed over the northeastern and northern MP (Figure 1B). The mean actual NPP showed obvious spatial heterogeneity, too (Figure 1C). Areas with mean actual NPP larger than 200 g C/(m^2·yr) were scattered in the northern and northeastern MP with good vegetation growth conditions. Areas with mean actual NPP lower than 100 g C/(m^2·yr) were mainly scattered in the regions with relatively scarce water resources and vegetation in the transition area of grassland and desert such as southwestern and western MP. We counted the different pixel values of grassland F_v (Figure 1a), F_b (Figure 1b), and NPP (Figure 1c) in the MP. The average F_v, F_b, and NPP values were 18.42%, 15.53%, and 61.41 g C/(m^2·yr), respectively, whereas the corresponding distribution rates of their peak value were 60–80%, 40–60%, and 150–200 g C/(m^2·yr). The F_v, F_b, and NPP of IM were 9.17%, 6.84%, and 24.82 g C/(m^2·yr), respectively. MG had higher values of the three indexes than IM. The corresponding distribution rates of IM and MG peak values were similar to the MP.

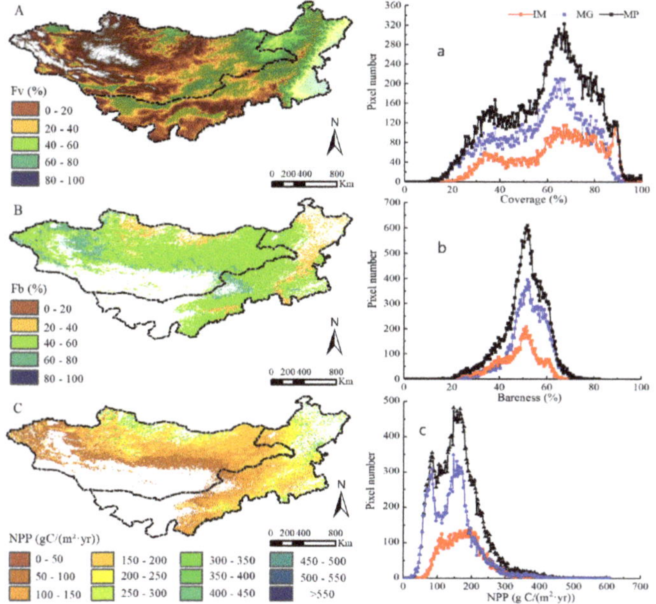

Figure 1. Spatial distribution of grassland F_v, F_b and NPP and statistics of the corresponding pixel number during 2000–2013. ((**A–C**) are the spatial distribution of F_v, F_b and NPP, respectively; (**a–c**) is the statistics of the corresponding pixel number).

In this study, grassland F_v in the MP exhibited an increasing trend from 2000 to 2013, with a 14-year cumulative increment of 0.18 (Figure 2). MG had a higher F_v value than IM (0.21 vs. 0.09). On the contrary, grassland F_b showed an overall decreasing trend in the MP, with the decreased rate of −0.08. The decrease rates of MG and IM were −0.09 and −0.05, respectively. NPP had the largest change rate compared with the two other indexes, with a 14-year cumulative increment of 0.43. MG had a higher increase value than IM (0.63 vs. 0.39).

2.2. Dynamic Analysis of Grassland

The changing trend and significance levels of grassland F_v, F_b, and NPP in the MP from 2000 to 2013 are shown in Figure 3. The growth rate of F_v occupied 60.51% of the MP grassland, mainly found in the east and central MG, east Xing'an, south Ordos, and central IM (Figure 3A). On the contrary, the regions of F_b exhibiting decreasing trends were extremely larger than that with increasing trends (92.64% vs. 7.36%), with the decreased rate of −0.0005/14a. The decreased regions were mainly found in the entire MP, typically occurring in the southwest and middle MP (Figure 3B). The NPP increasing areas occupied 79.54% of the MP grassland, mainly found in Kent Mountains and Hanggai Mountains in

MG and east Xing'an League in IM (Figure 3C). F_v with clear increases was distributed in the east Dornod, Hangai Mountains, and Kent Mountains in MG, and east Xing'an and south Ordos in IM (Figure 3a). F_b exhibited a significant decrease (SD) and an extremely significant decrease (ESD), accounting for 14.77% and 11.61% of the MP grassland, respectively (Figure 3b). The regions of NPP with a significant increase (SI) accounted for 4.27% of the MP grassland. The regions with significant increase were mainly distributed in the east Selenge in MG and east Xing'an in IM (Figure 3c).

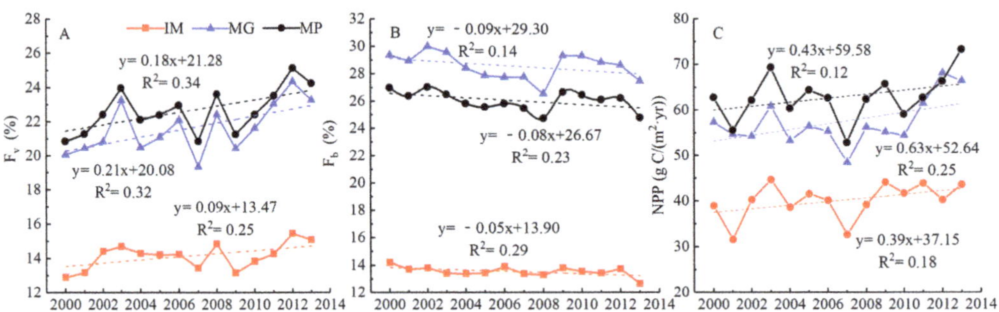

Figure 2. The statistics of the temporal distribution of grassland F_v, F_b, and NPP. ((**A–C**) are the statistics of F_v, F_b, and NPP during 2000–2013, respectively).

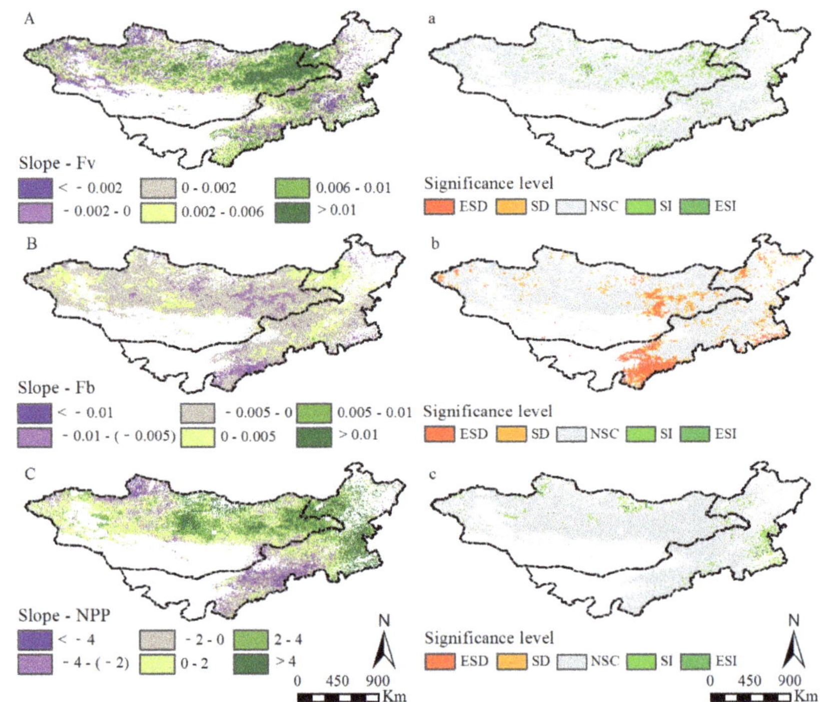

Figure 3. The changing trend and significance levels of grassland F_v, F_b and NPP. ((**A–C**) are the changing trend of grassland F_v, F_b and NPP, respectively. (**a–c**) are the corresponding significance levels. ESD (extremely significant decrease, slope < 0, $p < 0.01$); SD (significant decrease, slope < 0, $0.01 < p < 0.05$); NSC (no significant change, slope = 0, $p > 0.05$); SI (Significant Increase, Slope > 0, $0.01 < p < 0.05$); ESI (extremely significant increase, slope > 0, $p < 0.01$)).

2.3. Correlation Analysis of Grassland Indexes to scPDSI

The correlation coefficient of grassland indexes and scPDSI was analyzed because grassland dynamic is driven by global climate change (Figure 4). F_v and NPP were positively correlated with scPDSI, with a value of 0.12 and 0.85, respectively, whereas F_b accounted for -0.08. The areas with a positive correlation between F_v, NPP, and scPDSI were approximately 84.08 and 93.88%. On the contrary, a negative correlation between F_b and scPDSI accounted for 57.43%. The grassland regions (2.02%) showed a significant positive correlation ($p < 0.05$) between F_v and PSDI, mainly distributed in Baotou, Hohhot, and south Ulaan Chab in IM. The grassland areas (8.28%) showed significant negative correlation ($p < 0.05$), mainly distributed in the Kent mountain area of MG and Tong Liao, Chi Feng, and Xilin Gol of IM. In the regions with a positive correlation between NPP and PSDI, 19.57% of them showed a significant positive correlation ($p < 0.05$), mainly distributed over the west, north, and central MG, and central IM.

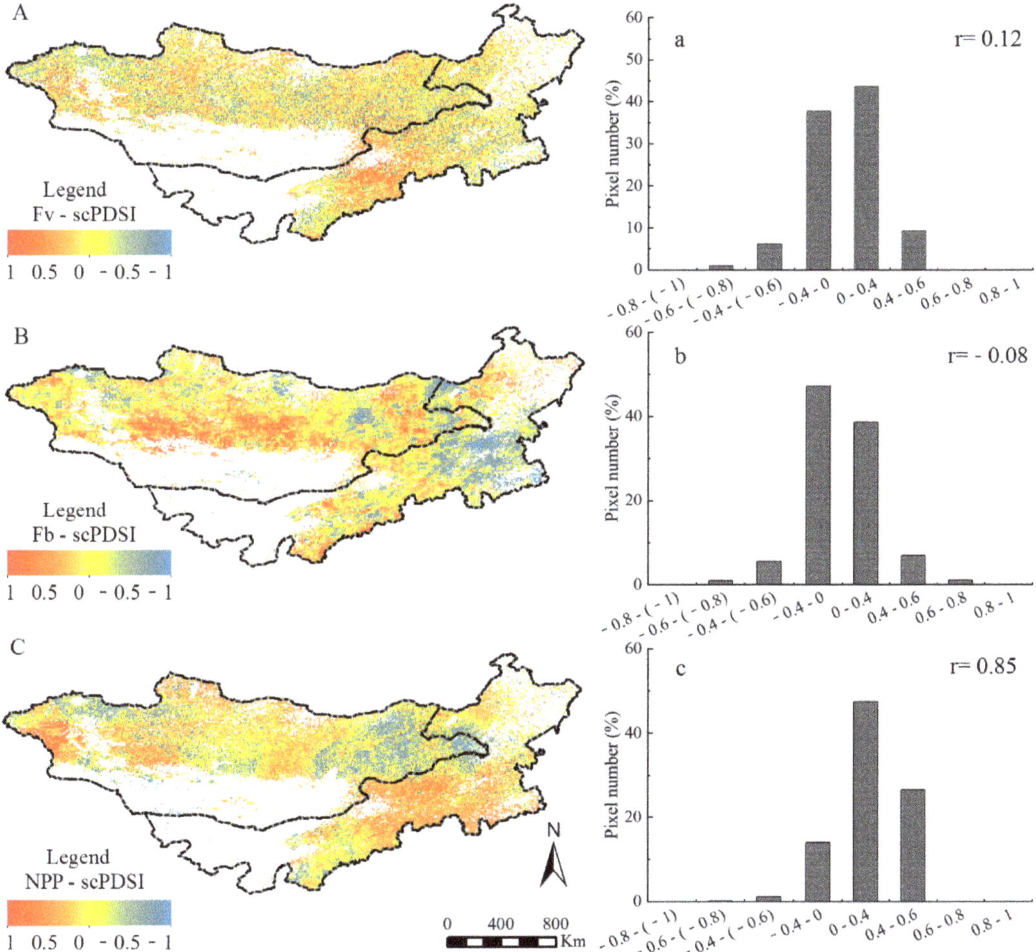

Figure 4. The spatial distribution and pixel frequency of correlation coefficient between scPDSI and three vegetation indicators ((**A–C**) are the correlation coefficient spatial distribution and (**a–c**) corresponding its pixel frequency).

2.4. Changes and Trends in Grassland Response to Drought

The area of grassland F_v, F_b, and NPP responding to scPDSI in the control response for grassland increase accounts for 36.55, 37.99, and 28.23% of the total area, respectively (Figure 5A–C). The control response to grassland increase from F_v, F_b, and NPP to scPDSI appear in similar areas, mainly concentrating on central, north, and west MG, and west IM. On the contrary, the control response to grassland decrease accounts for 10.09, 17.10, and 14.98% (Figure 5a–c). It is mainly concentrated on the south Sayan Mountains, south Hangai Mountains, and Dornod in MG, and northeast and south IM. The area of grassland F_v, F_b, and NPP responding to scPDSI in the counter response for grassland increase accounts for 44.73, 37.76, and 40.34% of the total area, respectively (Figure 6A–C). The counter response to grassland increase from F_v, F_b, and NPP to scPDSI appear in similar areas, mainly concentrating on the northeast and west MG and northeast and south IM. On the contrary, the counter response to grassland decreases accounts for 8.63, 7.15, and 16.45% (Figure 6a–c). It is mainly concentrated on central MG and west IM.

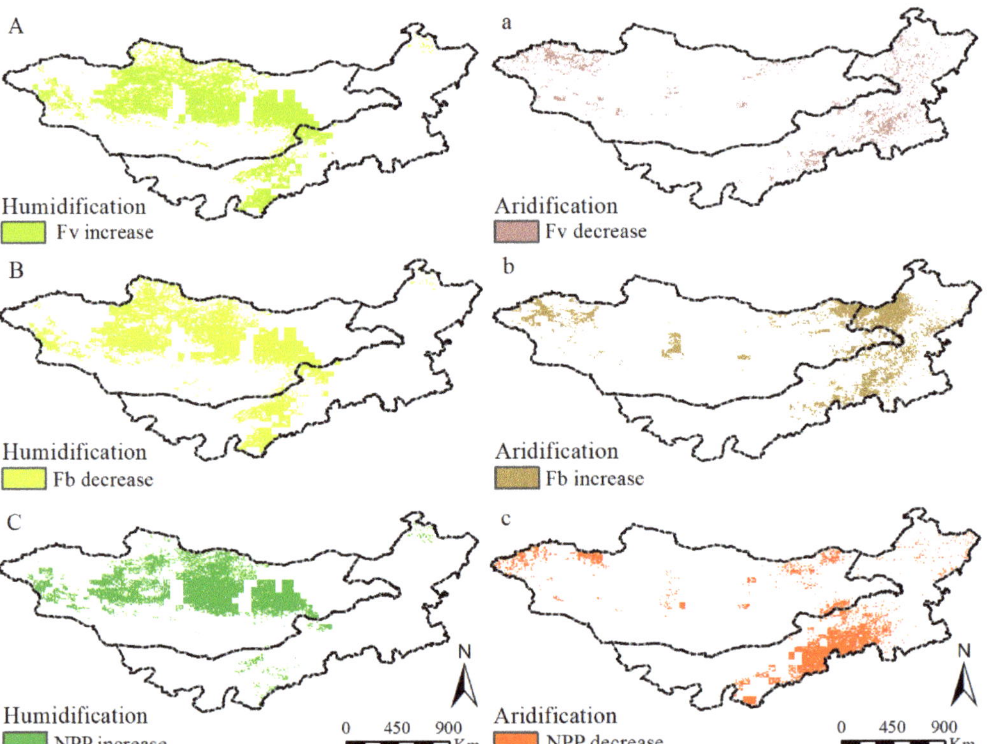

Figure 5. The control response of F_v, F_b, and NPP to PDSI changes in the Mongolian plateau from 2000 to 2013. ((**A–C**) are the correlation coefficient spatial distribution and (**a–c**) corresponding its pixel frequency).

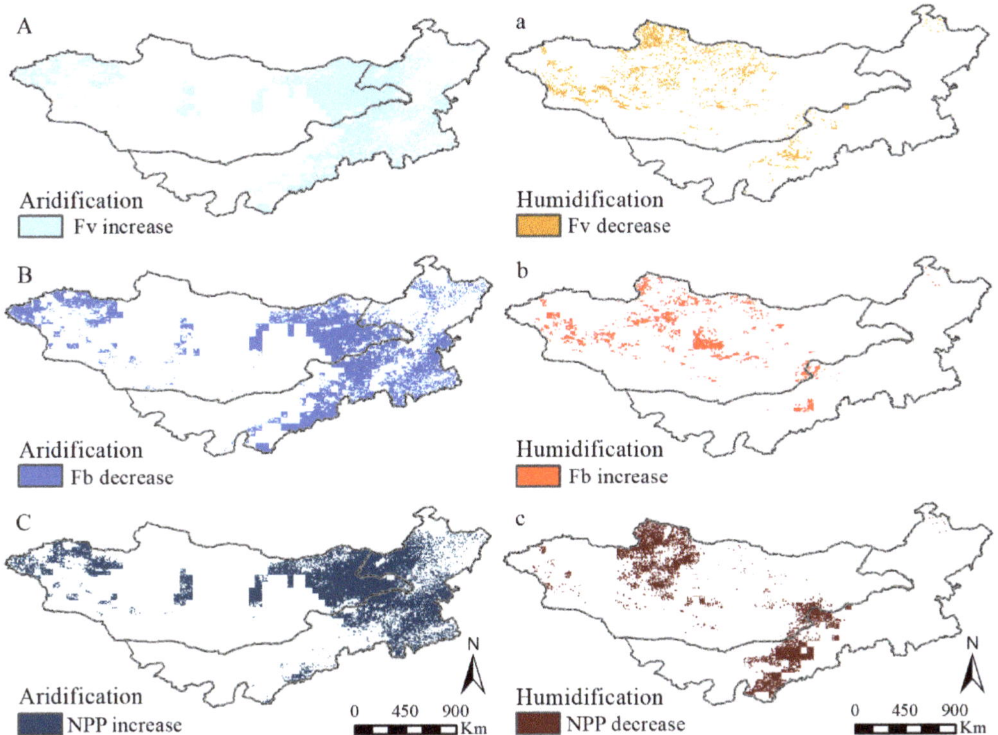

Figure 6. The counter response of F_v, F_b, and NPP to PDSI changes in the Mongolian plateau from 2000 to 2013. ((**A–C**) are the correlation coefficient spatial distribution and (**a–c**) corresponding its pixel frequency).

3. Discussion

3.1. Methodology

The current study used the slope-combined analysis based on multi-index to simulate grassland vegetation dynamic and monitor grassland response to droughts. The hypothesis is that grassland F_v and NPP dynamic are a positive feedback, whereas F_b is on the contrary. Previous studies applied single index, such as NDVI, F_v, and NPP, to simulate the grassland dynamic. However, many uncertainties remain due to the inversion model or the uncertainty of dataset itself [34]. The advantage of the current method is the reference of F_b index. Our findings show that 12.93% of the grassland in the MP experiences an increasing trend compared with 0.73% of the grassland that experienced a decreasing trend during the study period. Several studies about grassland NPP showed that grassland has an increasing trend in the similar area during the study period [35,36]. Similarly, studies on vegetation indexes, such as NDVI, F_v, and EVI, show an increasing trend of grassland vegetation [37–39]. Thus, the present studies confirmed that the grassland shows a recovery trend in the MP, which agrees with our findings.

3.2. Climate Factors on Grassland Vegetation Dynamic

In this study, we assessed the grassland dynamic on the basis of F_v, F_b, and NPP and their impact on droughts during 2001 to 2013. The results provided a new understanding of drought-driven grassland change in the MP. Climate variations, such as temperature and precipitation, influenced terrestrial vegetation directly. These climate factors regulated soil respiration, photosynthesis, growth status, and distribution [40]. Here, we calculated the temporal trends of temperature, precipitation, and radiation during the study period (Figure 7). The temperature in this study showed a downward trend (-0.03 °C), whereas precipitation and radiation showed an increasing trend (2.02 mm and 3.39 MJ/m^2) due to the short study period. Evidence shows that global warming is definitely occurring, and the climate in our study area tended to be wet and warm [41]. Typically, the combination of warmer temperature and higher precipitation concentration during the early growing season possibly increased NPP, partly by lengthening the growing season [11]. The related research shows that the carbon sequestration capacity of grassland ecosystem is enhanced by increased precipitation, which supports our findings [42,43]. Another study revealed that global warming helps to increase the productivity and carbon storage of grasslands in China [34].

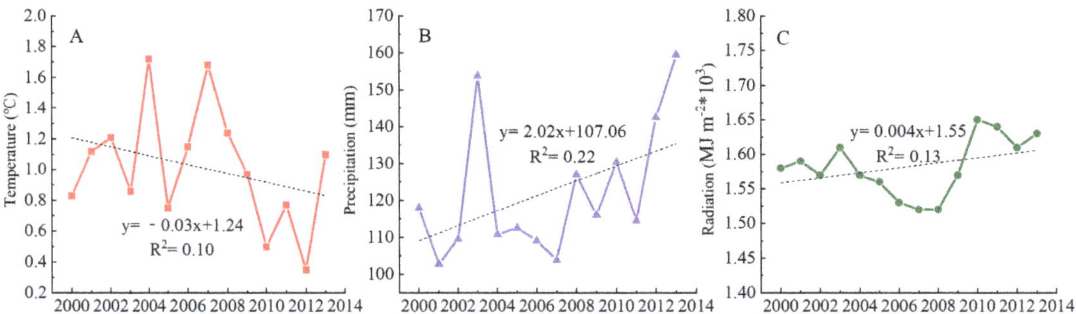

Figure 7. Temporal variations of meteorological variables during 2000–2013 ((**A–C**) are the annual mean temperature, annual cumulative precipitation, and annual cumulative solar radiation, respectively).

3.3. The Role of Ecological Policies in Grassland Restoration

As a limited resource, water is necessary for plant growth and development, especially in arid and semi-arid ecosystems [44]. Evidence showed that grasslands experience different degrees of drought in the MP (Figure 8A-1). Although drought associates with decreased precipitation, increased precipitation does not necessarily weaken the drought [45,46]. A slight reduction of drought is observed in the MP (-0.02), mainly concentrating in the western and eastern MG (Figure 8A-2). This finding is consistent with other studies that used SPI and SPEI to show drought [47–49]. We fitted the response from grassland F_v, F_b, and NPP to scPDSI, and the results showed a recovery trend (Figure 8B-1,B-2). Grassland increased regions are obviously larger than the decrease regions (12.93% vs. 0.73%), strongly confirming the recovery of grassland in the MP. The grassland increase regions with control response to drought mainly distributed in the central MP. Few human activities were found in these areas, and the vegetation growth was mainly affected by natural climatic factors [50]. The grassland increase regions with counter response to drought were in the eastern and western MG and northeast IM. This finding shows that other factors, such as human activities affecting local grassland restoration, are greater than the climate factors [51]. The distribution of grassland decrease regions with control response and counter response is minimal (0.44% vs. 0.29%). The grassland decrease regions with control response mainly distributed in south central IM and north MG. This finding reveals the grassland degradation in these areas are under complex influence factors, such as increase pressures from people and livestock populations [52].

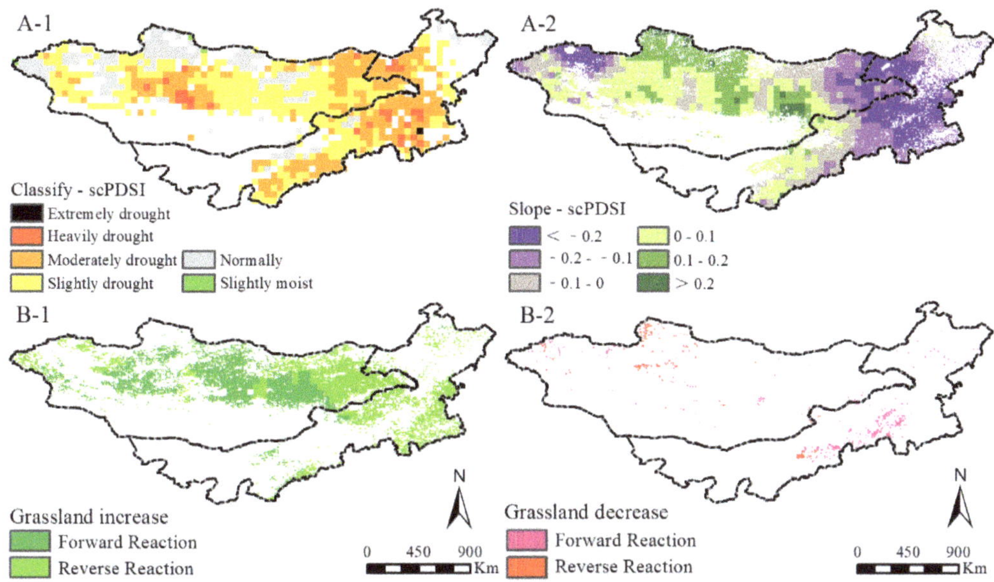

Figure 8. The spatial distribution (**A-1**) and changing trend (**A-2**) of scPDSI and the reaction of F_v, F_b, and NPP to scPDSI changes (**B-1,B-2**) in the MP from 2000 to 2013.

4. Materials and Methods

4.1. Study Area

The MP is located in Siberia in the north to the northern Yinshan in the south, from the Outer Xing'an Mountains in the east, to the Altai Mountains in the west (37°22′–53°20′ N and 87°43′–126°04′ E, Figure 9). The Altai Mountains in the southwest, the Kent Mountains in the north, and the outreach area of the Xing'an Mountains in the east are found in the study area, with the mean elevation of 1580 m (Figure 9a). The MP mainly includes Mongolia and Inner Mongolia autonomous region of China, with a total area of approximately 1.56×10^6 and 1.18×10^6 km^2, respectively. The MP climate is dry and lacks precipitation due to its long distance from the ocean. The annual average temperature and mean annual rainfall is 4.12 °C and 269 mm, respectively. The MP has a wide variety of regional climates, and most of them are from arid to humid from west to east. All regions are sensitive and vulnerable to drought. The grassland types mainly include grasslands, woody savannas, savannas, and shrub lands in descending order (Figure 9b).

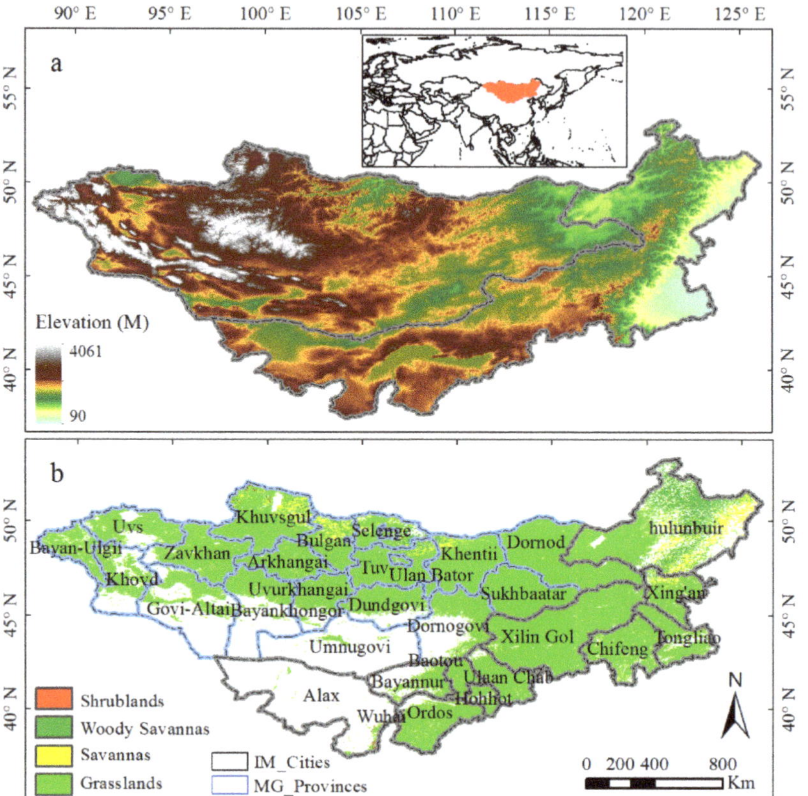

Figure 9. Elevation of the MP (**a**) and grassland types derived from MODIS land cover product (MCD12Q1, IGBP) by the year of 2013 (**b**).

4.2. Data Source and Processing

We obtained the global land cover maps of 2013 from the MODIS data products (MCD12Q1, http://modis-land.gsfc.nasa.gov/landcover.html/, accessed on 24 February 2020) with the International Geosphere-biosphere Programme (IGBP) classification system. The IGBP classification system defines 17 classes of primary land cover types. In this study, classes 1 to 5 were reclassified as forest, classes 6 to 9 were reclassified as shrubland, classes 12 and 13 were reclassified as farmland, classes 15–16 were reclassified as water, and class 17 was reclassified as city (Table 1).

The 0.05 degree monthly NDVI (normalized difference vegetation index) was offered from the Moderate Resolution Imaging Spectroradiometer (MODIS) data products MOD13C2 (http://ladsweb.nascom.nasa.gov/data/search.html, accessed on 4 March 2020). The 0.05 degree monthly NDII (normalized difference impervious index) was calculated by using a red band 1 from MOD13C2 and thermal infrared band 32 from MOD11C3. Both MOD13C2 and MOD11C3 image datasets were converted to Albers equal area conical projection and WGS-84 datum using the ArcGIS V9.3 software (ESRI, San Diego, CA, USA). To reduce the image noise from the atmospheric clouds, particles, shadows, etc., three-point smoothing was used to improve data quality.

Table 1. The reclassification of land use type according to the IGBP classification system.

Original Serial Number	Original Land Use Type	Serial Number after Reclassification	Land Use Type after Reclassification
1	Evergreen Needleleaf Forest		
2	Evergreen Broadleaf Forest		
3	Deciduous Needleleaf Forest	1	Forest
4	Deciduous Broadleaf Forest		
5	Mixed Forest		
6	Closed Shrublands		
7	Open Shrublands		
8	Woody Savannas	2	Shrubland
9	Savannas		
10	Grasslands	3	Grassland
11	Permanent Wetlands	4	Wetland
12	Croplands		
13	Cropland/Natural Vegetation Mosaic	5	Farmland
14	Barren or Sparsely Vegetated	6	Desert
15	Snow and Ice		
16	Water Bodies	7	Water
17	City and Built-up	8	City

We obtained the monthly meteorological data from the gridded datasets of Climatic Research Unit (CRU) TS 3.22 (http://crudata.uea.ac.uk/cru/data, accessed on 6 January 2019). These gridded datasets cover the global land surface (excluding Antarctica) at a 0.5° resolution and provide the best estimates for month-by-month variations in climate variables [53]. No measurement value is missing in the datasets. The scPDSI datasets were provided by the CRU. The scPDSI uses −0.99–0.99 as normal, and negative values indicate drought. Classification relevant to this research mainly includes extreme moist, heavy moist, moderate moist, slightly moist, slightly normal drought, moderate drought, heavy drought, and extreme drought (Table 2). In order to facilitate spatial statistics, meteorological data are resampled by ArcGIS 10.2 software with a resolution of 0.05 degree.

Table 2. Classification relevant of self-calibrating Palmer drought severity index (scPDSI).

scPDSI	Classification of Dry and Wet	scPDSI	Classification of Dry and Wet
≥4	Extreme moist	−1~−2	Slightly drought
3~4	Heavy moist	−2~−3	Moderate drought
2~3	Moderate moist	−3~−4	Heavy drought
1~2	Slightly moist	≤−4	Extreme drought
−1~1	Normally	-	-

4.3. Methods

4.3.1. Estimation of F_v

F_v is an index directly used to determine grassland health condition. We estimated the grassland coverage by using the *NDVI* data due to the significant linear correlation relationship between grassland coverage and *NDVI*. The calculated model is pixel dichotomy model. The specific calculation formula is as follows:

$$F_v = \frac{NDVI - NDVI_{\min}}{NDVI_{\max} - NDVI_{\min}} \times 100\% \tag{1}$$

where F_v is the grassland coverage (%), *NDVI* is the *NDVI* value of a single pixel, $NDVI_{\min}$ is the NDVI value of bare soil or areas without vegetation coverage, and $NDVI_{\max}$ is the NDVI value of pixels completely covered by vegetation. Theoretically, the $NDVI_{\min}$ value should be close to 0, and the $NDVI_{\max}$ value represents the maximum value of *NDVI* per unit pixel of total vegetation coverage. However, considering the influence of vegetation type, noise, terrain, image quality, and other factors, the $NDVI_{min}$ and $NDVI_{max}$ values will deviate from the actual values, which are generally represented by the maximum and minimum values within a certain confidence range. In this paper, $NDVI$ values near 2% and 98% of the cumulative percentage of $NDVI$ values in remote sensing images in the study area are selected as $NDVI_{min}$ and $NDVI_{max}$ values.

4.3.2. Estimation of F_b

F_b is a concept corresponding to F_v. It includes the reflectance and temperature characteristics of the object surface and is not a supplement to F_v. F_b enriches the research index of grassland ecosystem on the basis of remote sensing technology. In this study, we chose Wang's estimation formula for F_b [54], which is expressed as follows:

$$F_b = \frac{NDII - NDII_{min}}{NDII_{max} - NDII_{min}} \times 100\% \tag{2}$$

$$NDII = \frac{\lambda_R - \lambda_T}{\lambda_R + \lambda_T} \tag{3}$$

where F_b delineates the surface bareness fractions, $NDII$ is the normalized difference impervious index, $NDII_{min}$ refers to the minimum value of $NDII$ (high grassland coverage and low-temperature pixel), and $NDII_{max}$ represents the maximum value of $NDII$ (high temperature and reflectivity). λ_R is red band 1 from MOD13C2, and λ_T refers to thermal infrared band 32 from MOD11C3.

4.3.3. Estimation of NPP

NPP was extracted from the dataset at a spatial resolution of 1 km and calculated on the basis of the BIOME-BGC model [55]. The specific formula is as follows:

$$NPP = \sum_{t}^{365} PSNet - (R_m + R_g) \tag{4}$$

$$PSNet = GPP - R_{lr} \tag{5}$$

where NPP represents the actual NPP (g cm^{-2} year^{-1}), and $PSNet$ refers to the net photosynthesis. R_m is the annual maintenance respiration of live cells in woody tissue, and R_g delineates the annual growth respiration. GPP is the gross primary productivity from MOD17A2 datasets, and R_{lr} is the daily leaf and fine root maintenance respiration.

4.3.4. Grassland Dynamic Analysis

The grassland vegetation dynamic analysis is a significant ecological process of grassland health condition. We can assess grassland degradation or restoration by using F_v, F_b, and NPP as fundamental indicators. The slope was determined by using ordinary least squares regression, which is expressed as follows:

$$Slope_A = \frac{n \times \sum_{i=1}^{n} i \times (A)_i - (\sum_{i=1}^{n} i)(\sum_{i=1}^{n} (A)_i)}{n \times \sum_{i=1}^{n} i^2 - (\sum_{i=1}^{n} i)^2} \tag{6}$$

where A refers to grassland F_v, F_b, and NPP; i is the sequence number of the year (in this study, 1 is for the year 2000, 2 is for the year 2001, and so on); n represents the number of years, which is 14 in this study. A negative slope value shows a degradation trend, whereas a positive slope value shows a restoration trend. In this study, combined analysis of slopes (Table 3) was conducted to quantitatively evaluate the grassland response to drought.

Table 3. Scenarios to assess the role of F_v, F_b and NPP responding to scPDSI in the MP.

Change Direction	Grassland Conditions	scPDSI	F_v	F_b	NPP
control response	Grassland Restoration	Slope > 0	Slope > 0		
		Slope > 0		Slope < 0	
		Slope > 0			Slope > 0
	Grassland degradation	Slope < 0	Slope < 0		
		Slope < 0		Slope > 0	
		Slope < 0			Slope < 0
counter response	Grassland Restoration	Slope < 0	Slope > 0		
		Slope < 0		Slope < 0	
		Slope < 0			Slope > 0
	Grassland degradation	Slope > 0	Slope < 0		
		Slope > 0		Slope > 0	
		Slope > 0			Slope < 0

The significance of the variation tendency was determined in terms of F-test to represent the confidence level of variation. The calculation for statistics is expressed as follows:

$$F = U \times \frac{n-2}{Q} \tag{7}$$

$$U = \sum_{i=1}^{n} (\hat{y}_i - \bar{y})^2 \tag{8}$$

$$Q = \sum_{i=1}^{n} (y_i - \hat{y}_i)^2 \tag{9}$$

$$\hat{y}_i = Slope \times i + b \tag{10}$$

$$b = \bar{y} - Slope \times \bar{i} \tag{11}$$

where U represents the residual sum of the squares; Q is the regression sum; $\hat{y}_i$ refers to the regression value; y_i delineates the average data of year i; $\bar{y}$ is the mean value of F_v, F_b or NPP over n years; b refers to the intercept of the regression formula.

We classified the variation tendency into the following six levels on the basis of the F-test results: extremely significant decrease (ESD, slope < 0, $p < 0.01$); significant decrease (SD, slope < 0, $0.01 < p < 0.05$); no significant change (NSC, slope = 0, $p > 0.05$); Significant Increase (SI, Slope > 0, $0.01 < p < 0.05$); extremely significant increase (ESI, slope > 0, $p < 0.01$).

4.3.5. Correlation Analysis

The Pearson correlation coefficient was used to reflect the long-term dynamic of two variables in a given time n. The specific calculation formula is as follows:

$$r = \frac{n \times \sum_{i=1}^{n} (x_i \times y_i) - (\sum_{i=1}^{n} x_i)(\sum_{i=1}^{n} y_i)}{\sqrt{n \times (\sum_{i=1}^{n} x_i^2) - (\sum_{i=1}^{n} x_i)^2} \sqrt{n \times (\sum_{i=1}^{n} y_i^2) - (\sum_{i=1}^{n} y_i)^2}} \tag{12}$$

where r is the correlation coefficient, n refers to the sequential year, which is 14 in this study; x_i and y_i represent F_v, F_b or NPP and climatic factors, respectively. r refers to a description of linear correlation degrees between the two variables. The value of r ranges from -1 to 1. -1 and 1 are completely related, whereas 0 indicates irrelevant. The greater the absolute value of r, the stronger the correlation, but no causal relationship is found.

5. Conclusions

This study investigated the grassland vegetation dynamic on the basis of multi-index and its response to droughts in the MP during 2000–2013 in terms of the relations between F_v, F_b, NPP, and scPDSI. The spatial distribution of grassland F_v and NPP decreases from northeast to southwest, showing an increasing trend of 0.18 and 0.43, respectively. On the contrary, F_b increases from northeast to southwest, presenting a decreasing trend, with the value of -0.16. The grassland degradation condition of the MP shows a restoration trend during the study period. F_v and NPP shows a positive relationship with scPDSI, whereas F_b is exactly on the country. The areas with a positive correlation between F_v, NPP, and scPDSI are 84.08% and 93.88%. The grassland increase regions with control and counter response to drought account for 6.06% and 6.87%. However, the distribution of grassland decrease regions with control and counter response (0.44% vs. 0.29%) is minimal. The regions of grassland increase from control response mainly distribute in central MG, whereas the grassland increase regions with counter response are in the eastern and western MG and northeast IM. Such detailed analysis of grassland-related indexes and its responses to drought are useful to clarify the grassland condition, potential effect of drought, and is beneficial to help policymakers for develop proper measures for grassland protection.

Author Contributions: Conceptualization, Y.Z. and J.L.; Methodology, Z.W.; Software, Q.W.; Validation, Y.Z., Y.Y. and Z.W.; Formal Analysis, J.L. and Y.B.; Data Curation, Y.Z. and W.X.; Writing—Original Draft Preparation, Y.Z.; Visualization, Y.Z.; Supervision, J.L. and W.X.; Project Administration, J.L. and W.X. All authors have read and agreed to the published version of the manuscript.

Funding: This research was funded by the National Natural Science Foundation of China (42067069), the Technology Innovation Leading Program of Shaanxi Province (2021KJXX-53), the National Key Research and Development project (2018YFD0800201), the Natural Science Foundation of China (32060279) and the National Research Project of the State Ethnic Affairs Commission of China (2019-GMD-034). We are grateful to the editor and anonymous reviewers. We also appreciate the Oak Ridge National Laboratory Distributed Active Archive Center (ORNL DAAC) for sharing a series of original remote sensing dataset and Climatic Research Unit in University of East Anglia for sharing climate dataset.

Institutional Review Board Statement: Not applicable.

Informed Consent Statement: Not applicable.

Data Availability Statement: Not applicable.

Conflicts of Interest: The authors declare no conflict of interest.

References

1. Jiang, L.P.; Qin, Z.H.; Xie, W. Estimation of Grassland Ecosystem Services Value of China Using Remote Sensing Data. *J. Nat. Resour.* **2007**, *22*, 161–170.
2. Wei, Z.; Han, Y.; Liang, Z.; Yizhao, C.; Lu, H.; Weimin, J. Dynamics of grassland carbon sequestration and its coupling relation with hydrothermal factor of Inner Mongolia. *Ecol. Indic.* **2018**, *95*, 1–11.
3. Yanagawa, A.; Sasaki, T.; Jamsran, U.; Okuro, T.; Takeuchi, K. Factors limiting vegetation recovery processes after cessation of cropping in a semiarid grassland in Mongolia. *J. Arid. Environ.* **2016**, *131*, 1–5. [CrossRef]
4. Gang, C.; Wang, Z.; Zhou, W.; Chen, Y.; Groisman, P.Y. Assessing the Spatiotemporal Dynamic of Global Grassland Water Use Efficiency in Response to Climate Change from 2000 to 2013. *J. Agron. Crop Sci.* **2016**, *202*, 343–354. [CrossRef]
5. Soussana, J.F.; Loiseau, P.; Vuichard, N.; Ceschia, E.; Balesdent, J.; Chevallier, T.; Arrouays, D. Carbon cycling and sequestration opportunities in temperate grasslands. *Soil Use Manag.* **2010**, *20*, 219–230. [CrossRef]
6. Luo, Z.; Wu, W.; Yu, X.; Song, Q.; Yang, J.; Wu, J.; Zhang, H. Variation of Net Primary Production and Its Correlation with Climate Change and Anthropogenic Activities over the Tibetan Plateau. *Remote Sens.* **2018**, *10*, 1352. [CrossRef]
7. Yang, H.; Mu, S.; Li, J. Effects of ecological restoration projects on land use and land cover change and its influences on territorial NPP in Xinjiang, China. *Catena* **2014**, *115*, 85–95. [CrossRef]
8. Zhang, R.; Liang, T.; Guo, J.; Xie, H.; Feng, Q.; Aimaiti, Y. Grassland dynamics in response to climate change and human activities in Xinjiang from 2000 to 2014. *Sci. Rep.* **2018**, *8*, 1–11. [CrossRef]
9. Ying, Q.; Hansen, M.C.; Potapov, P.V.; Tyukavina, A.; Wang, L.; Stehman, S.V.; Moore, R.; Hancher, M. Global bare ground gain from 2000 to 2012 using Landsat imagery. *Remote Sens. Environ.* **2017**, *194*, 161–176. [CrossRef]
10. Zhang, Y.; Zhang, C.B.; Wang, Z.Q.; An, R.; Li, J.L. Comprehensive Research on Remote Sensing Monitoring of Grassland Degradation: A Case Study in the Three-River Source Region, China. *Sustainability* **2019**, *11*, 1845. [CrossRef]
11. Nanzad, L.; Zhang, J.; Tuvdendorj, B.; Yang, S.; Rinzin, S.; Prodhan, F.A.; Sharma, T.P.P. Assessment of Drought Impact on Net Primary Productivity in the Terrestrial Ecosystems of Mongolia from 2003 to 2018. *Remote Sens.* **2021**, *13*, 2522. [CrossRef]

12. Vicente-Serrano, S.M.; Gouveia, C.; Camarero, J.J.; Begueria, S.; Trigo, R.; L6Pez-Moreno, J.I.; Azorin-Molina, C.; Pasho, E.; Lorenzo-Lacruz, J.; Revuelto, J. Response of vegetation to drought time-scales across global land biomes. *Proc. Natl. Acad. Sci. USA* **2013**, *110*, 52–57. [CrossRef]
13. Dai, A. Drought under Global Warming: A Review. *Wiley Interdiscip. Rev. Clim. Chang.* **2011**, *2*, 45–65. [CrossRef]
14. Li, C.; Filho, W.L.; Yin, J.; Hu, R.; Wang, J.; Yang, C.; Yin, S.; Bao, Y.; Ayal, D.Y. Assessing vegetation response to multi-time-scale drought across inner Mongolia plateau. *J. Clean. Prod.* **2018**, *179*, 210–216. [CrossRef]
15. Di, L.; Scanlon, B.R.; Longuevergne, L.; Sun, A.Y.; Fernando, D.N.; Save, H. GRACE satellite monitoring of large depletion in water storage in response to the 2011 drought in Texas. *Geophys. Res. Lett.* **2013**, *40*, 3395–3401.
16. Kazemzadeh, M.; Malekian, A. Spatial characteristics and temporal trends of meteorological and hydrological droughts in northwestern Iran. *Nat. Hazards* **2016**, *80*, 1–20. [CrossRef]
17. Vicente-Serrano, S.M.; López-Moreno, J.I.; Beguería, S.; Lorenzo-Lacruz, J.; Azorin-Molina, C.; Morán-Tejeda, E. Accurate Computation of a Streamflow Drought Index. *J. Hydrol. Eng.* **2012**, *17*, 318–332. [CrossRef]
18. Zhang, Q.; Kong, D.; Singh, V.P.; Shi, P. Response of vegetation to different time-scales drought across China: Spatiotemporal patterns, causes and implications. *Glob. Planet. Chang.* **2017**, *152*, 1–11. [CrossRef]
19. Li, L.Y.; Tian, M.R.; Liang, H.; Chen, Y.M.; Qian, J.P. Spatial and Temporal Changes of Vegetation Coverage and Influencing Factors in Hulun Buir Grassland During 2000–2016. *J. Ecol. Rural. Environ.* **2018**, *34*, 584–591.
20. Ciais, P.; Reichstein, M.; Viovy, N.; Granier, A.; Ogee, J.; Allard, V.; Aubinet, M.; Buchmann, N.; Bernhofer, C.; Carrara, A.; et al. Europe-wide reduction in primary productivity caused by the heat and drought in 2003. *Nature* **2005**, *437*, 529–533. [CrossRef] [PubMed]
21. Zhao, M.; Running, S.W. Drought-Induced Reduction in Global Terrestrial Net Primary Production from 2000 through 2009. *Science* **2010**, *329*, 940–943. [CrossRef]
22. Bao, G.; Liu, Y.; Liu, N.; Linderholm, H.W. Drought variability in eastern Mongolian Plateau and its linkages to the large-scale climate forcing. *Clim. Dyn.* **2015**, *44*, 717–733. [CrossRef]
23. Yin, G.; Zengyun, H.U.; Chen, X.; Tashpolat, T. Vegetation dynamics and its response to climate change in Central Asia. *J. Arid. Land* **2016**, *8*, 375–388. [CrossRef]
24. Azimi, S.; Moghaddam, M.A. Modeling Short Term Rainfall Forecast Using Neural Networks, and Gaussian Process Classification Based on the SPI Drought Index. *Water Resour. Manag.* **2020**, *34*, 1369–1405. [CrossRef]
25. Karina, W.; Ursula, G.; Volker, H. Identifying Droughts Affecting Agriculture in Africa Based on Remote Sensing Time Series between 2000–2016: Rainfall Anomalies and Vegetation Condition in the Context of ENSO. *Remote Sens.* **2017**, *9*, 831.
26. Nobre, C.A.; Marengo, J.A.; Seluchi, M.E.; Cuartas, L.A.; Alves, L.M. Some Characteristics and Impacts of the Drought and Water Crisis in Southeastern Brazil during 2014 and 2015. *J. Water Resour. Prot.* **2016**, *8*, 252–262. [CrossRef]
27. Zhang, L.; Wu, P.; Zhou, T. Aerosol forcing of extreme summer drought over North China. *Environ. Res. Lett.* **2017**, *12*, 1–7. [CrossRef]
28. Mohammat, A.; Wang, X.; Xu, X.; Peng, L.; Yang, Y.; Zhang, X.; Myneni, R.B.; Piao, S. Drought and spring cooling induced recent decrease in vegetation growth in Inner Asia. *Agric. For. Meteorol.* **2013**, *178–179*, 21–30. [CrossRef]
29. Yi, L.; Ren, L.; Ma, M.; Yang, X.; Fei, Y.; Jiang, S. An insight into the Palmer drought mechanism based indices: Comprehensive comparison of their strengths and limitations. *Stoch. Environ. Res. Risk Assess.* **2016**, *30*, 119–136.
30. Wang, Q.; Yang, Y.; Liu, Y.Y.; Tong, L.J.; Zhang, Q.P.; Li, J.L. Assessing the Impacts of Drought on Grassland Net Primary Production at the Global Scale. *Sci. Rep.* **2019**, *9*, 1–8. [CrossRef]
31. Huang, L.; He, B.; Chen, A.; Wang, H.; Liu, J.; Lü, A.; Chen, Z. Drought dominates the interannual variability in global terrestrial net primary production by controlling semi-arid ecosystems. *Sci. Rep.* **2016**, *6*, 1–6.
32. Miao, B.; Li, Z.; Liang, C.; Wang, L.; Jia, C.; Bao, F.; Chao, J. Temporal and spatial heterogeneity of drought impact on vegetation growth on the Inner Mongolian Plateau. *Rangel. J.* **2018**, *40*, 113–128.
33. Xie, B.; Qin, Z.; Yang, W.; Chang, Q. Spatial and temporal variation in terrestrial net primary productivity on Chinese Loess Plateau and its influential factors. *Trans. Chin. Soc. Agric. Eng.* **2014**, *30*, 244–253.
34. Wang, Z.Q.; Chang, J.F.; Peng, S.S.; Piao, S.L.; Ciais, P.; Betts, R. Changes in productivity and carbon storage of grasslands in China under future global warming scenarios of 1.5 degrees C and 2 degrees C. *J. Plant Ecol.* **2019**, *12*, 804–814. [CrossRef]
35. Dukes, J.S.; Chiariello, N.R.; Cleland, E.E.; Moore, L.A.; Shaw, M.R.; Thayer, S.; Tobeck, T.; Mooney, H.A.; Field, C.B. Responses of grassland production to single and multiple global environmental changes. *PLoS Biol.* **2005**, *3*, e319. [CrossRef]
36. Wang, J.; Brown, D.G.; Chen, J. Drivers of the dynamics in net primary productivity across ecological zones on the Mongolian Plateau. *Landsc. Ecol.* **2013**, *28*, 725–739. [CrossRef]
37. John, R.; Chen, J.; Ou-Yang, Z.-T.; Xiao, J.; Becker, R.; Samanta, A.; Ganguly, S.; Yuan, W.; Batkhishig, O. Vegetation response to extreme climate events on the Mongolian Plateau from 2000 to 2010. *Environ. Res. Lett.* **2013**, *8*, 1–14. [CrossRef]
38. Wang, W.; Wang, W.J.; Jun-Sheng, L.I.; Hao, W.U.; Chao, X.U.; Liu, X.F. Remote Sensing Analysis of Impacts of Extreme Drought Weather on Ecosystems in Southwest Region of China Based on Normalized Difference Vegetation Index. *Res. Environ. Sci.* **2010**, *23*, 1447–1455.
39. Xu, L.; Samanta, A.; Costa, M.H.; Ganguly, S.; Nemani, R.R.; Myneni, R.B. Widespread decline in greenness of Amazonian vegetation due to the 2010 drought. *Geophys. Res. Lett.* **2011**, *38*. [CrossRef]

40. Matthews, H.D.; Weaver, A.J.; Meissner, K.J.; Gillett, N.P.; Eby, M. Natural and anthropogenic climate change: Incorporating historical land cover change, vegetation dynamics and the global carbon cycle. *Clim. Dyn.* **2004**, *22*, 461–479. [CrossRef]

41. Shi, Y. A preliminary study of signal, impact and foreground of climatic shift from warm-dry to warm-humid in Northwest China. *J. Glaciol. Geocryol.* **2002**, *24*, 219–226.

42. Mu, S.; Zhang, C.; Zhou, W.; Li, J.; Sun, Z.; Gang, C. Spatial-temporal dynamics of grassland coverage and its response to climate change in China during 1982–2010. *Acta Geogr. Sin.* **2014**, *69*, 15–30.

43. Vandandorj, S.; Gantsetseg, B.; Boldgiv, B. Spatial and temporal variability in vegetation cover of Mongolia and its implications. *J. Arid. Land* **2015**, *7*, 450–461. [CrossRef]

44. Zhang, B.; Zhu, J.J.; Liu, H.M.; Pan, Q.M. Effects of extreme rainfall and drought events on grassland ecosystems. *Chin. J. Plant Ecol.* **2014**, *38*, 1008–1018.

45. Hua, L.; Ma, Z.; Zhong, L. A comparative analysis of primary and extreme characteristics of dry or wet status between Asia and North America. *Adv. Atmos. Sci.* **2011**, *28*, 352–362. [CrossRef]

46. Yin, Y.; Liu, H.; Liu, G.; Hao, Q.; Wang, H. Vegetation responses to mid-Holocene extreme drought events and subsequent long-term drought on the southeastern Inner Mongolian Plateau, China. *Agric. For. Meteorol.* **2013**, *178–179*, 3–9. [CrossRef]

47. Zhou, Y.; Ning, L.I.; Zhong-Hui, J.I.; Xiao-Tian, G.U.; Fan, B.H. Temporal and Spatial Patterns of Droughts Based on Standard Precipitation Index (SPI) in Inner Mongolia during 1981–2010. *J. Nat. Resour.* **2013**, *28*, 1694–1706.

48. Jiang, Y.; Wang, R.; Peng, Q.; Wu, X.; Cheng, L. The relationship between drought activity and vegetation cover in Northwest China from 1982 to 2013. *Nat. Hazards* **2018**, *92*, 1–19. [CrossRef]

49. Zhang, X.; Pan, X.; Xu, L.; Wei, P.; Yin, Z.; Shao, C. Analysis of spatio-temporal distribution of drought characteristics based on SPEI in Inner Mongolia during 1960–2015. *Trans. Chin. Soc. Agric. Eng.* **2017**, *33*, 190–199.

50. Mu, S.; Yang, H.; Li, J.; Chen, Y.; Gang, C.; Zhou, W.; Ju, W. Spatio-temporal dynamics of vegetation coverage and its relationship with climate factors in Inner Mongolia, China. *J. Geogr. Sci.* **2013**, *23*, 231–246. [CrossRef]

51. Zhang, Y.; Wang, Q.; Wang, Z.; Yang, Y.; Li, J. Impact of human activities and climate change on the grassland dynamics under different regime policies in the Mongolian Plateau. *Sci. Total Environ.* **2020**, *698*, 1–10. [CrossRef]

52. Kemp, D.R.; Guodong, H.; Xiangyang, H.; Michalk, D.L.; Fujiang, H.; Jianping, W.; Yingjun, Z. Innovative grassland management systems for environmental and livelihood benefits. *Proc. Natl. Acad. Sci. USA* **2013**, *110*, 8369–8374. [CrossRef]

53. Harris, I.; Jones, P.D.; Osborn, T.J.; Lister, D.H. Updated high-resolution grids of monthly climatic observations—The CRU TS3.10 Dataset. *Int. J. Climatol.* **2014**, *34*, 623–642. [CrossRef]

54. Wang, Z.; Zhang, Y.; Yang, Y.; Zhou, W.; Gang, C.; Zhang, Y.; Li, J.; An, R.; Wang, K.; Odeh, I. Quantitative assess the driving forces on the grassland degradation in the Qinghai—Tibet Plateau, in China. *Ecol. Inform.* **2016**, *33*, 32–44. [CrossRef]

55. White, M.A.; Thornton, P.E.; Running, S.W.; Nemani, R.R. Parameterization and Sensitivity Analysis of the BIOME–BGC Terrestrial Ecosystem Model: Net Primary Production Controls. *Earth Interact.* **2000**, *4*, 1–84. [CrossRef]

Article

The Influence of Climate Warming and Humidity on Plant Diversity and Soil Bacteria and Fungi Diversity in Desert Grassland

Yi Zhang [1], Yingzhong Xie [1,2], Hongbin Ma [1,2], Juan Zhang [1], Le Jing [1], Yutao Wang [1] and Jianping Li [1,2,*]

1 College of Agriculture, Ningxia University, Yinchuan 750021, China; 13995378736@163.com (Y.Z.);
   xieyz@nxu.edu.cn (Y.X.); ma_hb@nxu.edu.cn (H.M.); 18435121531@163.com (J.Z.); jinglevip@163.com (L.J.);
   18234171762@163.com (Y.W.)
2 State Key Laboratory Cultivation Base for Northwest Degraded Ecosystem Recovery and Reconstruction,
   Yinchuan 750021, China
* Correspondence: lijianpingsas@nxu.edu.cn

Abstract: Our study, which was conducted in the desert grassland of Ningxia in China (E 107.285, N 37.763), involved an experiment with five levels of annual precipitation 33% (R33), 66% (R66), 100% (CK), 133% (R133), 166% (R166) and two temperature levels (inside Open-Top Chamber (OTC) and outside OTC). Our objective was to determine how plant, soil bacteria, and fungi diversity respond to climate change. Our study suggested that plant α-diversity in CK and TCK were significantly higher than that of other treatments. Increased precipitation promoted root biomass (RB) growth more than aboveground living biomass (ALB). R166 promoted the biomass of Agropyron mongolicum the most. In the fungi communities, temperature and precipitation interaction promoted α-diversity. In the fungi communities, the combination of increased temperature and natural precipitation (TCK) promoted β-diversity the most, whose distance was determined to be 25,124 according to PCA. In the bacteria communities, β-diversity in CK was significantly higher than in other treatments, and the distance was determined to be 3010 according to PCA. Soil bacteria and fungi α- and β-diversity, and ALB promoted plant diversity the most. The interactive effects of temperature and precipitation on C, N, and P contents of plants were larger than their independent effects.

Keywords: precipitation changes; increased temperatures; desert grassland; plant diversity; plant biomass; soil bacteria diversity; soil fungi diversity

Citation: Zhang, Y.; Xie, Y.; Ma, H.; Zhang, J.; Jing, L.; Wang, Y.; Li, J. The Influence of Climate Warming and Humidity on Plant Diversity and Soil Bacteria and Fungi Diversity in Desert Grassland. *Plants* **2021**, *10*, 2580. https://doi.org/10.3390/plants10122580

Academic Editor: Emilia Fernández Ondoño

Received: 16 October 2021
Accepted: 22 November 2021
Published: 25 November 2021

Publisher's Note: MDPI stays neutral with regard to jurisdictional claims in published maps and institutional affiliations.

1. Introduction

One of the most severe environmental problems facing mankind is undoubtedly climate change, which is dominated by climate warming [1]. During 1982–2012, high latitudes experienced greater temperature increases than middle latitudes, land temperatures increased faster than ocean temperatures, and the average global temperature increased by 0.85 °C compared with the same time period (1981–2017). In the context of global climate change, extreme weather events frequently occur, and precipitation is unevenly distributed [2].

Vegetation is the link connecting soil, the atmosphere, and water. As an important part of the terrestrial ecosystem, it plays an "indicator" role in global climate change. The vegetation index reflects surface vegetation characteristics and vegetation cover information [3]. Therefore, to a large extent it represents the ecological quality of a certain area [4].

Plant biodiversity includes four components: genetic diversity, species diversity, intraspecific genetic diversity, and ecosystem diversity. Species diversity is the manifestation of biodiversity in species. It is a simple measure of biodiversity [5], counting only the number of different species in a given area. Plant species diversity is the basic unit that constitutes a plant community and forms the core part of biodiversity. Plant species diversity mainly includes plant species composition, richness, uniformity, interspecies relationships, and their changes in time and space [6]. A sustained high level of plant diversity can improve ecosystem stability and its resistance to changes in habitat. Plant species diversity is an important indicator for analyzing the structure, function, succession, and ecological restoration stages of plant communities [7].

Species alpha (α) diversity refers to the diversity of species in the same location or community. It is caused by differences in niches among species [8]. Its measurement is divided into four categories: species richness, relative abundance models of species, the ecological diversity index, and the evenness index [9]. Biodiversity is reflected by plant diversity and its functional characteristics, and it can also explain the interaction between vegetation and the environment [10]. The composition and development of plant communities can be well understood by studying biodiversity. This work has theoretical and practical significance for maintaining a greater variety of organisms [11]. Plant community diversity generally refers to differences in composition, structure, function, and dynamics of plant communities and is a foundational level of research among all levels of biodiversity [12]. In biodiversity conservation practice, the status of communities or ecosystems is often evaluated based on the diversity index. The measurement of community alpha diversity can be divided into four categories: (1) the Shannon–Wiener diversity index reflects the hierarchical characteristics of a biological community; the higher the value, the higher the community diversity; (2) the Margalef species richness index reflects the number of plant species in the community; (3) the Pielou evenness index refers to the distribution ratio of the number of individuals in the community; (4) the Simpson dominance index reflects the distribution of numerically abundant species within a community. The four commonly used α-diversity indexes incorporate two measurements—the number (richness) of species and the uniformity of species [13].

Soil microbes degrade and detoxify environmental pollutants and play an important role in maintaining soil quality and ecosystem stability [14]. The diversity and variability within the soil microorganism community reflects its diverse responses and adaptations to the environment [15].

As an important part of the soil, soil microorganisms are often hailed as the "converter" of nutrient cycling, the "regulator" of terrestrial ecosystem stability, and the "purifier" of environmental pollution [16]. The soil microorganism community can be used to monitor changes in the structure and function of grassland ecosystems following fluctuations in water availability. Changing precipitation patterns have been one of the hot issues in global climate research in recent years, and grassland ecosystems are widely distributed, with most of them located in ecologically fragile zones [17]. They are susceptible to disturbance and global climatic imbalances, especially those in areas with restricted water such as arid and semi-arid grassland ecosystems. Soil microorganisms act as catalysts for soil nutrient cycling and transformation, which can promote absorption [18].

We conducted a study in the Ningxia Hui Autonomous Region of China. We used an Open-Top Chamber (OTC) to simulate increased temperature and artificial shelters and sprinklers to simulate precipitation changes. We systematically studied the changes in temperature and precipitation and the interaction of the two factors:

(i)   Dynamic changes with regards to about the α-diversity, biomass, organic carbon, total nitrogen, and total phosphorus of plants in the desert steppe ecosystem;

(ii)   Dynamic changes with regards to the α- and β-diversity of the soil fungi and soil bacteria;

(iii) The synergistic relationship between the α-diversity, biomass, organic carbon, total nitrogen, total phosphorus of plants, and the α- and β-diversity of the soil fungi and soil bacteria. The research results provide a reliable theoretical basis for the formulation of reasonable response strategies for desert grasslands.

2. Results

2.1. Plant Importance Value

Under the changing precipitation condition and the interaction of the precipitation changed and the temperature increased, the main value of the Agropyron mongolicum, Lespedeza bicolor, and Polygala tenuifolia were all higher than the other plants (Figure 1).

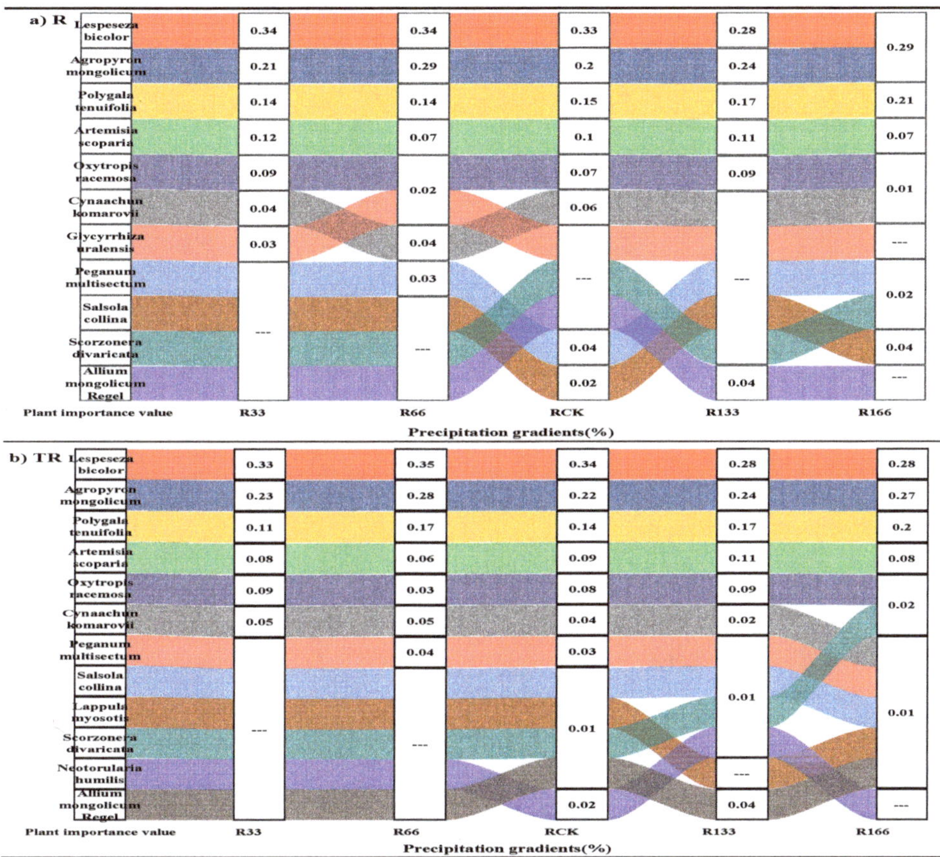

Figure 1. The main values of plants in the study sites. (**a**) The main values of plants under precipitation changing treatment (R); (**b**) The main values of plants under the interaction of precipitation changing and temperature increasing treatment (TR). Five levels of rainfall (R) were used: 33% (R33), 66% (R66), 100% (CK), 133% (R133), and 166% (R166) of the annual average. The first two rainfall conditions were obtained by using two rainout shelters with two manipulated rainfall doses: 97 mm (R33) and 194 mm (R66). For the three other rainfall conditions, we artificially increased rainfall pot in unsheltered plots using a watering: 295 mm (CK), 392 mm (R133), and 490 mm (R166). The temperature consisted of two levels: the actual temperature (CK) and the interaction between rainfall and the temperature, which was increased by about 2 °C (T) with the OTC (Open-Top Chamber) in each plot. TR33 was the first site of interaction between 33% precipitation (R33) and the temperature increase of about 2 °C (T); the marks of TR66, TCK, TR133, TR166 were the same. R33 was the first site of 33% precipitation, and the marks of R66, CK, R133, R166 were the same.

2.2. Plant α-Diversity

The plant α-diversity had no significant differences among the different precipitation and temperature treatments (Table 1).

Table 1. Grassland plant α-diversity index under different precipitation and temperature treatments.

	Shannon–Wiener	Pielou	Margalef	Simpson
	F = 6.12 $p > 0.05$	F = 4.84 $p > 0.05$	F = 6.34 $p > 0.05$	F = 3.71 $p > 0.05$
R33	1.64 ± 0.15 a	0.80 ± 0.03 a	3.17 ± 0.76 a	0.76 ± 0.05 a
R66	1.65 ± 0.07 a	0.75 ± 0.04 a	3.75 ± 0.39 a	0.75 ± 0.03 a
CK	1.80 ± 0.11 a	0.83 ± 0.04 a	3.30 ± 0.35 a	0.79 ± 0.03 a
R133	1.70 ± 0.07 a	0.85 ± 0.05 a	2.84 ± 0.21 a	0.77 ± 0.04 a
R166	1.68 ± 0.11 a	0.73 ± 0.07 a	3.70 ± 0.39 a	0.76 ± 0.03 a
	F = 6.09 $p > 0.05$	F = 4.79 $p > 0.05$	F = 5.98 $p > 0.05$	F = 3.69 $p > 0.05$
TR33	1.65 ± 0.14 a	0.82 ± 0.04 a	3.18 ± 0.74 a	0.77 ± 0.03 a
TR66	1.67 ± 0.06 a	0.76 ± 0.03 a	3.76 ± 0.38 a	0.76 ± 0.02 a
TCK	1.81 ± 0.10 a	0.84 ± 0.03 a	3.31 ± 0.37 a	0.80 ± 0.02 a
TR133	1.71 ± 0.08 a	0.86 ± 0.06 a	2.85 ± 0.23 a	0.78 ± 0.03 a
TR166	1.69 ± 0.10 a	0.75 ± 0.06 a	3.71 ± 0.40 a	0.77 ± 0.02 a

Mean ± SE followed by lowercase letters in each column indicates significant differences between the variance percentage of precipitation, according to LSD test ($p < 0.05$). Five levels of rainfall (R) were used: 33% (R33), 66% (R66), 100% (CK), 133% (R133), and 166% (R166) of the annual average. The first two rainfall conditions were obtained by using two rainout shelters with two manipulated rainfall doses: 97 mm (R33) and 194 mm (R66). For the three other rainfall conditions, we artificially increased rainfall in unsheltered plots using a watering pots: 295 mm (CK), 392 mm (R133), and 490 mm (R166). The temperature consisted of two levels: the actual temperature (CK) and the interaction between rainfall and the temperature, which was increased by about 2 °C (T) with the OTC (Open-Top Chamber) in each plot. TR33 was the first site of interaction between 33% precipitation (R33) and the temperature increase of about 2 °C (T), and the marks of TR66, TCK, TR133, TR166 were the same. R33 was the first site of 33% precipitation, and the marks of R66, CK, R133, R166 were the same.

2.3. The Number of Species

Under the interaction of rising precipitation levels and increasing temperature conditions, the number of species was highest in CK, which was significantly higher than R33 ($p < 0.05$). With increased precipitation, the number of species was highest under R166, which was significantly higher than other precipitation treatments ($p < 0.05$). The difference between the natural temperature and the temperature increases was obvious ($p < 0.05$) (Figure 2).

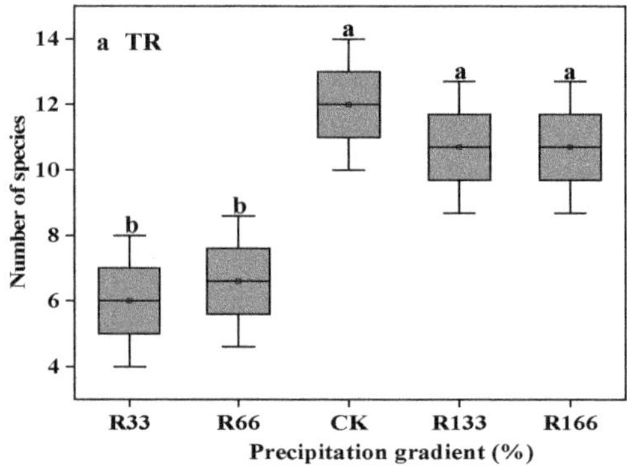

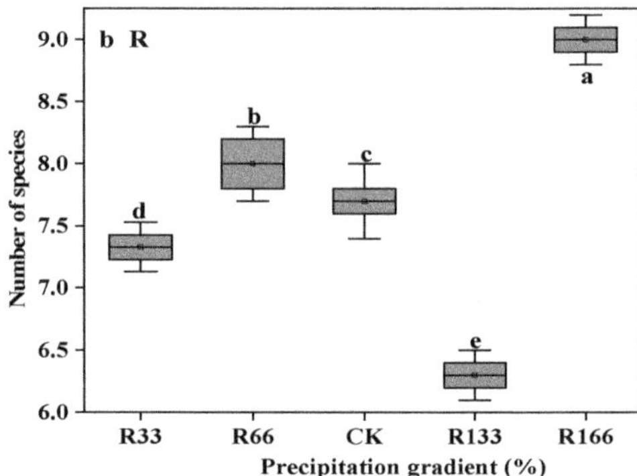

Figure 2. *Cont.*

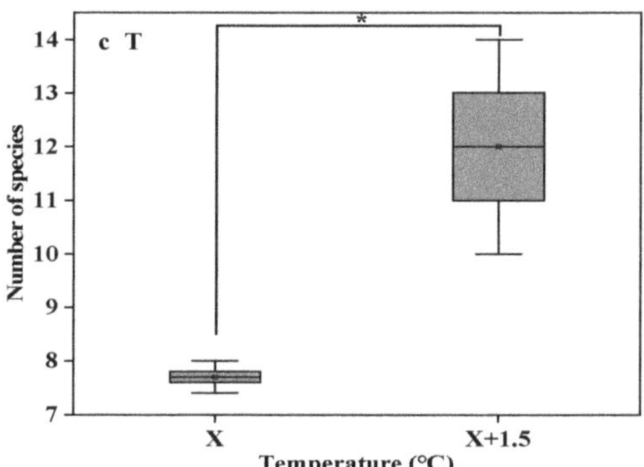

Figure 2. The number of species in the study sites. (**a**) The number of species under precipitation changing treatment (R); (**b**) The number of species under interaction of precipitation changing and temperature increasing treatment (TR). (**c**) The number of species under temperature increasing treatment (T). Five levels of rainfall (R) were used: 33% (R33), 66% (R66), 100% (CK), 133% (R133), and 166% (R166) of the annual average. The first two rainfall conditions were obtained by using two rainout shelters with two manipulated rainfall doses: 97 mm (R33) and 194 mm (R66). For the three other rainfall conditions, we artificially increased rainfall in unsheltered plots using a watering pot: 295 mm (CK), 392 mm (R133), and 490 mm (R166). The temperature consisted of two levels: the actual temperature (CK) and the interaction between rainfall and the temperature, which was increased by about 2 °C (T) with the OTC (Open-Top Chamber) in each plot. TR33 was the first site of interaction between 33% precipitation (R33) and the temperature increase of about 2 °C (T), R33 was the first site of 33% precipitation, and other marks were the same. Values indicate the mean ± SE; different letters represent a significant difference according to LSD test ($p < 0.05$). * represents a significant difference according to t-test ($p < 0.05$).

2.4. The Biomass of Plants and Dominant Plant Species

The RB was significantly higher than the ALB. The differences of the ALB were not significant under the interaction of the changed precipitation and increased temperature conditions ($p < 0.05$). The root biomass was highest in R166 but lowest in R33, and their difference were very significant ($p < 0.05$). The difference of the ALB under the natural temperature and the increased temperature were not significant, the same as the RB ($p < 0.05$) (Figure 3).

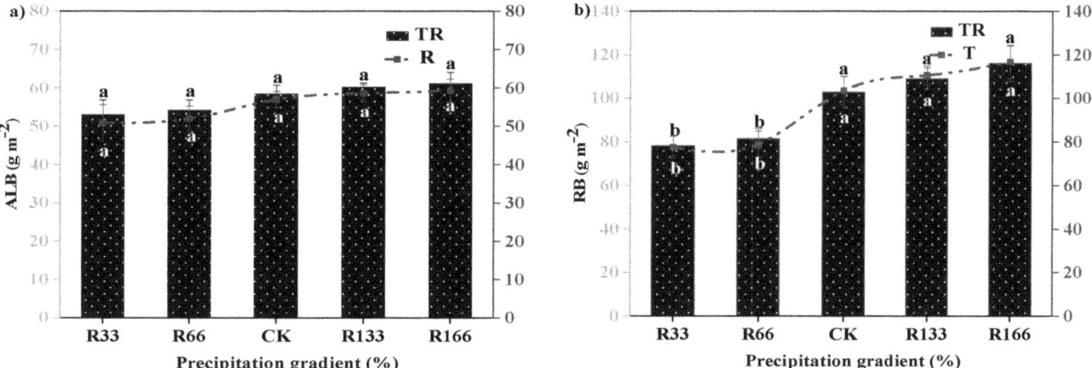

Figure 3. *Cont.*

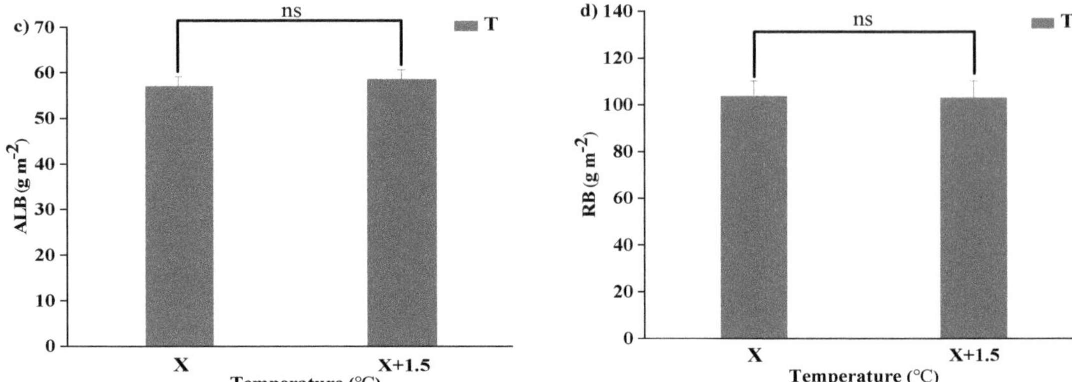

Figure 3. Variations of aboveground plant living biomass (ALB) and plant root biomass (RB) of vegetation in the study sites. (**a**) Aboveground plant living biomass (ALB) under precipitation changing (R) and the interaction of the precipitation changing and temperature increasing (TR). (**b**) Root biomass (RB) under precipitation changing (R) and the interaction of the precipitation changing and temperature increasing (TR). (**c**) Aboveground plant living biomass (ALB) under temperature increasing (T). (**d**) Root biomass (RB) under temperature increasing (T). Five levels of rainfall (R) were used: 33% (R33), 66% (R66), 100% (CK), 133% (R133), and 166% (R166) of the annual average. The first two rainfall conditions were obtained by using two rainout shelters with two manipulated rainfall doses: 97 mm (R33) and 194 mm (R66). For the three other rainfall conditions, we artificially increased rainfall in unsheltered plots using a watering pot: 295 mm (CK), 392 mm (R133), and 490 mm (R166). The temperature consisted of two levels: the actual temperature (CK) and the interaction between rainfall and the temperature increased by about 2 °C (T) with the OTC (Open-Top Chamber) in each plot. TR33 was the first site of interaction between 33% precipitation (R33) and the temperature, which was increased by about 2 °C (T), and the marks of TR66, TCK, TR133, TR166 are the same. R33 was the first site of 33% precipitation, and the marks of R66, CK, R133, R166 were the same. Values indicate the mean ± SE, and different letters represent a significant difference according to LSD test ($p < 0.05$). ns represents a no significant difference according to *t*-test ($p < 0.05$).

When precipitation was increased, Agropyron mongolicum had the highest RB under R166, and had the lowest RB under R33, the RB difference between the precipitation gradients was significant ($p < 0.05$). The ALB of Agropyron mongolicum was highest under the R166, which was significantly higher than other precipitation treatments, the same as the total biomass of Agropyron mongolicum ($p < 0.05$). The RB, ALB, and total biomass of Lespedeza bicolor were highest under R33, which were significantly higher than other precipitation gradients ($p < 0.05$). The RB, ALB, and total biomass of Polygala tenuifolia were the highest under natural precipitation, which were significantly higher than other precipitation treatments ($p < 0.05$) This phenomenon with regards to the Agropyron mongolicum, Lespedeza bicolor, and Polygala tenuifolia under the warming and precipitation interaction was the same ($p < 0.05$). With the increases temperature, the differences of the RB, ALB, and total biomass of Agropyron mongolicum and Lespedeza bicolor were significant, but the temperature increases had no obvious impact on the Polygala tenuifolia ($p < 0.05$) (Figure 4).

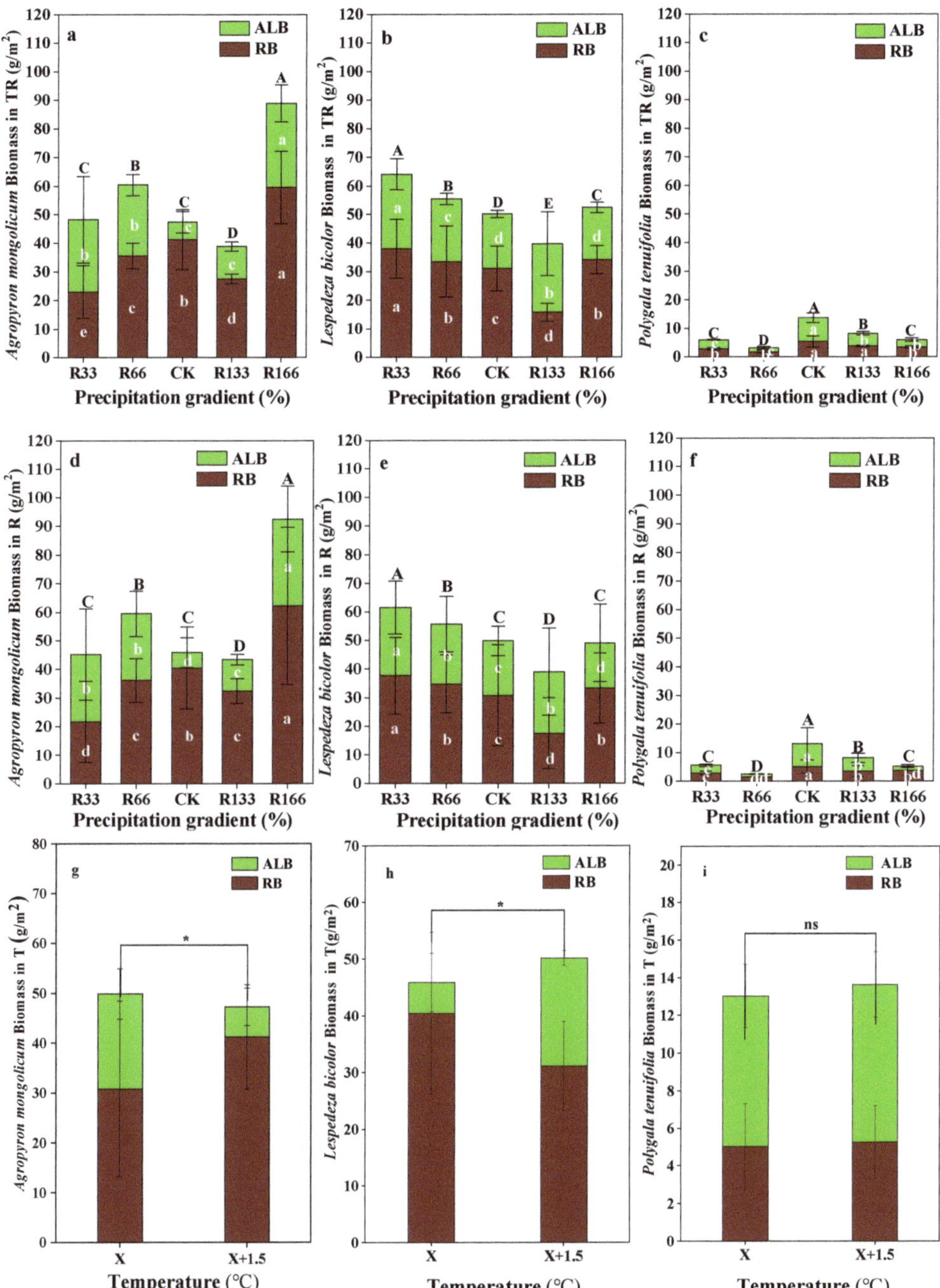

Figure 4. Variations of aboveground plant living biomass (ALB) and plant root biomass (RB) of dominant species in the study sites. (**a**) Aboveground plant living biomass (ALB) and plant root biomass (RB) of Agropyron mongolicum under the precipitation changing and temperature increasing(TR); (**b**) Aboveground plant living biomass (ALB) and plant root biomass (RB) of Lespedeza bicolor under the precipitation changing and temperature increasing(TR); (**c**) Aboveground plant

living biomass (ALB) and plant root biomass (RB) of Polygala tenuifolia under the precipitation changing and temperature increasing(TR); (**d**) Aboveground plant living biomass (ALB) and plant root biomass (RB) of Agropyron mongolicum under the precipitation changing (R); (**e**) Aboveground plant living biomass (ALB) and plant root biomass (RB) of Lespedeza bicolor under the precipitation changing (R); (**f**) Aboveground plant living biomass (ALB) and plant root biomass (RB) of Polygala tenuifolia under the precipitation changing (R); (**g**) Aboveground plant living biomass (ALB) and plant root biomass (RB) of Agropyron mongolicum under the temperature increasing (T); (**h**) Aboveground plant living biomass (ALB) and plant root biomass (RB) of Lespedeza bicolor under the temperature increasing (T); (**i**) Aboveground plant living biomass (ALB) and plant root biomass (RB) of Polygala tenuifolia under the temperature increasing (T). Five levels of rainfall (R) were used: 33% (R33), 66% (R66), 100% (CK), 133% (R133), and 166% (R166) of the annual average. The first two rainfall conditions were obtained by using two rainout shelters with two manipulated rainfall doses: 97 mm (R33) and 194 mm (R66). For the three other rainfall conditions, we artificially increased rainfall in unsheltered plots using a watering pot: 295 mm (CK), 392 mm (R133), and 490 mm (R166). The temperature consisted of two levels: the actual temperature (CK) and the interaction between rainfall and the temperature increased by about 2 °C (T) with the OTC (Open-Top Chamber) in each plot. TR33 was the first site of interaction between 33% precipitation (R33) and the temperature, which was increased by about 2 °C (T), and the marks of TR66, TCK, TR133, TR166 were the same. R33 was the first site of 33% precipitation, and the marks of R66, CK, R133, R166 were the same. Values indicate the mean ± SE; different letters represent a significant difference according to LSD test ($p < 0.05$). ns represents a nonsignificant difference according to *t*-test ($p < 0.05$).

2.5. The Organic Carbon, Total Nitrogen, and Total Phosphorus of Plants and Dominant Plant Species

The differences in plant organic carbon, plant nitrogen, and plant phosphorus content were not significant under the changing precipitation condition and the interaction of the changing precipitation and increasing temperature conditions ($p < 0.05$), but the plant organic carbon, plant nitrogen, and plant phosphorus content under the precipitation condition changed less than under the interaction of the temperature and precipitation conditions. With the increases in temperature, the differences of the plant organic carbon, plant total nitrogen, and plant total phosphorus were also not obvious ($p < 0.05$) (Figure 5).

For the interaction of the temperature and precipitation conditions, plant organic carbon was the lowest at R166 for *Agropyron mongolicum*, which was significantly lower than other treatments ($p < 0.05$). Plant organic carbon was the highest at R66 for Lespedeza bicolor and was highest at R133 for Polygala tenuifolia, which were both significantly higher than other treatments ($p < 0.05$). Plant total nitrogen and plant total phosphorus were not significant among the different treatments ($p < 0.05$). The plant organic carbon, total nitrogen, and total phosphorus for *Agropyron mongolicum*, Lespedeza bicolor, and Polygala tenuifolia at the precipitation changed was lower. With the increased temperature, the differences of plant organic carbon, plant total nitrogen, and plant total phosphorus for *Agropyron mongolicum*, *Lespedeza bicolor*, and *Polygala tenuifolia* were not significant ($p < 0.05$) (Figure 6).

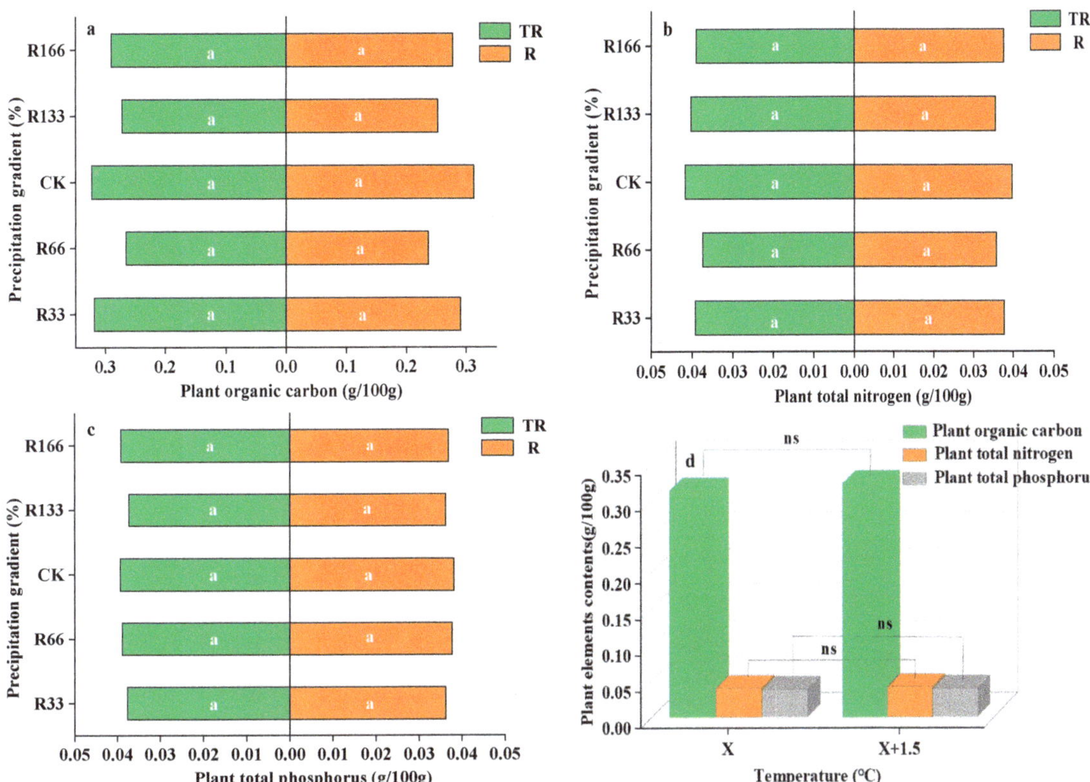

Figure 5. Percentages of plant organic carbon, plant total nitrogen, and total phosphorus of vegetation in the study sites. (**a**) Plant organic carbon under precipitation changing (R) and the interaction of the precipitation changing and temperature increasing (TR); (**b**) Plant total nitrogen under precipitation changing (R) and the interaction of the precipitation changing and temperature increasing (TR); (**c**) Plant total phosphorus under precipitation changing (R) and the interaction of the precipitation changing and temperature increasing(TR); (**d**) Plant organic carbon, plant total nitrogen, plant total phosphorus under temperature increasing (T). Five levels of rainfall (R) were used: 33% (R33), 66% (R66), 100% (CK), 133% (R133), and 166% (R166) of the annual average. The first two rainfall conditions were obtained by using two rainout shelters with two manipulated rainfall doses: 97 mm (R33) and 194 mm (R66). For the three other rainfall conditions, we artificially increased rainfall in unsheltered plots using a watering pot: 295 mm (CK), 392 mm (R133), and 490 mm (R166). The temperature consisted of two levels: the actual temperature (CK) and the interaction between rainfall and the temperature, which was increased by about 2 °C (T) with the OTC (Open-Top Chamber) in each plot. TR33 is the first site of interaction between 33% precipitation (R33) and the temperature increase of about 2 °C (T), and the marks of TR66, TCK, TR133, TR166 were the same. R33 was the first site of 33% precipitation, and the marks of R66, CK, R133, R166 were the same. Values indicate the mean ± SE, and different letters represent a significant difference according to LSD test ($p < 0.05$). ns represents a nonsignificant difference according to *t*-test.

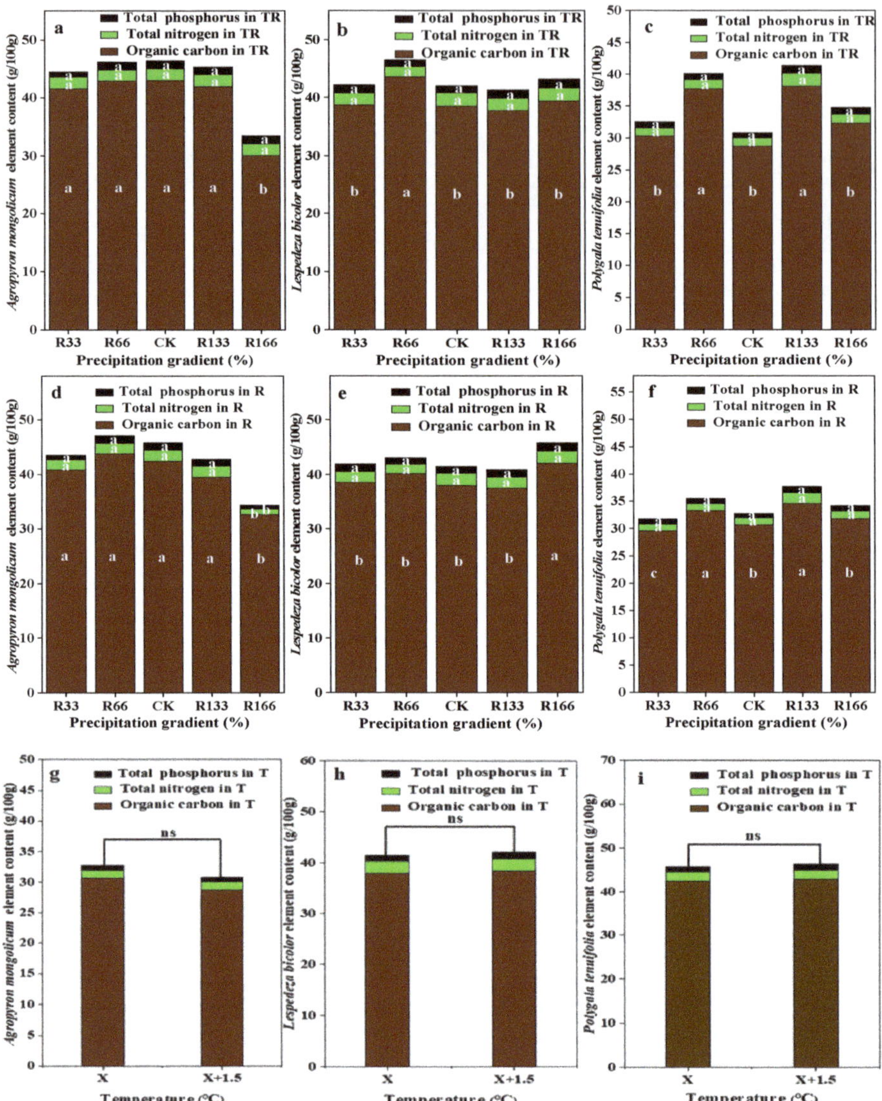

Figure 6. Percentage of plant organic carbon, plant total nitrogen, and total phosphorus of dominant species in the study sites. (**a**) The organic carbon, total nitrogen, total phosphorus of Agropyron mongolicum under the interaction of the precipitation changing and temperature increasing (TR); (**b**) The organic carbon, total nitrogen, total phosphorus of Lespedeza bicolor under the interaction of the precipitation changing and temperature increasing(TR); (**c**) The organic carbon, total nitrogen, total phosphorus of Polygala tenuifolia under the interaction of the precipitation changing and temperature increasing(TR); (**d**) The organic carbon, total nitrogen, total phosphorus of Agropyron mongolicum under the precipitation changing (R); (**e**) The organic carbon, total nitrogen, total phosphorus of Lespedeza bicolor under the precipitation changing (R); (**f**) The organic carbon, total nitrogen, total phosphorus of Polygala tenuifolia under the precipitation changing (R); (**g**) The organic carbon, total nitrogen, total phosphorus of Agropyron mongolicum under the temperature increasing (T); (**h**) The organic carbon, total nitrogen, total phosphorus of Lespedeza bicolor under the temperature increasing (T); (**i**) The organic carbon, total nitrogen, total phosphorus of Polygala tenuifolia under the temperature increasing (T). Five levels of rainfall (R) were used: 33% (R33), 66% (R66), 100% (CK), 133% (R133), and 166% (R166) of the annual average. The first two rainfall

conditions were obtained by using two rainout shelters with two manipulated rainfall doses: 97 mm (R33) and 194 mm (R66). For the three other rainfall conditions, we artificially increased rainfall in unsheltered plots using a watering pot: 295 mm (CK), 392 mm (R133), and 490 mm (R166). The temperature consisted of two levels: the actual temperature (CK) and the interaction between rainfall and the temperature, which was increased by about 2 °C (T) with the OTC (Open-Top Chamber) in each plot. TR33 was the first site of interaction between 33% precipitation (R33) and the temperature increase of about 2 °C (T), and the marks of TR66, TCK, TR133, TR166 are the same. R33 was the first site of 33% precipitation, and the marks of R66, CK, R133, R166 are the same. Values indicate the mean ± SE, and different letters represent a significant difference according to LSD test ($p < 0.05$).

2.6. Soil Microorganism α-Diversity

In the fungi communities, α-diversity gradually decreased under the interaction of the increasing temperature and precipitation conditions. However, when only precipitation increased, α-diversity first decreased and then increased.

In the bacteria communities, under the control of the increasing temperature and precipitation conditions, the α-diversity did not show any obvious patterns.

In both the fungi and bacteria communities, α-diversity increased with increased temperature (Figure 7).

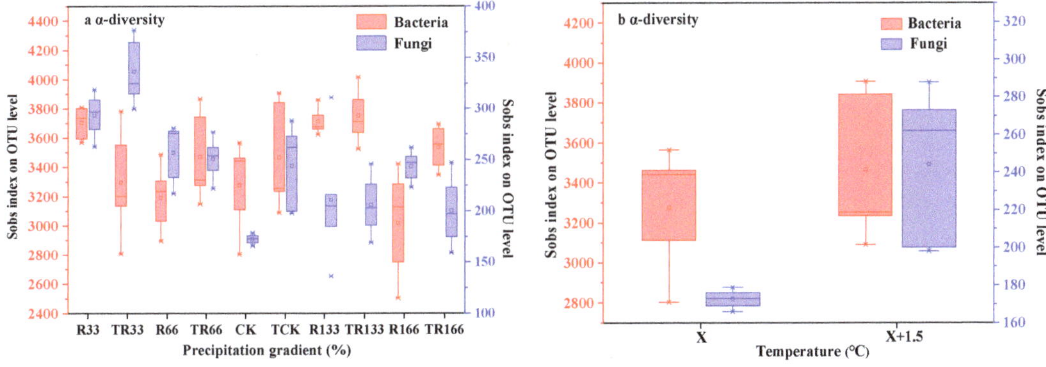

Figure 7. Soil microbial α-diversity of (**a**) fungi and (**b**) bacteria in the study sites by principal component analysis (PCA). Sobs index was the observed richness. Five levels of rainfall (R) were used: 33% (R33), 66% (R66), 100% (CK), 133% (R133), and 166% (R166) of the annual average. The first two rainfall conditions were obtained by using two rainout shelters with two manipulated rainfall doses: 97 mm (R33) and 194 mm (R66). For the three other rainfall conditions, we artificially increased rainfall in unsheltered plots using a watering pot: 295 mm (CK), 392 mm (R133), and 490 mm (R166). The temperature consisted of two levels: the actual temperature (CK) and the interaction between rainfall and the temperature, which was increased by about 2 °C (T) with the OTC (Open-Top Chamber) in each plot. TR33 was the first site of interaction between 33% precipitation (R33) and the temperature increase of about 2 °C (T), and the marks of TR66, TCK, TR133, TR166 were the same. R33 was the first site of 33% precipitation, and the marks of R66, CK, R133, R166 were the same.

2.7. Soil Bacteria and Fungi β-Diversity

In the fungi communities, the distance between each sample point was the farthest under TCK, and the distance was 25124 according to PCA; therefore, the corresponding β-diversity was the highest under TCK. In the bacteria communities, the distance between each sample point was the farthest under CK; therefore, the corresponding β-diversity was the highest under CK, and the distance was 3010 according to PCA (Figure 8).

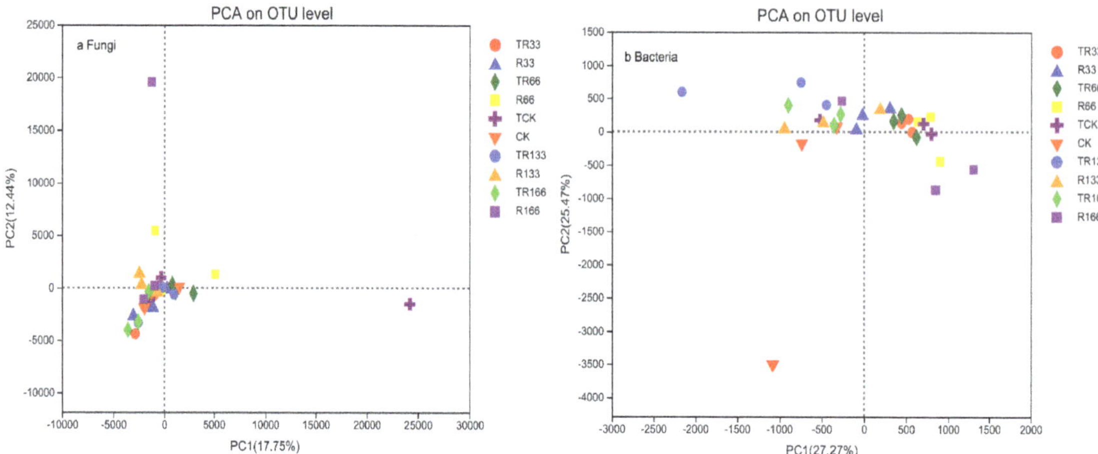

Figure 8. Soil microorganism β-diversity of (**a**) fungi and (**b**) bacteria in the study sites by principal component analysis (PCA). Five levels of rainfall (R) were used: 33% (R33), 66% (R66), 100% (CK), 133% (R133), and 166% (R166) of the annual average. The first two rainfall conditions were obtained by using two rainout shelters with two manipulated rainfall doses: 97 mm (R33) and 194 mm (R66). For the three other rainfall conditions, we artificially increased rainfall in unsheltered plots using a watering pot: 295 mm (CK), 392 mm (R133), and 490 mm (R166). The temperature consisted of two levels: the actual temperature (CK) and the interaction between rainfall and the temperature, which was increased by about 2 °C (T) with the OTC (Open-Top Chamber) in each plot. TR33 was the first site of interaction between 33% precipitation (R33) and the temperature increase of about 2 °C (T), and the marks of TR66, TCK, TR133, TR166 were the same. R33 was the first site of 33% precipitation, and the marks of R66, CK, R133, R166 were the same.

2.8. The Relationship between Grassland Plant Diversity, Biomass, Soil Bacteria, and Fungi α- and β-Diversity

The Shannon–Wiener diversity index was positively correlated with the ALB, α-diversity, and β-diversity but negatively correlated with the RB. The Pielou evenness index was positively correlated with α-diversity and β-diversity but negatively correlated with the ALB and RB. The Margalef species richness index was positively correlated with the RB and ALB. The Simpson dominance index was positively correlated with α-diversity and β-diversity but negatively correlated with the RB and ALB. The RB was greatly negatively correlated with β-diversity (Figure 9).

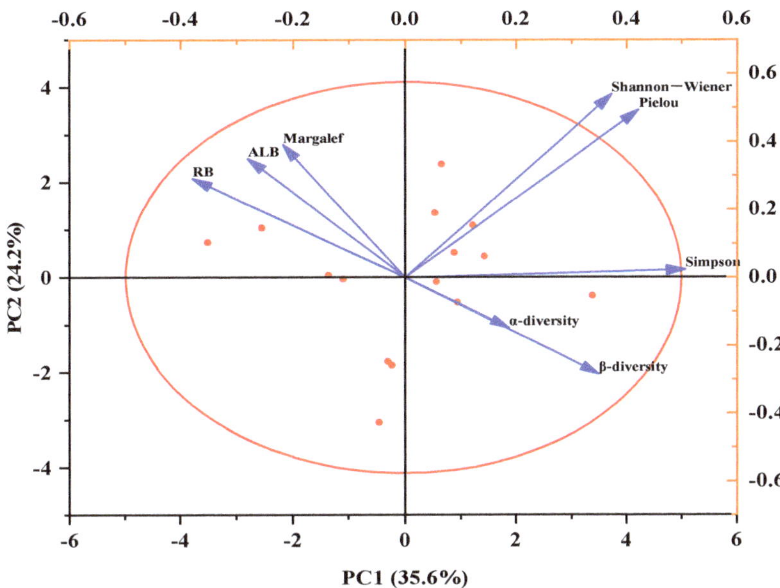

Figure 9. Principal components analysis (PCA) plots showing the influence of the Shannon–Wiener, Pielou, Margalef, and Simpson indexes, and the above-living biomass (ALB), root biomass (RB), α-diversity, and β-diversity, which represented effects of different temperature (recorded as CK and T) and variation in precipitation (recorded as R33, R66, CK, R133, R166).

3. Discussion

3.1. Effects of Precipitation Changes and Temperature on Plant Main Value

Under the changing precipitation condition and the interaction of the changing precipitation and the increasing temperature conditions, the main values of *Agropyron mongolicum*, *Lespedeza bicolor*, and *Polygala tenuifolia* were all higher than the other plants, so we made sure that the three plants were dominant plants; the reason might have been because the root system of *Lespedeza bicolor* forms vertical and horizontal networks in the soil layer, helping it to make full use of the water and nutrients therein. Furthermore, the proportion of woody and sclerenchyma cell tissue in Mongolia wheatgrass is large, and the roots of *Polygala tenuifolia* are strong and can absorb water better. Thus, these characteristics of the three plants give them better drought resistance than other species.

3.2. Effects of Precipitation Changes and Temperature on Plant α-Diversity

Species diversity directly affects ecosystem function and stability, which is the foundation of human survival and development [19]. The distribution pattern of species diversity and its influencing factors have become the core problem of ecology and biogeography research [20].

The Shannon–Wiener and Simpson in CK and TCK are significantly higher than other treatments. This might have been because the sparse surface vegetation, loose soil structure, and low water-holding capacity of desert steppe. Precipitation directly affects the soil moisture content. The decrease of precipitation leads to the drought of topsoil and thus directly reduces the effective moisture content in soil. The change of soil moisture content indirectly affects soil nutrients and indirectly affects the absorption, transportation, and utilization of nutrients by plants by limiting the normal activities of rhizosphere microorganisms, resulting in the reduction of plant species. When precipitation increases, soil erosion occurs, and the reduction of carbon, nitrogen, and other nutrients in soil limits the growth of plants. The distribution pattern of species diversity and its influencing factors have become the core problem of ecology and biogeography research.

3.3. Effects of Changing Precipitation and Increasing Temperature on the Number of Species

With increased precipitation, the number of species was highest under R166, possibly because plant roots need to absorb more water to grow in desert grasslands. This finding agrees with a previous study that found fine roots may have complex responses to the higher amount of precipitation predicted for the future [21].

3.4. Effects of Changing Precipitation and Increasing Temperature on Plants and Dominant Species Biomass

This study found that increased precipitation promoted the growth of RB more than ALB, but the effect of rising temperature on RB was not clear. When a plant is subjected to drought stress, it will reduce ALB and increase underground RB. Some studies have found that reducing drought stress caused by a lack of precipitation will prompt plant to allocate more biomass to the underground RB, so that the root system can better absorb water and nutrients in deep soil [22]. Plant biomass increases with rising precipitation, probably because increased precipitation can effectively supplement soil moisture and promote plant growth and development.

With increasing precipitation, R166 was found to promote the ALB and total biomass of *Agropyron mongolicum* the most, and R33 promoted the ALB, RB, and total biomass of *Lespedeza bicolor* the most. The ALB, RB, and total biomass of *Polygala tenuifolia* was highest under natural precipitation. Under rising temperatures, with increased precipitation, R166 was found to promote the total biomass of *Agropyron mongolicum* the most, and R33 promoted the total biomass of *Lespedeza bicolor* the most. *Polygala tenuifolia* had the highest biomass under natural precipitation, possibly because *Agropyron mongolicum* is a graminoid plant and needs to absorb more water than *Polygala tenuifolia* (*Leguminosae* plant) and *Lespedeza bicolor* (*Lespedeza bicolor* plant).

3.5. Effects of Changing Precipitation and Increasing Temperature on Plant and Dominant Species Organic Carbon, Total Nitrogen, and Total Phosphorus

Under rising temperature with increasing precipitation, the differences in plant nutrients were not obvious. TR promoted plant organic carbon, nitrogen, and phosphorus content more than R, possibly because higher temperatures promote root respiration, and the root absorbs more elements from soil through active transpiration. This accords with a previous study that indicated that roots were the main source of SR, which contributed >70% of CO_2 emissions [23].

3.6. Effects of Changing Precipitation and Increasing Temperature on Soil Bacteria and Fungi Diversity

According to a recent study, compared with soil bacteria and fungi biomass and activity, community structure is more sensitive to warming, with seasonal changes in temperature found to have a significant impact on soil bacteria and fungi communities [24]. Our research found that changes in precipitation and increased temperature promoted fungi communities but did not have a significant effect on the bacteria communities. In fungi communities, TCK promoted the most β-diversity, but in the bacteria communities, CK promoted the most β-diversity. Increasing temperatures may provide a more suitable growth environment for fungi by affecting the availability of plant litter components and nutrients. Therefore, increasing temperature may be beneficial to the growth of fungi and may inhibit bacteria growth, thereby changing bacteria and fungi community structure. For desert grassland, an increase in temperature will also change soil temperature and moisture levels, which may further change the diversity of soil bacteria and fungi communities.

3.7. Effects of Precipitation Changes and Temperature on the Relationship between Grassland Vegetation Diversity, Biomass, and Soil Bacteria and Fungi Diversity

Soil bacteria and fungi α- and β-diversity and ALB all promoted plant diversity, but RB had a limited function. The soil bacteria and fungi communities were able to promote the decomposition of soil nutrients, offering more nutrients for plant to absorb and allowing more different kinds of plants to grow. This is consistent with previous research [25].

4. Materials and Methods

4.1. Study Site

The study area was located in the desert steppe of the Sidunzi ecological field station of Ningxia in China (37°47′ N 107°25′ E), which is on the southern edge of the Mu Us Sandy land and the yellow transition zone from the soil plateau to the Ordos platform (Figure 10). The natural conditions are relatively poor, characterized by drought, low rainfall, and strong winds and storms, and the region has typical temperate continental monsoon climate. Its annual average temperature is 8.1 °C, its monthly average temperature is −13.0–22.7 °C, its extreme maximum temperature was 34.9 °C, its extreme minimum temperature was −24.2 °C, its annual average frost-free period is 162 days, and its annual average precipitation is less than 300 mm. Its zonal soil structure is loose, and its soil fertility is low. Its zonal vegetation is typical of a desert steppe, and its dominant plants are Agropyron mogolicum Keng, Lespedeza potaninii Vass, and Polygala tenuifolia Willd. Due to the influence of climatic conditions and human activities, the grassland in the region has been degraded in large areas for a long time.

Figure 10. Location of the Sidunzi Village of Ningxia Observatory on the Loess Plateau.

4.2. Experimental Design

We completed all control experiment devices from June 2018 to March 2019. We started our experiment in May 2019 and collected soil and plant samples from July 2019.

According to meteorological monitoring of the study site from 1981 to 2017, its annual average precipitation, ground temperature, and air temperature all showed rising trends (Figure 11). Artificial rain-collecting greenhouses and sprinkler irrigation techniques were used to achieve 66% and 133% precipitation gradients and to ensure the precipitation treatment was kept within the 37-year average precipitation and fluctuation extremes. We established two temperature increase gradients to reflect the steady increases in ground temperature and air temperature recorded by meteorological monitoring.

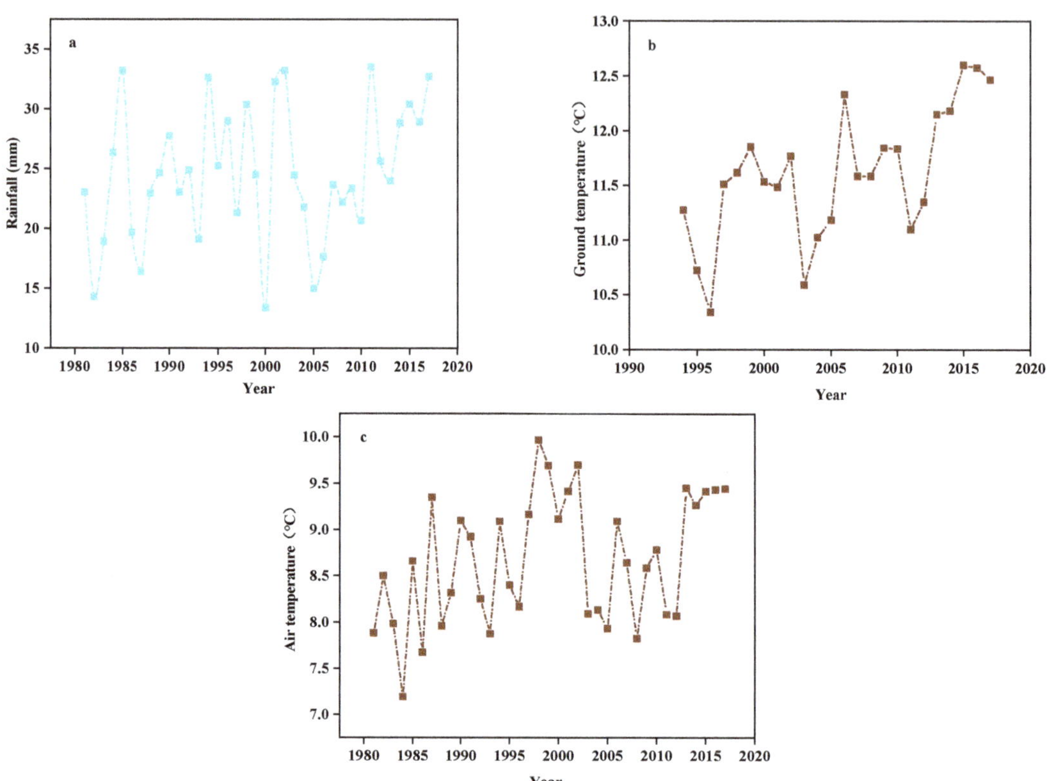

Figure 11. The (**a**) Rainfall, (**b**) Ground temperature, and (**c**) Air temperature from 1981 to 2017.

The designed rainout shelter was completed in November 2018, and these shelters were randomly built (Figure 12). The rainfall gradient was constructed with artificial shelters and sprinklers, and the rainout shelters were made of polycarbonate material, which can allow 90% of photosynthetic effective radiation to pass through it. A two-factor completely randomized experimental design was used based on rainfall and temperature factors. Five levels of rainfall were used: 33% (R33), 66% (R66), 100% (CK), 133% (R133), and 166% (R166) of the annual average. The first two rainfall conditions were obtained by using two rainout shelters with two manipulated rainfall doses: 97 mm (R33) and 194 mm (R66). For the three other rainfall conditions, we artificially increased rainfall in unsheltered plots using a watering pot: 295 mm (CK), 392 mm (R133), and 490 mm (R166). The temperature consisted of two levels: the actual temperature and the interaction between the rainfall and the temperature, which was increased by about 1.5 °C with the OTC (Open-Top Chamber) in each plot [26]. The OTC was made of acrylic transparent board material, which can allow 90% of photosynthetic effective radiation to pass through it. The area of each plot was 6 × 6 m, and each treatment (n = 5) was repeated three times, for a total of 15 plots (temperature treatments are included in the precipitation treatments) (Figure 3). On the 15th and 30th of each month, R33 and R66 of the natural rainfall during the 1st–15th and the 16th–30th of the month, respectively, were collected from the actual rainfall and then evenly replenished to the plots containing R133 and R166 by a watering pot.

Figure 12. Rain shelter construction and Open-Top Chamber (OTC) arrangements of the subplots at the study sites. Five levels of rainfall (R) were used: 33% (R33), 66% (R66), 100% (CK), 133% (R133), and 166% (R166) of the annual average. The first two rainfall conditions were obtained by using two rainout shelters with two manipulated rainfall doses: 97 mm (R33) and 194 mm (R66). For the three other rainfall conditions, we artificially increased rainfall in unsheltered plots using a watering pot: 295 mm (CK), 392 mm (R133), and 490 mm (R166). The temperature consisted of two levels: the actual temperature (CK) and the interaction between rainfall and the temperature, which was increased by about 2 °C (T) with the OTC (Open-Top Chamber) in each plot.

4.3. Collection of Soil Microorganism Samples

In each plot, including the inner OTC, we collected 0–10 cm of soil from each sample plot. We removed the impurities in the soil, including plants, moss, visible roots, litter, and visible soil animals, and then wiped the sampler with alcohol-soaked cotton. After the alcohol had completely evaporated, we used the soil in the sample to soak the sampler. This step needed to be repeated each time the sample changed. Three points sampled from the same quadrant were mixed as one soil sample. We placed mixed soil into a 10 mL centrifuge tube and then transferred it to a −80 °C refrigerator for determination of soil microbes. We used the Operational Taxonomic Units (OTU) lever to determine the soil microorganism taxonomic group. The Operational Taxonomic Unit (OTU) is an operational definition used to classify groups of closely related individuals. In the context of numerical taxonomy, the most abundant sequence type was selected to represent each OUT [27]. In our research, OTUs are in the absence of traditional systems of biological classification (which are available for macroscopic organisms), pragmatic proxies for "species" (microbial). For several years, OTUs have been the most commonly used units of diversity, especially when analyzing small subunit 16S for prokaryotes (as is the case of this work bacteria) or 18S (fungi) marker gene sequence datasets [28].

4.4. Collection of Plant Samples

The measurement of community α-diversity can be divided into the Shannon–Wiener diversity index, Margalef species richness index, Pielou evenness index, and Simpson dominance index. These four commonly used α-diversity indexes incorporate two measurements, the number (richness) of species, and the uniformity of species.

The richness index mainly measures the number of species within a certain spatial range to express the richness of organisms; the evenness index is a single statistic that combines the richness index and the evenness index; and the diversity index is based on the number of species to reflect a community's diversity, which can describe the disorder and uncertainty of an individual species. An increase in the number of species in the community represents an increase in the community's complexity. The greater the index value, the greater the amount of information contained in the community [9]. The importance value (IV) is calculated using the relative density, relative frequency, relative coverage, relative height, and relative biomass, according to the following formula:

$$\text{Importance value (IV)} = (\text{Relative density} + \text{relative coverage} + \text{relative frequency} + \text{relative height} + \text{relative biomass})/5$$

We calculated plant diversity in terms of the number of species in each plot(s), the relative importance of the species in the plot (Pi), and the number of individuals in all species (N), according to the following formula:

Shannon–Wiener diversity index: $H = -\sum_{i=1}^{s} \text{PilnPi}$
Margalef species richness index: $R = (S - 1)/\log_{10} N$
Pielou evenness index: $E = H/\ln S$
Simpson dominance index: $C = 1 - \sum(\text{Pi})^2$

Dominant plant species: We measured the plant relative biomass, relative height, relative cover, relative frequency, and relative density to determine the importance values, and then used importance values to determine the dominant species.

Plant carbon, nitrogen, and phosphorus: Plant organic carbon and total carbon were measured with a total organic carbon (TOC) analyzer (CS Analysis Instrument, Naples, FL, USA), and total nitrogen and total phosphorus were measured with a $HCLO_4$-H_2SO_4 digestion-flow injection instrument (model Skalar-SAN++, Delft, The Netherlands).

Litter: We picked up the litter on the ground by hand in the sample squares that cut off the ground plants and carefully removed the fine soil particles attached to the litter and put them in the envelopes according to the sample squares. We dried samples at 65 °C to constant weight, weighed them, and recorded the dry weight data.

Plant height: The natural height of different plants in each sample box was measured 5 times, respectively. If there were not 5 plants, the plants outside the sample box were selected for measurement.

Plant coverage: The acupuncture method was used. A 1 m^2 square sample rope was placed on the ground and divided into an average of 100 grids. Plants were acupunctured in order every 10 cm from top to bottom and from left to right with a 2 mm needle. If the needle contacted with the plant, it was counted as 1, and it was not counted if there was no contact. The coverage of each plant in the sample was not more than 100. If two plants occurred simultaneously during acupuncture, the total coverage was reduced by 1, and if three plants occurred simultaneously, the total coverage was reduced by 2. (Note: Total coverage is the sum of coverage of each plant.)

Plant frequency: Rounds were thrown 10–15 times in the plot, and the total number of times that each plant appeared was the frequency of the plant. (Note: In each round thrown, as long as the plant appeared, regardless of the number of plants, the frequency was 1). The total frequency was the sum of the frequency of each plant.

Plant density: The kinds of plants and the number of times each plant appeared in the sample box were recorded.

Plant biomass: We measured plant biomass in a 1 m^2 quadrat that was randomly selected in each plot at the end of July 2019. We dug all plants in each plot out from the soil. We then cut the aboveground living plant and plant roots separately while sorting them according to species and placed them into respective envelopes. These species were then taken into the laboratory and dried at 65 °C in the oven for 48 h; the aboveground living plant biomass (ALB) and plant root biomass (RB) were then calculated.

4.5. Statistical Analysis

We used repeated-measures ANOVA to examine the differences in the plant α-diversity index, the number of species and plants and the dominant species of the biomass, the organic carbon, total nitrogen, and total phosphorus of the plants and the dominant plant species under different precipitation levels and the interaction between precipitation and temperature by SPSS 21.0. t-tests were used to examine the differences in the plant α-diversity index, the number of species and plants and the dominant species of biomass, and the organic carbon, total nitrogen, and total phosphorus of plants and dominant plant species under different temperatures by SPSS 21.0. The soil microorganism β-diversity was analyzed by principal component analysis (PCA) using R. We used principal components analysis (PCA) to examine the relationships between the grassland plant diversity, biomass, and soil bacteria and fungi α- and β-diversity by Origin 2021.

5. Conclusions

In this study, plant α-diversity in CK and TCK under the altered precipitation were significantly higher than other treatments. Under the interaction of the increasing precipitation and the rising temperature conditions, R166 promoted the number of species the most. Increasing precipitation was found to promote the growth of RB more than ALB, but the effect of rising temperatures on RB was not clear. The changing precipitation and increasing temperature factors, and the interaction of the two factors, all had no significant impact on the biomass, organic carbon, total nitrogen, and total phosphorus of plants. R166 promoted the ALB, RB, and total biomass of Agropyron mongolicum the most. The TR treatment promoted plant organic carbon, nitrogen, and phosphorus content more than R. In the fungi communities, under rising temperature, increasing precipitation promoted α-diversity, but α-diversity did not obviously vary in the bacteria communities. In the fungi communities, TCK promoted the most β-diversity, but in the bacteria communities, CK promoted the most β-diversity. Soil bacteria and fungi α- and β-diversity, and ALB promoted plant diversity the most.

Author Contributions: Conceptualization, J.L.; investigation, Y.Z., L.J., J.Z. and Y.W.; writing—original draft preparation, Y.Z.; writing—review and editing, J.L.; project administration, Y.X., J.L. and H.M.; funding acquisition, Y.X. and J.L. All authors have read and agreed to the published version of the manuscript.

Funding: This research was funded by the Ningxia key research and development program (2020BEG03046), the funder is Jian-Ping Li, and the Top Discipline Construction Project of Pratacultural Science (NXYLXK2017A01), the funder is Yingzhong Xie.

Institutional Review Board Statement: Not applicable.

Informed Consent Statement: Not applicable.

Data Availability Statement: Not applicable.

Acknowledgments: We thank the team from Majorbio Bio-Pharm Technology Co., Ltd. (Shanghai, China) for technical advice relating to our microbial analysis.

Conflicts of Interest: The authors declare no conflict of interest.

References

1. Zhang, X.K.; Du, X.D.; Hong, J.T.; Du, Z.Y.; Lu, X.Y.; Wang, X.D. Effects of climate change on the growing season of alpine grassland in Northern Tibet, china. *Glob. Ecol. Conserv.* **2020**, *23*, e01126. [CrossRef]
2. Zhang, R.; Guo, J.; Liang, T.; Feng, Q. Grassland vegetation phenological variations and responses to climate change in the Xinjiang region, China. *Quat. Int.* **2019**, *513*, 56–65. [CrossRef]
3. Panitsa, M.; Iliopoulou, N.; Petrakis, E. Citizen Science, Plant Species, and Communities' Diversity and Conservation on a Mediterranean Biosphere Reserve. *Sustainability* **2021**, *13*, 9925. [CrossRef]
4. Damasceno, G.; Fidelis, A. Abundance of invasive grasses is dependent on fire regime and climatic conditions in tropical savannas. *J. Environ. Manag.* **2020**, *271*, 111016. [CrossRef]
5. Garske, B.; Ekardt, F. Economic policy instruments for sustainable phosphorus management: Taking into account climate and biodiversity targets. *Environ. Sci. Eur.* **2021**, *33*, 1–20. [CrossRef]
6. Maršík, P.; Zunová, T.; Vaněk, T.; Podlipná, R. Metazachlor effect on poplar—Pioneer plant species for riparian buffers. *Chemosphere* **2021**, *274*, 129711. [CrossRef]
7. Noroozi, J.; Moser, D.; Essl, F. Diversity, distribution, ecology and description rates of alpine endemic plant species from Iranian mountains. *Alp. Bot.* **2016**, *126*, 1–9. [CrossRef]
8. Shah, S.Q.; Khan, N.; Ahmad, S.U.; Bashir, Z.; Mahmood, T. The relationship of synonymous codon usage bias analyses of stress resistant and reference genes in significant species of plants. *Pak. J. Bot.* **2021**, *53*, 841–846. [CrossRef]
9. Keke, U.N.; Arimoro, F.O.; Ayanwale, A.V.; Odume, O.N.; Edegbene, A. Weak relationships among macroinvertebrates beta diversity (β), river status, and environmental correlates in a tropical biodiversity hotspot. *Ecol. Indic.* **2021**, *129*, 107868. [CrossRef]
10. Múgica, L.; Canals, R.M.; Emeterio, L.S.; Peralta, J. Decoupling of traditional burnings and grazing regimes alters plant diversity and dominant species competition in high-mountain grasslands. *Sci. Total Environ.* **2021**, *790*, 147917. [CrossRef]
11. Manolis, E.N.; Manoli, E.N. Raising awareness of the Sustainable Development Goals through Ecological Projects in Higher Education. *J. Clean. Prod.* **2021**, *279*, 123614. [CrossRef]
12. Kudo, G. Dynamics of flowering phenology of alpine plant communities in response to temperature and snowmelt time: Analysis of a nine-year phenological record collected by citizen volunteers. *Environ. Exp. Bot.* **2020**, *170*, 170. [CrossRef]
13. Wan, Z.Q.; Yan, Y.L.; Chen, Y.L.; Gu, R.; Gao, Q.Z. Ecological responses of Stipa steppe in Inner Mongolia to experimentally increased temperature and preccipitation. 2: Plant species diversity and sward characteristics. *Rangel. J.* **2018**, *40*, 147–152. [CrossRef]
14. McGuire, K.; Treseder, K. Microbial communities and their relevance for ecosystem models: Decomposition as a case study. *Soil Biol. Biochem.* **2010**, *42*, 529–535. [CrossRef]
15. van Bruggen, A.; Semenov, A. In search of biological indicators for soil health and disease suppression. *Appl. Soil Ecol.* **2000**, *15*, 13–24. [CrossRef]
16. Li, H.; Qiu, Y.; Yao, T.; Han, D.; Gao, Y.; Zhang, J.; Ma, Y.; Zhang, H.; Yang, X. Nutrients available in the soil regulate the changes of soil microbial community alongside degradation of alpine meadows in the northeast of the Qinghai-Tibet Plateau. *Sci. Total Environ.* **2021**, *792*, 148363. [CrossRef]
17. Zhumanova, M.; Wrage-Mönnig, N.; Jurasinski, G. Long-term vegetation change in the Western Tien-Shan Mountain pastures, Central Asia, driven by a combination of changing precipitation patterns and grazing pressure. *Sci. Total Environ.* **2021**, *781*, 146720. [CrossRef]
18. Bagriacik, B.; Sani, Z.K.; Uslu, F.M.; Yigittekin, E.S.; Dincer, S. An experimental approach to microbial carbonate precipitation in improving the engineering properties of sandy soils. *Ann. Microbiol.* **2021**, *71*, 1–13. [CrossRef]
19. Motevalli-Haghi, S.F.; Ozbaki, G.M.; Hosseini-Vasoukolaei, N.; Nikookar, S.H.; Dehghan, O.; Yazdani-Charati, J.; Siahsarvie, R.; Dehbandi, R.; Fazeli-Dinan, M.; Enayati, A. Rodent Species Diversity and Occurrence of Leishmania in Northeastern Iran. *Pol. J. Ecol.* **2021**, *69*, 57–70. [CrossRef]
20. Escaravage, V.; Herman, P.; Merckx, B.; Wlodarska-Kowalczuk, M.; Amouroux, J.; Degraer, S.; Grémare, A.; Heip, C.; Hummel, H.; Karakassis, I.; et al. Distribution patterns of macrofaunal species diversity in subtidal soft sediments: Biodiversity–productivity relationships from the MacroBen database. *Mar. Ecol. Prog. Ser.* **2009**, *382*, 253–264. [CrossRef]
21. Li, X.; Zhang, C.; Zhang, B.; Wu, D.; Zhu, D.; Zhang, W.; Ye, Q.; Yan, J.; Fu, J.; Fang, C.; et al. Nitrogen deposition and increased precipitation interact to affect fine root production and biomass in a temperate forest: Implications for carbon cycling. *Sci. Total Environ.* **2021**, *765*, 144497. [CrossRef] [PubMed]
22. Li, S.-X.; Wang, Z.-H.; Malhi, S.; Li, S.-Q.; Gao, Y.-J.; Tian, X.-H. Nutrient and Water Management Effects on Crop Production, and Nutrient and Water Use Efficiency in Dryland Areas of China. *Adv. Agron.* **2009**, *102*, 223–265. [CrossRef]
23. Shi, P.; Qin, Y.; Liu, Q.; Zhu, T.; Li, Z.; Li, P.; Ren, Z.; Liu, Y.; Wang, F. Soil respiration and response of carbon source changes to vegetation restoration in the Loess Plateau, China. *Sci. Total Environ.* **2021**, *707*, 135507. [CrossRef] [PubMed]
24. Sunnemann, M.; Alt, C.; Kostin, J.E.; Lochner, A.; Reitz, T.; Siebert, J.; Schadler, M.; Eisenhauer, N. Low-intensity land-use enhances soil microbial activity, biomass and fungal-to-bacterial ratio in current and future climates. *J. Appl. Ecol.* **2021**, *39*, 1–12. [CrossRef]
25. Benizri, E.; Nguyen, C.; Piutti, S.; Slezack-Deschaumes, S.; Philippot, L. Additions of maize root mucilage to soil changed the structure of the bacterial community. *Soil Biol. Biochem.* **2007**, *39*, 1230–1233. [CrossRef]

26. Wan, Z.W.; Hu, G.Z.; Chen, Y.; Chao, C.; Gao, Q.Z. Ecologial responses of Stipa steppe in Inner Mongolia to experimentally increased temperature and precipitation: 1: Background and experimental design. *Rangel. J.* **2018**, *40*, 143–146. [CrossRef]
27. Baldrian, P.; Větrovský, T.; Lepinay, C.; Kohout, P. High-throughput sequencing view on the magnitude of global fungal diversity. *Fungal Divers.* **2021**, 1–9. [CrossRef]
28. Du, Y.; Ke, X.; Dai, L.; Cao, G.; Zhou, H.; Guo, X. Moderate grazing increased alpine meadow soils bacterial abundance and diversity index on the Tibetan Plateau. *Ecol. Evol.* **2020**, *10*, 8681–8687. [CrossRef]

Article

The Effect of Soil Water Deficiency on Water Use Strategies and Response Mechanisms of *Glycyrrhiza uralensis* Fisch

Kechen Song [1], Haiying Hu [1,2,*], Yingzhong Xie [1,2,*] and Li Fu [3]

[1] College of Agriculture, Ningxia University, Yinchuan 750021, China; nxdxskc@163.com
[2] Breeding Base for State Key Laboratory of Land Degradation and Ecological Restoration of North-Western China, Ningxia University, Yinchuan 750021, China
[3] College of Politics and History, Ningxia Normal University, Yinchuan 750021, China; fuli39501793@163.com
* Correspondence: haiying@nxu.edu.cn (H.H.); xieyz@nxu.edu.cn (Y.X.); Tel.: +86-951-2061351

Abstract: We aimed to investigate the water use strategies and the responses to water shortages in *Glycyrrhiza uralensis*, which is a dominant species in the desert steppe. Water stress gradients included control, mild, moderate, and severe. The time intervals were 15, 30, 45, and 60 d. Our study suggested that with the aggravation of water stress intensity, the total biomass of *Glycyrrhiza uralensis* gradually decreased and allometric growth was preferred to underground biomass accumulation. From 30 d and mild to moderate water stress, the water potential (WP) of leaves decreased considerably compared to the CK. The relative water content (EWC) decreased over time and had a narrow range of variation. Proline (PR) was continuously increased, then declined at 45–60 d under severe and more severe water stress. The $\delta^{13}C$ values increased in all organs, showed roots > stems > leaves. The net photosynthetic rate (Pn) and transpiration rate (Tr) decreased to varying degrees. The instantaneous water use efficiency (WUEi) and limiting value of stomata (Ls) increased continuously at first and decreased under severe water stress. Meanwhile, severe water stress triggered the most significant changes in chloroplast and guard cell morphology. In summary, *Glycyrrhiza uralensis* could maintain water content and turgor pressure under water stress, promote root biomass accumulation, and improve water use efficiency, a water-conservation strategy indicating a mechanism both avoidable dehydration and tolerable drought.

Keywords: water stress; *Glycyrrhiza uralensis*; biomass allocation; water use efficiency; drought resistance

Citation: Song, K.; Hu, H.; Xie, Y.; Fu, L. The Effect of Soil Water Deficiency on Water Use Strategies and Response Mechanisms of *Glycyrrhiza uralensis* Fisch. *Plants* **2022**, *11*, 1464. https://doi.org/10.3390/plants11111464

Academic Editor: James A. Bunce

Received: 19 April 2022
Accepted: 25 May 2022
Published: 30 May 2022

Publisher's Note: MDPI stays neutral with regard to jurisdictional claims in published maps and institutional affiliations.

1. Introduction

In the desert steppe, water is the most important factor limiting plant growth. Different plants adopt water use strategies to adapt to changes of water conditions [1]. A plant's drought adaptability is closely related to their water use strategies [2,3]. Zhang et al. believes that drought resistance and the ability to use water effectively are essential mechanisms for revealing a plant's drought resistance under long-term water stress conditions [4]. Therefore, the adaptation of plants to arid habitats requires the evolution of their water use strategies.

A previous study reported that plants could adjust the distribution of biomass among different organs to adapt to water stress, showing that the biomass of various organs of plants decreases due to water stress, while the root to shoot ratio (R/S) increases [5]. The biomass allocation strategy is also directly influenced by the ability of plants to adapt to their environment. Plants allocate resources to the most needed organs after environmental changes to effectively obtain scarce resources [6,7]. Plants obtain water by increasing root biomass when water is deficient. When soil moisture is sufficient, plants promote photosynthetic capacity by increasing the distribution ratio of above-ground biomass. However, above-ground biomass and underground biomass accumulation are not synchronized [8]. Therefore, plant organ biomass allocation results from a balance between reproduction and survival and a trade-off strategy for plants to adapt to their environment.

Precipitation can change Ci/Ca (intercellular CO_2 concentration/atmospheric CO_2 concentration) by changing the stomatal conductance of leaves, which then affects the $\delta^{13}C$ value of plants [9]. The $\delta^{13}C$ values of C_3 plants in arid and desert areas ranged from $-20‰$ to $-35‰$ with decreasing precipitation. In extreme arid areas, the $\delta^{13}C$ values ranged from $-20‰$ to $-26‰$, and precipitation has a significant negative correlation with the $\delta^{13}C$ values [10]. It has been suggested that C_3 plant tissues under water stress generally have higher $\delta^{13}C$ values, while those under non-water stress have lower $\delta^{13}C$ values [11], but most of the previous results were from field experiments. Liu Ying et al. studied the effects of different drought stress conditions on $\delta^{13}C$ values of the C_4 plant Leymus chinensis and found that the $\delta^{13}C$ values were significantly positively correlated with water use efficiency (WUE) [12]. It is feasible to determine water use efficiency of *L. chinensis* by the $\delta^{13}C$ value. However, some earlier studies suggested that $\delta^{13}C$ values have growth stages and organ specificity when used to study WUE and biomass in plants [13].

Glycyrrhiza uralensis is the dominant species in the natural restoration of the desert steppe. It is not only a high-quality pasture and a Chinese herbal medicine, but it also has a higher ability to survive and compete, making it important in the study of grassland restoration. Hu found that *Glycyrrhiza uralensis* has the highest $\delta^{13}C$ value and photosynthetic rate among four dominant species of the desert steppe, including *Glycyrrhiza uralensis*, *Lespedeza potaninii*, *Stipa breviflora*, and *Agropyron mongolicum*, under various precipitation conditions [14,15]. This research focuses on *Glycyrrhiza uralensis*. It uses a water control experiment to investigate the responses of biomass allocation, water use efficiency, and physiological and morphological characteristics to water stress to uncover the water use strategies and drought resistance mechanisms in *Glycyrrhiza uralensis*. It gives a scientific basis for choosing species to restore the grassland to its natural state.

2. Results

2.1. Biomass Allocation and the Root–Shoot Ratio of Glycyrrhiza uralensis under Water Stress

Under the same level of water stress, the total biomass of *Glycyrrhiza uralensis* increased first, then decreased; the highest biomass was obtained after 45 d of treatment. The root and leaf biomass accumulated quickly within 30–45 d, with root biomass at 45 d and 60 d significantly higher than at 15 and 30 d ($p < 0.01$). The root–shoot ratio (R/S) of *Glycyrrhiza uralensis* gradually increased over time, the R/S values under different water stress all showed 60 d > 45 d > 30 d > 15 d.

Under the same time, the total biomass of *Glycyrrhiza uralensis* decreased gradually as the degree of water stress increased, but the stem biomass remained unchanged. The R/S increased significantly ($p < 0.05$). The total biomass of the T4 treatment was the lowest, and the R/S of the T4 treatment was the highest at 15, 30, 45 and 60 d. These results indicated that drought stress resulted in more allocation of biomass to the roots and less allocation to the stems and leaves (Figure 1).

2.2. Growth Relationship between Above-Ground and Underground Biomass of Glycyrrhiza uralensis under Water Stress

Under different water stress treatments, there was an extremely significant correlation between the below-ground biomass (BGB) and the above-ground biomass (AGB) of *Glycyrrhiza uralensis* ($p < 0.01$), and the allometric growth relationship was biased toward under-ground biomass accumulation (Figure 2). Each treatment had a correlation growth index α greater than one. T2 had the highest α index ($\alpha = 1.9$), the CK had the lowest, and T3 had the best fitting effect ($R^2 = 0.8592$).

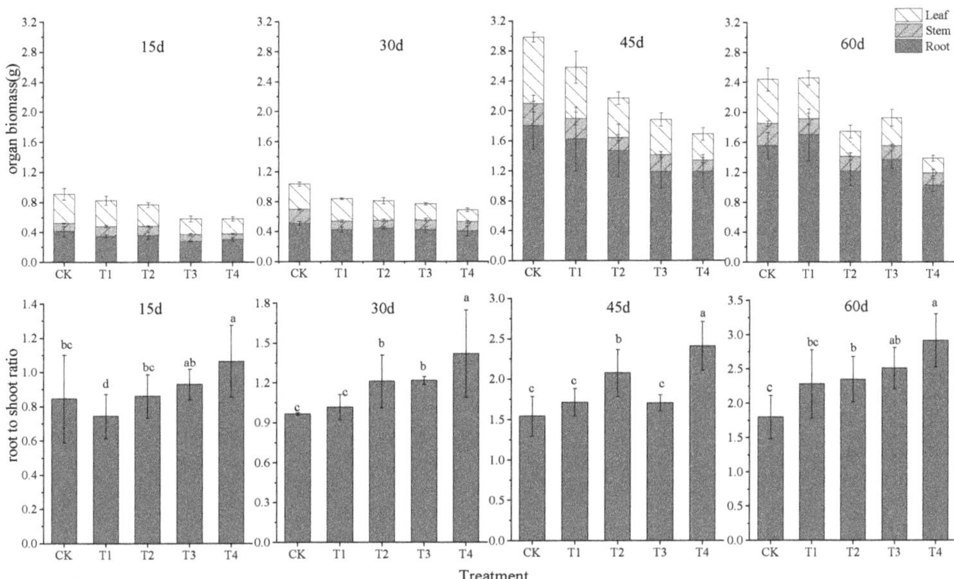

Figure 1. The biomass distribution in organs of *Glycyrrhiza uralensis* under different water stress treatments. Different lowercase letters indicate significant differences among treatments at 5% level.

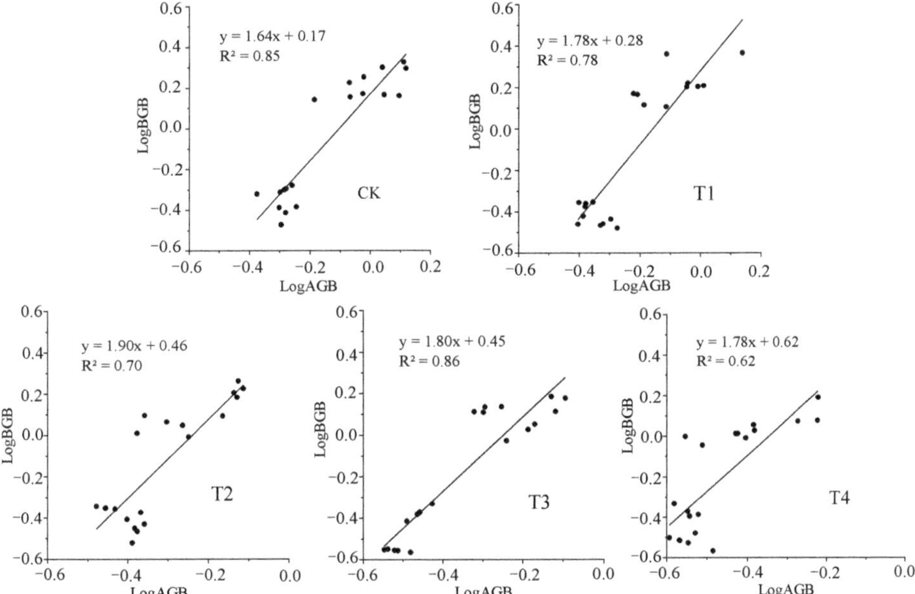

Figure 2. Growth relationship between above-ground and underground biomass of *Glycyrrhiza uralensis* in different water stress treatments. AGB and BGB represent the above-ground biomass and below-ground biomass. The biomass data are logarithmic transformed, its power function is converted to the form of $\log = \log\beta + \alpha\log X$, where α is the slope of linear regression, and $\log\beta$ is the intercept of linear regression.

2.3. Effects of Water Stress on Water Potential of Glycyrrhiza uralensis Leaves

The leaf water potential (WP) of *Glycyrrhiza uralensis* decreased to varying degrees compared to the CK under water stress (Figure 3). With the increase in the water stress degree, WP remained unchanged at the early stages (15 d), decreased first, increased at the middle stages (30 and 45 d), and gradually increased at the late stages (60 d). Except for 15 d, the CK and T1 treatments had a significantly higher WP than T2, T3, and T4 treatments ($p < 0.05$).

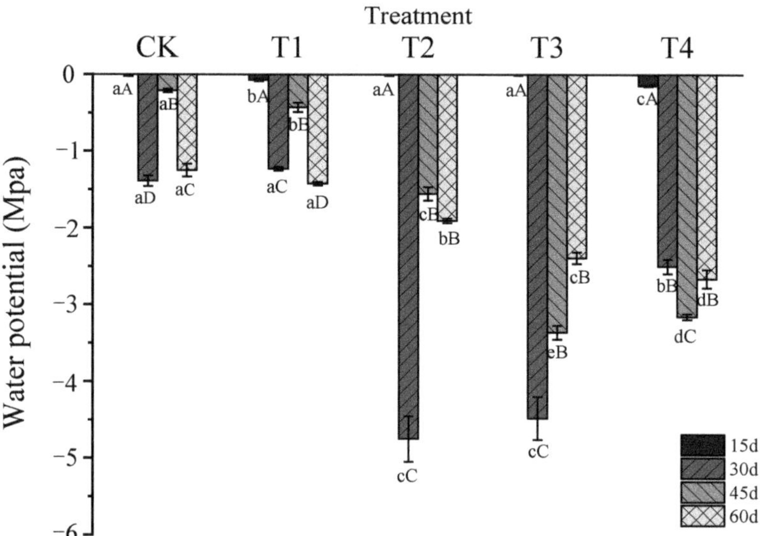

Figure 3. Changes of water potential in leaves of *Glycyrrhiza uralensis* under different water stress treatments. Different lowercase letters indicate a significant difference among treatments at a 5% level. Different capital letters indicate a significant difference at a 5% level between different times.

The CK and T1 treatments fluctuated over time. The WP of the T2 and T3 treatment decreased rapidly at 30 d and then increased rapidly, the lowest at 30 d and the highest at 15 d. The WP at 15 d, 45 d, and 60 d was significantly higher than that at 30d, and the WP at 15 d was significantly higher than that at 45 d and 60 d ($p < 0.05$). The WP of the T4 treatment decreased rapidly and increased slowly, with the lowest at 45 d and the highest at 15 d. The WP at 15 d was significantly higher than that at 30 d, 45 d, and 60 d, and the WP at 30 d and 65 d was significantly higher than that at 45 d ($p < 0.05$).

2.4. Effects of Water Stress on Relative Water Content and Proline Content of Glycyrrhiza uralensis Leaves

The relative water content (RWC) of *Glycyrrhiza uralensis* leaves was significantly affected by both the degree of water stress and time passage (Table 1). The RWC continued to decrease slowly as the degree of water stress increased ($p < 0.05$), while proline content (PR) continued to increase significantly ($p < 0.05$). Compared with the CK, the RWC of T1, T2, T3, and T4 treatments decreased by 8.21%, 11.79%, 21.26%, 21.99%, and the PR of T1, T2, T3, and T4 treatments increased by 44.60%, 111.07%, 174.30%, and 246.91%, respectively.

Table 1. The effects of different water stress treatments on RWC and proline content of *Glycyrrhiza uralensis*. Note: Different lowercase letters indicate a significant difference among treatments or times at a 5% level.

Treatment	Relative Water Content (RWC)	Proline Content (PR)
CK	85.48 ± 1.59a	70.46 ± 9.90e
T1	78.46 ± 1.95b	101.89 ± 15.39d
T2	72.40 ± 2.17c	148.72 ± 18.24c
T3	67.31 ± 2.25d	193.28 ± 19.29b
T4	66.68 ± 5.75d	244.43 ± 13.54a
15 d	67.81 ± 3.20c	176.51 ± 16.90b
30 d	78.84 ± 2.24a	201.02 ± 21.84a
45 d	78.25 ± 1.55a	119.35 ± 17.16c
60 d	71.36 ± 1.54b	110.14 ± 14.60c

The RWC at 30 and 45 d was significantly higher than that at 15 and 60 d ($p < 0.05$) as the stress time increased. The PR of *Glycyrrhiza uralensis* leaves increased at first, then decreased. At 30 d, the PR was significantly higher than at 15, 45, and 60 d; at 15 d, PR was significantly higher than at 45, and 60 d ($p < 0.05$).

2.5. Effects of Water Stress on $\delta^{13}C$ Values in Different Organs of Glycyrrhiza uralensis

In different organs of *Glycyrrhiza uralensis*, the $\delta^{13}C$ values were as follows: $\delta^{13}C_{root}$ > $\delta^{13}C_{stem}$ > $\delta^{13}C_{leave}$, the $\delta^{13}C$ values of different organs were significantly different ($p < 0.05$).

The $\delta^{13}C$ values of all organs showed an upward trend as the degree of water stress increased. The $\delta^{13}C_{root}$ of T2, T3, and T4 treatments was significantly higher than the CK and the T1 treatment. The $\delta^{13}C_{stem}$ of the T4 treatment was significantly higher than the CK, T1, and T2 treatments, and the $\delta^{13}C_{leave}$ was significantly higher than the CK, T1, T2, and T3 treatments ($p < 0.05$) (Figure 4).

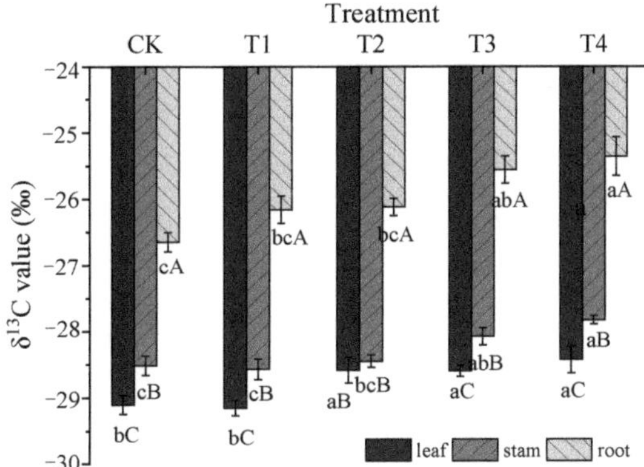

Figure 4. The composition of $\delta^{13}C$ value in organs of *Glycyrrhiza uralensis* under different water stress conditions. Different lowercase letters indicate a significant difference among treatments at the 5% level. Different capital letters indicate a significant difference at a 5% level between different organs.

2.6. Effects of Water Stress on the Gas Exchange Parameters of Glycyrrhiza uralensis

As demonstrated in Table 2, the Pn, Tr, Ci, and Gs of *Glycyrrhiza uralensis* leaves decreased with the increasing degree of water stress, the CK had the highest, and the CK and the T1 treatment were significantly higher than the T3 and T4 treatments ($p < 0.05$). The WUEi and Ls increased continuously at first, and then decreased significantly in the

T4 treatment, with the WUEi of T1, T2, and T3 treatments being significantly higher than the CK and T4 treatment ($p < 0.05$).

Table 2. Effects of water stress on gas exchange parameters of *Glycyrrhiza uralensis*. Different lowercase letters indicate a significant difference at a 5% level between different times.

Index	CK	T1	T2	T3	T4
Net photosynthetic rate (Pn)	19.74 ± 0.77a	17.61 ± 0.37b	16.84 ± 0.65b	10.14 ± 0.92c	3.62 ± 0.24d
Transpiration rate (Tr)	17.16 ± 0.56a	11.39 ± 0.20b	10.42 ± 0.44b	6.10 ± 0.46c	4.27 ± 0.45d
Intercellular CO_2 concentration (Ci)	356.90 ± 2.41a	334.15 ± 2.31b	319.93 ± 2.05c	293.90 ± 5.39d	315.64 ± 5.79c
Stomatal conductance (Gs)	1.07 ± 0.07a	0.50 ± 0.23b	0.39 ± 0.02c	0.18 ± 0.01d	0.10 ± 0.01d
Instantaneous water use efficiency (Wuei)	0.34 ± 0.02b	0.67 ± 0.03a	0.84 ± 0.04a	1.70 ± 0.14a	3.18 ± 0.33c
Limiting value of stomata (Ls)	1.15 ± 0.04d	1.55 ± 0.02c	1.63 ± 0.09b	1.66 ± 0.06a	0.87 ± 0.08bc

2.7. Effects of Water Stress on Chloroplast and Stomatal Ultrastructure

Under normal water conditions, the chloroplasts of *G. uralensis* mesophyll cells were long and semi-circular, distributed close to the cell edge, with a complete structure and a clear membrane structure. Among them, the crenellations of thylakoids were tight and smooth, and the stromal lamellae were more evenly distributed and arranged, and there were a small number of mesophyll granules. The ultrastructure of chloroplasts changed to varying degrees as the degree of water stress increased (Figure 5a). The T4 treatment caused the chloroplast to become shorter and swollen to varying degrees, and the membrane structure of the chloroplasts gradually blurred. The number of osmiophilic granules increased and accumulated. Starch grains appeared, and their volume gradually increased. The number of granalamellae decreased, and the structure was fuzzy in the T4 treatment.

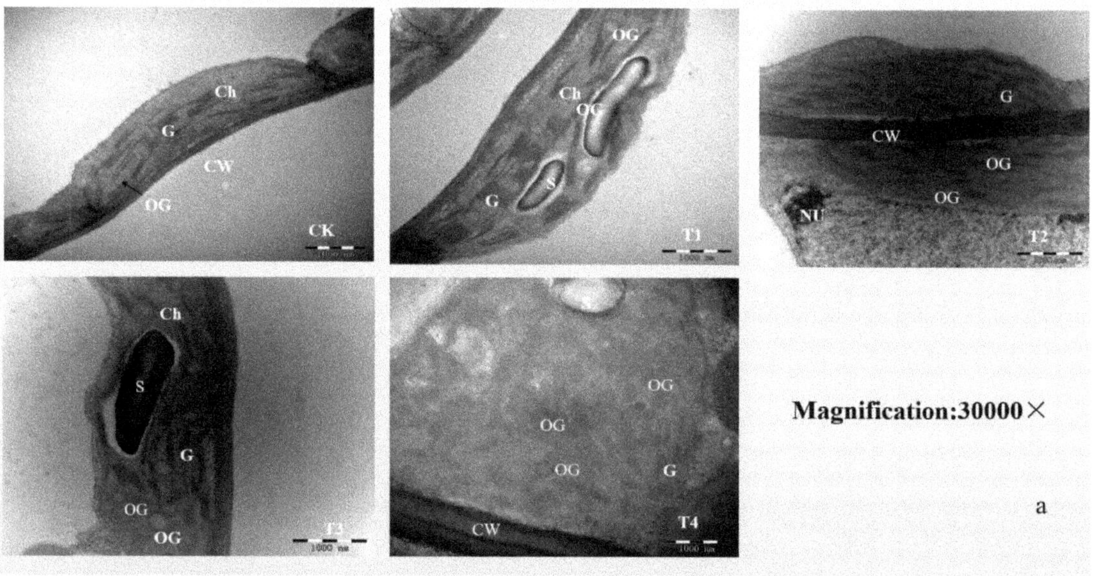

Figure 5. *Cont.*

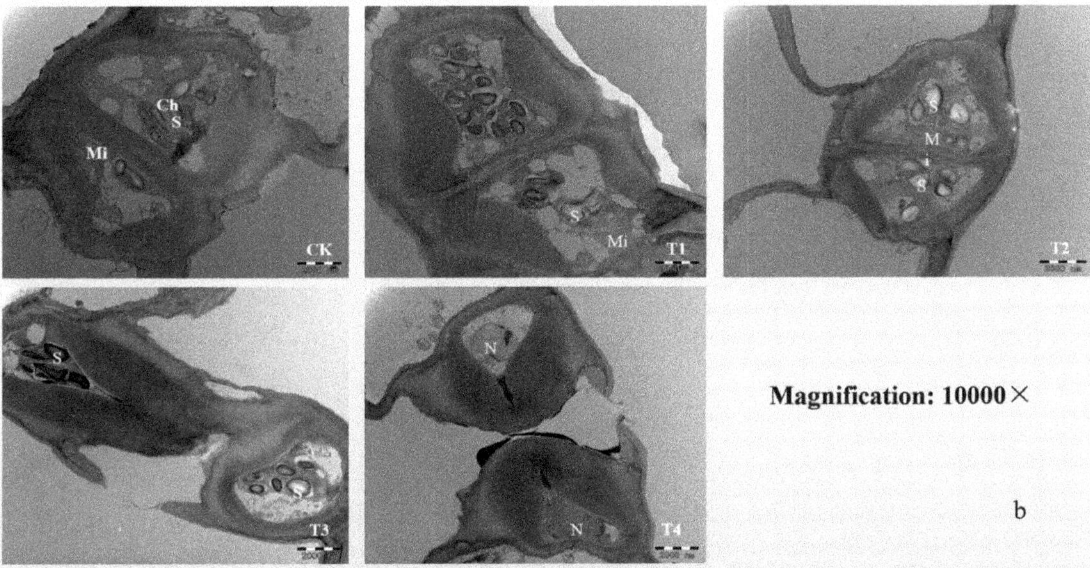

Figure 5. (**a**,**b**): Effects of water stress on chloroplast and stomatal ultrastructures of *Glycyrrhiza uralensis*. Ch, Mi, G, S, OG, P, CW, and N represent the chloroplast, mitochondrion, granalamellae, starch grain, osmiophilic granule, plastoglobulis, cell wall, and nucleus, respectively.

The guard cells in the leaves of *Glycyrrhiza uralensis* were found to have a typical reniform equithick wall, and the two guard cells were arranged symmetrically. Under normal water conditions, the thickness of the upper cell wall of guard cells was larger than that of the lower cell wall. The chloroplast, mitochondria, nuclear stomatal cavity, and other structures are clear and distinguishable, with normal morphology and more starch grains.

The guard cells became smaller as the degree of water stress increased, with uneven cell wall thickening (Figure 5b—T3, T4), protoplast shrinking volume (Figure 5b—T2, T3, T4), and starch grains gradually disintegrating. The stomatal cavity was shaped like a slender wine cup and severely deformed (Figure 5b–T4).

2.8. Analysis of Different Organ Biomass and Related Physiological Indexes of Glycyrrhiza uralensis

The biomasses of *Glycyrrhiza uralensis* organs were negatively correlated with R/S, $\delta^{13}C_{root}$, and PR, and positively correlated with Tr and Gs ($p < 0.05$) (Figure 6). The $\delta^{13}C$ values of each organ were positively correlated with the R/S and PR and negatively correlated with the leaf biomass, Pn, Tr, and Gs ($p < 0.05$). The R/S and PR were negatively correlated with the Pn, Ti, Ci, and Gs ($p < 0.05$). The RWC and WP were positively correlated with the Tr, Ci, and Gs and negatively correlated with the Ls ($p < 0.05$).

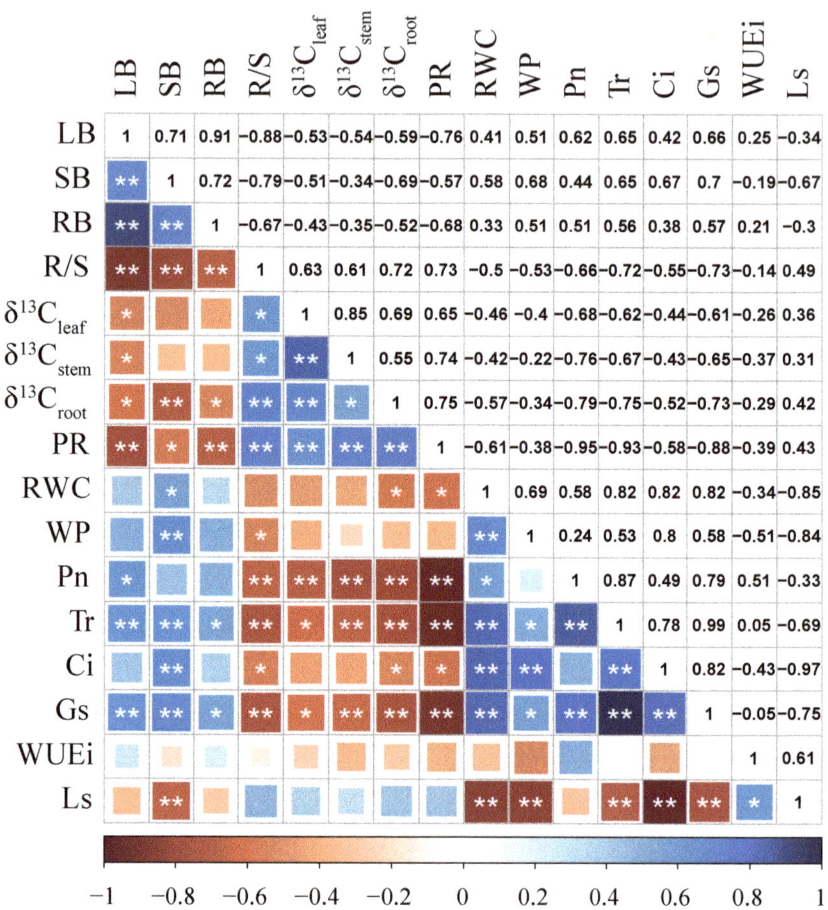

Figure 6. Correlation coefficient between different organ biomasses and related physiological indexes of *Glycyrrhiza uralensis*. $\delta^{13}C_{leaf}$, $\delta^{13}C_{stem}$, and $\delta^{13}C_{root}$ represent the $\delta^{13}C$ values of leaf, stem, and root; LB, SB, and RB represent the leaf biomass, stem biomass, and root biomass; R/S, PR, RWC, WP, Pn, Tr, Ci, Gs, WUEi, and Ls represent the root to shoot ratio, proline, relative water content, water potential, net photosynthetic rate, transpiration rate, intercellular CO_2 concentration, stomatal conductance, instantaneous water use efficiency, and limiting value of stomata, respectively. * and ** indicate significant correlation at 0.05 and 0.01 levels, respectively.

2.9. The Relationship between the Biomass of Different G.uralensis Organs and the Photosynthetic Physiological Indexes under Water Stress

Redundancy analysis revealed that RDA1 axis could explain 85.82% of the changes in organ biomass, which mainly reflected the changes in physiological regulatory factors. The proline content had the greatest influence on the changes of organ biomass, amounting to 54.3% of the explanations and 55.6% of the contribution rate ($p < 0.01$) (Figure 7). In addition, the Pn and $\delta^{13}C$ roots were important factors affecting the changes in organ biomass, they accounted for 9.9% of the explanation, contributed 10.1%, and the $\delta^{13}C_{root}$ had a significant influence ($p < 0.05$) (Table 3).

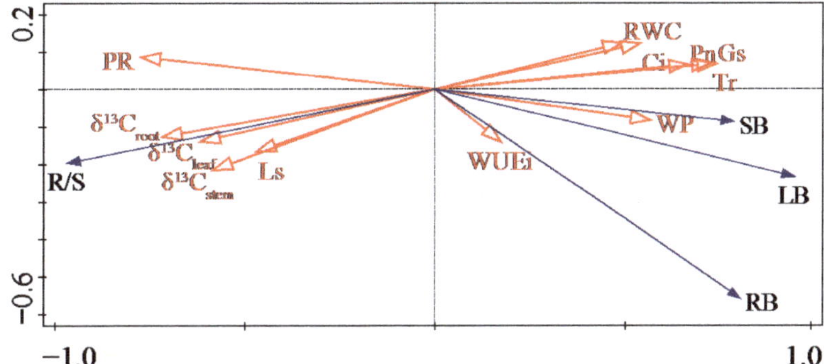

Figure 7. Results by redundancy analysis between biomass and physiological indexes of *Glycyrrhiza uralensis*. $\delta^{13}C_{leaf}$, $\delta^{13}C_{stem}$, and $\delta^{13}C_{root}$ represent the $\delta^{13}C$ values of leaf, stem, and root; LB, SB, and RB represent the leaf biomass, stem biomass, and root biomass; R/S, PR, RWC, WP, Pn, Tr, Ci, Gs, WUEi and Ls represent the root to shoot ratio, proline, relative water content, water potential, net photosynthetic rate, transpiration rate, intercellular CO_2 concentration, stomatal conductance, instantaneous water use efficiency, and limiting value of stomata, respectively.

Table 3. Results by redundancy analysis ordination with the first two axes and Monte Carlo permutation test.

Name	Explains %	Contribution %	Pseudo-F	*p*
Proline (PR)	54.3	55.6	15.4	0.002
Net photosynthetic rate (Pn)	9.9	10.1	3.3	0.072
$\delta^{13}C$ values of root ($\delta^{13}C_{root}$)	9.9	10.1	4.2	0.022
Stomatal conductance (Gs)	6.2	6.4	3.2	0.062
Water potential (WP)	2.8	2.8	1.5	0.25
Instantaneous water use efficiency (WUEi)	1.5	1.5	0.8	0.506
Transpiration rate (Tr)	0.8	0.8	0.4	0.72
$\delta^{13}C$ values of stem ($\delta^{13}C_{stem}$)	1.5	1.6	0.7	0.546
$\delta^{13}C$ values of leaf ($\delta^{13}C_{leaf}$)	5	5.1	3.1	0.084
Limiting value of stomata (Ls)	1.7	1.7	1.1	0.386
Intercellular CO_2 concentration (Ci)	0.9	0.9	0.5	0.648
Relative water content (RWC)	3.2	3.3	2.7	0.14
Statistic	Axis 1	Axis 2	Axis 3	Axis 4
Eigenvalues	0.8582	0.0754	0.0426	0.0003
Explained variation (cumulative)	85.82	93.36	97.62	97.65
Pseudo-canonical correlation	0.9959	0.9214	0.9671	0.9563

3. Discussion

3.1. Effects of Water Stress on Biomass Allocation of Glycyrrhiza uralensis

In response to water stress, changes in plant biomass and its distribution ratios to different organs reflect its adaptation methods and abilities to the ecological environment [16]. Meanwhile, there are stable allometric relationships between plant organ biomass and plant metabolism, this change in growth relationships may be the mechanism of drought tolerance in plants [17]. In the present study, the above-ground biomass and total biomass of *Glycyrrhiza uralensis* decrease to varying degrees, and the R/S significantly increased under severe water stress. Thus, the proportion of underground biomass was increased by slowing down the growth of the above-ground part, falling leaves, and other adaptive characteristics. This may be because plants often allocate more biomass to the ground to cope with extreme drought [18]. Under water deficit conditions, *Glycyrrhiza uralensis* increases the available soil water by increasing the root–shoot ratio and thereby boosting drought tolerance [19]. We also found a correlation between underground biomass and above-ground biomass under different water stresses ($p < 0.05$). The allometric growth content was greater than one, which further indicates that *Glycyrrhiza uralensis* allocates more resources to enhance the ability of water acquisition by the roots.

3.2. $\delta^{13}C$ Value Composition of Glycyrrhiza uralensis Organs under Water Stress

In this study, the $\delta^{13}C$ values of the roots of *Glycyrrhiza uralensis* were significantly higher than the leaves and stems ($p < 0.01$), and the $\delta^{13}C$ values of different organs were as follows: $\delta^{13}C_{roots} > \delta^{13}C_{stems} > \delta^{13}C_{leaves}$. Similarly, Gao also confirmed that the ^{13}C abundance of fine roots, thick roots, and thick branches was significantly higher than the leaves [20]. This may be because roots accumulate ^{13}C more easily than leaves and stems. Photosynthetic organs usually contain a low $\delta^{13}C$ value, and most of the stems of herbaceous plants are also green photosynthetic organs. Therefore, the isotopic fractionation metabolism of the stems of herbaceous plants is closer to the leaves than that of wood and roots [21,22]. This indicates that the ^{13}C fractionation produced by the leaves and stems was smaller than the roots. The main reason for plants ^{13}C fractionation may be the heterogeneity of the chemical composition of their organs, and organs with high cellulose content are more likely to enrich ^{13}C [23].

With water stress intensifies, plant species always have improved their $\delta^{13}C$ values, and WUE will increase with increasing $\delta^{13}C$ values [11,12]. The *Glycyrrhiza uralensis* in this study also conforms to this feature. However, we found that the leaf $\delta^{13}C$ value of seedling was significantly lower than that of uncultivated *Glycyrrhiza uralensis* [14,15], indicating that wild plants or cultivated perennials have higher leaf WUE than young plants. This indicates that $\delta^{13}C$ values of plants have organ and growth stages specificity under water stress [13].

3.3. Responses of Water Physiological Characteristics of Glycyrrhiza uralensis to Water Stress

Li suggested two types of drought-resistant plants: the first one has a high-water potential, a delayed dehydration, and drought tolerance, and the second one has a low water potential, tolerant dehydration, and drought tolerance [24]. The first type of plant delays the occurrence of dehydration through limited water loss, maintaining water absorption capacity. The second type of plant tolerates dehydration by increasing the water absorption capacity, maintaining turgor pressure, and reducing water loss. Yang and Song argued that under water stress, plants with less evident changes in water potential have a stronger ability to maintain normal turgor pressure [25,26]. They can maintain a specific amount of water and nutrient transport capacity. In this study, the leaf WP of *Glycyrrhiza uralensis* showed a downward trend as the intensity of water stress increased, dropping sharply under moderate water stress at the earlier times and increasing under severe stress at the late stages of stress. This demonstrates that *Glycyrrhiza uralensis* could regulate osmotic potential by decreasing the leaf WP, enhancing water absorption, resisting tissue dehydration, and being sensitive to drought-resistance.

Meanwhile, *Glycyrrhiza uralensis* accumulated PR in a large amount under water stress at the earlier times and decreased at the late stages of stress. Although *Glycyrrhiza uralensis* decreased the RWC with the intensity of water stress but stabilized it throughout all stress times. This showed that *Glycyrrhiza uralensis* maintained a certain turgor pressure to resist dehydration through the coordination of different physiological activities [27]. Based on the classification proposed by Li, we categorized this plant as close to the second type of drought-resistance [24].

3.4. Photosynthetic Physiological Properties of Glycyrrhiza uralensis Respond to Water Stress

The photosynthetic capacity of plant mesophyll cells is determined to some extent by the mesophyll cell's ultrastructural characteristics [28]. In this study, water stress had a significant effect on the ultrastructure of epidermal cells of *Glycyrrhiza uralensis* leaves. Under severe water stress, the grana of chloroplasts underwent changes such as bending, swelling, and disarrangement, and osmophilic grains (lipid bodies) significantly increased and aggregated. Some chloroplast envelopes ruptured or even disintegrated, and inclusions flowed out. These results are consistent with the findings of various previous studies [29,30]. The guard cells of *Glycyrrhiza uralensis* were severely deformed, the cell wall was inhomogeneously thickened, the organelles and solutes decreased, and the cells were damaged. At this point, osmotic substances in guard cells were expelled or consumed, leading to water outflow and stomatal closure [31]. It was further confirmed in the present study that *Glycyrrhiza uralensis* was sensitive to water stress and regulated stomata by osmotic substance content and morphological changes.

In general, the response mechanism of photosynthesis is different under different water stress conditions. the decrease of Pn and Ci under mild water stress is the result of stomatal restriction. Under severe water stress, the photosynthetic apparatus and the photosynthetic enzyme system of plants were destroyed, and the non-stomatal restriction was the main reason for the decrease of Pn. Due to the decrease of the photosynthetic enzyme activity of mesophyll cells, the discrimination of $\delta^{13}C$ was weakened, which resulted in the increase of the $\delta^{13}C$ value, but the WUEi decreased [20,32]; our data also confirmed this feature of *Glycyrrhiza uralensis*. Therefore, under mild water stress, the Pn and Ci of plants decrease, and their absorption of $^{12}CO_2$ is relatively reduced, but their $\delta^{13}C$ value and water use efficiency increase due to stomatal restriction [33,34]. In this study, the change patterns of Pn, Tr, Gs, and Ci of *Glycyrrhiza uralensis* belonged to stomatal restriction under the T1–T3 treatments. Under severe water stress, the decreased range of Pn was higher than Tr, which led to the WUEi and Ls decreasing significantly, but the $\delta^{13}C$ value increased. Thus, a non-stomatal restriction may play a dominant role under severe water stress when combined with changes in morphological and water physiological characteristics.

3.5. Water Use Strategy and Adaptation Mechanism of Glycyrrhiza uralensis in Response to Water Stress

Many studies have shown that plants adopt a survival adaptation strategy in arid habitats by improving the WUE and coordinating related physiological and metabolic functions to maintain the plant's maximum water absorption capacity or minimum water loss to maintain survival [35,36]. Gulías found that neither *F. arundinacea* nor *D. glomerata* cultivars showed a trade-off between high WUE and biomass production, indicating that these plants have both the characteristics of efficient water use and productivity retention [2], whereas the other types of plants usually sacrifice above-ground biomass for efficient water use [37]. Here, we found that the $\delta^{13}C$ values of *Glycyrrhiza uralensis* organs were significantly positively correlated with PR and root–shoot ratio and significantly negatively correlated with leaf biomass and gas exchange parameters. The change in water use efficiency was consistent with the change of root biomass and its reaction direction. Redundancy analysis further verified these results; physiological water regulation of *Glycyrrhiza uralensis* was the main component in the response vector to water stress. So, we concluded that *Glycyrrhiza uralensis* maintains water content and turgor pressure, promotes root biomass accumulation by increasing the root WUEi, resists tissue dehydration with a higher RWC, and maintains a specific photosynthetic yield, which is a water conservation strategy.

As the above analysis showed, in response to water stress, there was a significant trade-off between the biomass accumulation and water use efficiency of *Glycyrrhiza uralensis*. *Glycyrrhiza uralensis* responded positively to drought stress by increasing the root-to–shoot ratio and WUE while reducing biomass accumulation, regulating water status to match soil water supply capacity by osmosis and stomatal regulation, and maintaining relative humidity at a stable RWC, so as to reduce the risk of hydraulic imbalance. However, in order to ensure the safety of the plant hydraulic system and maintain the viability of the plant, the adaptation mode will change continuously or transitively with the change of drought degree [38].

4. Materials and Methods

4.1. Seed Collection and Nursery

The mature seeds of *Glycyrrhiza uralensis* were collected from the desert steppe natural restoration area in Gaoshawo, Yanchi County, Ningxia. Seedlings were grown in plugs in April and transplanted when the seedlings of *Glycyrrhiza uralensis* grew to more than 5 cm.

4.2. Soil Treatment and Its Physical and Chemical Properties

The desert soil was collected from the field sampling site at Gaoshawo, Yanchi County, and was transported to the agricultural experiment base (greenhouse) of the College of Agriculture, Ningxia University, China. The temperature inside the greenhouse was 28 °C/16 °C (day/night), with natural light. The soil was packed into self-made PVC plastic tubes with a diameter of 20 cm and a height of 40 cm after three days of sun exposure and screening out weeds and stones. Each tube contained 7.5 kg of soil, and the bottom of the tube was sealed with hard gauze. The field water holding capacity (FC) was 20.44 ± 2.74%, the volume weight was 1.45 ± 0.06 g·cm^{-3}, and the saturation moisture capacity was 28.58%, close to the field sampling site. Bao's method was used to investigate the physical and chemical properties of soil [39]: soil organic matter was 5.80 g·kg^{-1}, total nitrogen was 0.38 g·kg^{-1}, total phosphorus was 0.22 g·kg^{-1}, total kaliun was 19.19 g·kg^{-1}, alkali-hydrolyzable nitrogen was 27.33 mg·kg^{-1}, rapidly available phosphorus was 5.27 mg·kg^{-1}, available potassium was 113.31 mg·kg^{-1}, total salt was 0.058 g·kg^{-1}, and pH values were 9.45.

4.3. Design of Experiment

The experiment adopted a randomized block design with two factors (water gradient × times), and the water gradient treatment was divided into 5 gradients, each with three blocks.

Control (CK) for normal water supply, maintaining field water capacity between 70–80%. Treatment 1 (T1) was mild water stress, and field water capacity was between 60–70%. Treatment 2 (T2) was moderate water stress, and the field water capacity was 40–60%. Treatment 3 (T3) was relatively severe water stress, and field water capacity was 30–40%. Treatment 4 (T4) was severe water stress, and field water capacity was 20–30%.

After the seedlings were transplanted and new roots grew, five plants with the same growth were kept in each pot. The soil moisture was balanced regularly (2–3 d), and the field water capacity was maintained around 70–80%. After 15 d of growth, the moisture was being controlled. A TDR soil moisture meter (Mini Trase with soil-moisture TDR Technology, USA) was used to measure soil volumetric moisture content in each pot daily to ensure the soil's water content in each treatment reaches the set level. The weight of each pot was measured every two days with an electronic scale. According to the soil moisture content and water consumption in each pot, the additional water contents were calculated to ensure the set range. The surface of each pot was covered with polyethylene plastic particles to prevent water evaporation.

4.4. Test Items

(1) Plant biomass of different organs

During different periods (15, 30, 45, and 60 d) of water stress, *Glycyrrhiza uralensis* was harvested and returned to the laboratory. The roots, stems, and leaves were put into paper bags and dried in an oven at 65 °C for 48 h, the dry weight of these plant components was measured. Each treatment was repeated five times.

Root to shoot ratio (R/S) = aboveground biomass/underground biomass.

$$\text{Aboveground biomass} = \text{dry weight of stem biomass (SB, g)} + \text{dry weight of leaf biomass (LB, g)}. \tag{1}$$

$$\text{Underground biomass} = \text{roots biomass dry weight (RB, g)}. \tag{2}$$

The linear correlation between the above-ground and underground biomass of *Glycyrrhiza uralensis* was analyzed using the correlation growth relationship model ($Y = \beta X^{\alpha}$). In the formula, Y is the underground biomass (BGB), β is the scaling constant, X is the above-ground biomass (AGB), α is allometric growth index, $\alpha = 1$ is the isometric growth relationship, and $\alpha \neq 1$ is the allometric growth relationship. After the logarithmic transformation of biomass data, its power function is converted to the form of $\log = \log\beta + \alpha\log X$, where α is the slope of linear regression, and $\log\beta$ is the intercept of linear regression.

(2) Relative leaf water content (RWC), free proline content (PR), and the leaf water potentials (WP)

During different periods (15, 30, 45, and 60 d) of water stress, fully expanded, healthy leaves were collected, while old leaves and new leaves were avoided. After sampling, a part of each plant was quickly determined for the fresh weight (FW) of the leaves, put in a container with distilled water (plastic bag), and let to stand for more than 12 h in a refrigerator at 4 °C. Then the absorbent paper was used to absorb water on the surface of the leaves and measure the weight of the leaves, which were the saturated weight (TW) of the leaves. Then the leaves were dried in an oven at 65 °C for 48 h, and the dry weight (DW) of the leaves was measured.

$$RWC\ (\%) = (FW - DW)/(TW - DW) \times 100.$$

The free proline content was determined by the acidic-ninhydrin method.

The situ determination method was used to measure the water potential of plant leaves in the early morning, and the Wescor psychrometer (PSYPRO water potential measurement system) was used to suite the C-52 sample room. During the determination, the plant leaves were clamped in the plant in-situ leaf chamber, balanced for 1 h, and the leaf chamber was completely sealed with plasticine. The leaves of the tested plant completely covered the measurement chamber. After 1 h, the probe was connected to the PSYPRO host to measure the leaf water potential.

(3) Carbon stable isotope composition ($\delta^{13}C$) of different organs.

Plant leaves of each treatment at the end of water stress were sampled, 15 leaves of five plants (3 leaves each plant) as one sample. Fully expanded, healthy, and southern leaves were selected from the sample plants. The plant leaves were dried at 60 °C for 72 h and then ground into powder and passed through a 100-mesh sieve for carbon isotope analyses, which were conducted at the Huake Jingxin Stable Isotope Laboratory of Tsinghua University, Shenzhen, China. The $^{13}C/^{12}C$ isotope composition was determined from 1-mg samples with a DELTAV Advantage isotopic ratio mass spectrometer (Thermo Fisher Scientific, Inc., USA). After high temperature combustion in the element analyzer, the samples generated CO_2. The mass spectrometer detected the ^{13}C and ^{12}C ratio of CO_2 and compared them with the international standard (Pee Dee Belnite or PDB, a shell fossil in the ocean, with a ^{13}C content of 1.124%), then calculated the $\delta^{13}C$ value of the samples. The $\delta^{13}C$ values were expressed in parts per thousand (‰) and expressed as follows:

$$\delta^{13}C\ (‰) = \left(R_{sample}/R_{standard} - 1\right) \times 1000,$$

where R is the molar ratio of the heavy to light isotopes in the sample relative to the appropriate standards. The Pee Dee Belemnite carbonate was used as the standard for C. The accuracies of analyses were $< \pm 0.1‰$.

(4) The gas exchange parameters.

The gas exchange parameters of flag leaves were measured between 9:00 and 11:00 using a potable photosynthesis analyzer (LI-6400 by Li-Cor, USA) at the end stage of water stress. The gas exchange measurements were the average of three readings within 15 s. Three leaves lied on the equal node in each pot were determined, averaging each parameter of three leaves.

$$Instantaneous\ water\ use\ efficiency\ (WUEi) = Pn/Tr,$$

$$\text{stomatal limit value (Ls)} = (1 - Ci/Ca) \times 100,$$

where Pn is net photosynthetic rate, Tr is transpiration rate, Ci is intercellular CO_2 concentration, Ca is atmospheric CO_2 concentration

(5) Observations of the leaf ultrastructure.

After 60 d of water stress treatment, the upper functional leaves of each treatment (five plants with basically the same growth status) were sampled at 8:00–9:00 in the morning under sunny weather. Six plants with basically the same growth status were selected from each treatment, and two upper functional leaves were taken from each plant. Afterwards, it was refrigerated at 4 °C and brought back to the laboratory. A sharp double-sided blade was used to crosscut the rectangular pieces of approximately 1 cm × 0.5 cm along the middle of the main veins of the leaves. The pieces were immediately fixed at 4 °C for 3 h with a 4% glutaraldehyde prefixative solution, then washed with 0.1 m sodium dimethyl arsenate three times, and the washing solution was replaced at an interval of 2 h. Then it was fixed with a 1% post-fixative solution of osmium at 4 °C for 2 h, washed twice with 0.1 m sodium dimethyl arsenate at an interval of 15 min, and dehydrated in gradient alcohol at room temperature, permeated and embedded with epoxy resin Epon812, and polymerized for 48–72 h in an incubator at 60 °C.

The embedded plant leaves were made into semi-thin slices with a thickness of 2–4 µm on leica UC-6 ultra-thin slicing machine. The test material was positioned under an optical microscope to determine the structural parts to be observed. After localization, leica UC-6 ultrathin slicing machine was used to slice the ultrathin slices with a thickness of 70–80 nm. The ultrathin slices were observed and photographed by JEM-2100HC transmission electron microscope after dual staining with uranium dioxy acetate and lead citrate.

4.5. Statistical Analyses

The original data were collated and displayed in Excel 2010. IBM SPSS Statistics 23 software was used to perform single or double factorial analysis of variance. Duncan's method for multiple comparisons was adopted when there were significant differences. Correlation analysis (CA) and redundancy analysis (RDA) was performed on the data by R software and Origin 2018.

5. Conclusions

Under water stress, more biomass was allocated to the underground part of *Glycyrrhiza uralensis*, resulting in a significant increase in the R/S ratio. The biomass of above-ground and underground parts showed allelic growth biased to the accumulation of underground biomass. The $\delta^{13}C$ values in all organs of *Glycyrrhiza uralensis* increased, the WP and PR of leaves significantly responded to mild and moderate water stress from 30 d and the RWC was exhibited a lower range of change. *Glycyrrhiza uralensis* exhibited sensitive responses to water shortages and maintained a certain turgor pressure to resist dehydration, which is a water conservation strategy. Meanwhile, the WUEi and Ls increased continuously and decreased significantly under severe water stress. The most significant morphological changes in chloroplast and guard cells began at T3 treatment. So, a non-stomatal restriction may play a dominant role under severe water stress.

We summarized that *Glycyrrhiza uralensis* could maintain water content and turgor pressure under water stress, promote root biomass accumulation, and improve water use efficiency, which was a water-conservation strategy showing a mechanism for both drought tolerance and avoidance.

Author Contributions: Conceptualization, H.H. and Y.X.; investigation, H.H., K.S. and L.F.; software, K.S.; writing—original draft preparation, K.S.; writing—review and editing, H.H.; funding acquisition, H.H. All authors have read and agreed to the published version of the manuscript.

Funding: This study was funded by Natural Science Foundation of Ningxia Province (2020AAC03080), the funder is Haiying Hu, and the First-class Discipline Construction Project (Grassland Science Discipline) for the high school in Ningxia (NXYLXK2017A01), the funder is Yingzhong Xie.

Institutional Review Board Statement: Not applicable.

Informed Consent Statement: Not applicable.

Data Availability Statement: Not applicable.

Acknowledgments: We would like to thank Jianping Li and Wang Xing at Ningxia University for the helpful discussion and data analysis.

Conflicts of Interest: The authors declare no conflict of interest.

Abbreviations

Abbreviation	Full Name
$\delta^{13}C_{leaf}$	$\delta^{13}C$ values of leaf
$\delta^{13}C_{stem}$	$\delta^{13}C$ values of stem
$\delta^{13}C_{root}$	$\delta^{13}C$ values of root
LB	leaf biomass
SB	stem biomass
RB	root biomass
R/S	root to shoot ratio
PR	proline
RWC	relative water content
WP	water potential
Pn	net photosynthetic rate
Tr	transpiration rate
Ci	intercellular CO_2 concentration
Gs	stomatal conductance
WUEi	instantaneous water use efficiency
Ls	limiting value of stomata

References

1. Bacelar, E.L.V.A.; Moutinho-Pereira, J.M.; José, M.; Brito, B.M.C.; Carlos, M. Water use strategies of plants under drought conditions. In *Plant Responses to Drought Stress*; Springer: Berlin, Germany, 2012; pp. 145–170.
2. Gulías, J.; Seddaiu, G.; Cifre, J.; Salis, M.; Ledda, L. Leaf and plant water use efficiency in cocksfoot and tall fescue accessions under differing soil water availability. *Crop Sci.* **2012**, *52*, 2321. [CrossRef]
3. Yin, H.; Tariq, A.; Zhang, B.; Lv, G.; Zeng, F.; Graciano, C.; Santos, M.; Zhang, Z.; Wang, P.; Mu, S. Coupling Relationship of Leaf Economic and Hydraulic Traits of *Alhagisparsifolia* Shap. in a Hyper-Arid Desert Ecosystem. *Plants* **2021**, *10*, 1867. [CrossRef] [PubMed]
4. Zhang, K.; Tian, C.; Li, C. Influence of saline soil and sandy soil on growth and mineral constituents of common annual halophytes in Xinjiang. *Acta Ecol. Sin.* **2012**, *32*, 3069–3076. [CrossRef]
5. Enquist, B.; Niklas, K. Global allocation rules for patterns of biomass partitioning in seed plants. *Science* **2002**, *295*, 1517–1520. [CrossRef]
6. Chen, G.; Zhao, W.; He, S.; Xiao, F. Biomass allocation and allometrec relationship in aboveground components of Salix psammophila branches. *J. Desert Res.* **2016**, *36*, 357–363.
7. Zhang, J.; Zuo, X.; Zhao, X.; Ma, J.; Medina-Roldán, E. Effects of rainfall manipulation and nitrogen addition on plant biomass allocation in a semiarid sandy grassland. *Sci. Rep.* **2020**, *10*, 9026. [CrossRef]
8. Wilcox, K.R. Contrasting above-and be-lowground sensitivity of three Great Plains grasslands to altered rainfall regimes. *Glob. Change Biol.* **2014**, *21*, 335–344. [CrossRef]
9. Shen, F.; Fan, H.; Wu, J.; Liu, W.; Lei, X. Review on carbon isotope composition ($\delta^{13}C$) and its relationship with water use efficiency at leaf level. *J. Beijing For. Univ.* **2017**, *39*, 114–124.
10. Zhang, C.J.; Chen, F.H.; Jin, M. Study on modern plant C^{13} in western China and its significance. *Chin. J. Geochem.* **2003**, *22*, 97–106.
11. Leidi, E.O.; López, M.; Forham, J. Variation in carbon isotope discrimination and other traits related to drought tolerance in upland cotton cultivars under dry land conditions. *Field Crops Res.* **1999**, *61*, 109–123. [CrossRef]
12. Liu, Y.; Li, P.; Shen, B.; Fen, C.; Liu, Q.; Zhang, Y. Effects of drought stress on Bothriochloa ischaemum water-use efficiency based on stable carbon isotope. *Acta Ecol. Sin.* **2017**, *37*, 3055–3064.
13. Saranga, Y.; Flash, I.; Paterson, A.H. Carbon isotope ratio in cotton varies with growth stage and plant organ. *Plant Sci.* **1999**, *142*, 47–56. [CrossRef]

14. Hu, H.; Li, H.; Ni, B. Characteristic of typical vegetation community and water use efficiency of dominant plants in desert steppe of Ningxia. *J. Zhejiang Univ. Agric. Life Sci.* **2019**, *45*, 460–471.

15. Hu, H.; Zhu, L.; Li, H. Seasonal changes in the water-use strategies of three herbaceous species in a native desert steppe of Ningxia, China. *J. Arid Land* **2021**, *13*, 109–122. [CrossRef]

16. Lv, S.J.; Liu, H.M.; Wu, Y.L.; Wei, Z.; Nie, Y. Effects of grazing on spatial distribution relationships between constructive and dominant species in Stipa breviflora desert steppe. *Chin. J. Appl. Ecol.* **2014**, *25*, 3469–3474.

17. Lozano, Y.M.; Aguilar-Trigueros, C.A.; Flaig, I.C.; Rillig, M.C. Root trait responses to drought are more heterogeneous than leaf trait responses. *Funct. Ecol.* **2020**, *34*, 2224–2235. [CrossRef]

18. Huang, D.Q.; Yu, L.; Zhang, Y.S.; Zhao, X. Above-ground Biomass and its relationship to soil moisture of Natural Grassland in the Northern Slopes of the Qilian Mountains. *Acta Prataculture Sin.* **2011**, *20*, 20–27.

19. Karcher, D.E.; Richardson, M.D.; Hignight, K.; Rush, D. Drought tolerance of tall fescue populations selected for high root/shoot ratios and summer survival. *Crop Sci.* **2008**, *48*, 771–777. [CrossRef]

20. Gao, H.H.; Wang, J.J.; Zhou, P.S.; Yu, W.; Kang, H. Carbon and Nitrogen allocation in organs of Quercusvariabilisseedlings by ^{13}C and ^{15}N tracer technigue. *J. Shaihai Jiaotong Univ. Agric. Sci.* **2017**, *35*, 67–73.

21. Xin, D.; Yang, W.R.; Peng, G.H.; Dan, S.; Liao, F.; Han, S. Stable isotope technique: An advanced technology in ascertaining plant-water relations. *Agric. Sci. Technol.* **2014**, *15*, 338–343.

22. Badeck, F.W.; Tcherkez, G.; Salvador, N.; Clément, P.; Jaleh, G. Post-photosynthetic fractionation of stable carbon isotopes between plant organs—A widespread phenomenon. *Rapid Commun. Mass Spectrom.* **2005**, *19*, 1381–1391. [CrossRef] [PubMed]

23. Farquhar, G.D.; Ehleringer, J.R.; Hubick, K.T. Carbon isotope discrimination and photosynthesis. *Annu. Rev. Plant Biol.* **1989**, *40*, 503–537. [CrossRef]

24. Li, J.Y. Mechanism: Of Drought Toleranee in Plants. *J. Beijing For. Univ.* **1991**, *13*, 92–100.

25. Yang, X.; Fu, H.; Zhang, H.; Zhao, J. Effect of soil water stress on leaf water potential and biomass of Zygophyllum xanthoxylum during seedling stage. *Acta Prataculture Sin.* **2006**, *15*, 37–41.

26. Song, M.; Du, G.; Feng, C.; Zhu, L. Research of Diunal Variation of Water Potential of Four Dominant Plants in Hulunbeier Typical Steppe. *Chin. J. Grassl.* **2012**, *34*, 114–119.

27. Liu, C.; Wang, W.; Cui, J.; Li, S.; Liu, C.; Wang, W.; Cui, J.; Li, S. Effects of Drought Stress on Photosynthesis Characteristics and Biomass Allocation of *Glycyrrhiza uralensis*. *J. Desert Res.* **2006**, *2*, 142–145.

28. Zheng, M.N.; Li, X.L.; Wan, L.Q.; Feng, H.; Xi, C. Effect of Water Stress on Ultrastructures of Chloroplast and Mitochondria and Photosynthesis in Six Gramineous Grass Species. *Acta Agrestia Sincia* **2009**, *17*, 643–649.

29. Liu, Y.; Yue, X.; Chen, G.L. Effects of water stress on ultrastructure and membrane lipid peroxidation of leaf and root cells of *Glycyrrhiza uralensis*. *Acta Prataculture Sin.* **2010**, *19*, 79–86.

30. Zhang, Z.; Liu, J.; Fu, X.; Zhao, B.; Li, L.; Liu, J.; Yang, H. Effect of Drought Stress on Stomata and Ultrastrucure of Mesophyll cells of Oat Leaf. *J. Triticeae Crops* **2017**, *37*, 1216–1223.

31. Zhou, X.; Zhang, H. Differential Responses in Stomatal Guard Cells of Different Drought Tolerant Poplars to Water Stress. *J. Beijing For. Univ.* **1999**, *21*, 1–6.

32. Gimenez, C.; Mitchell, V.G.; Lawlor, D.W. Regulation of photosynthetic rate of two sunflower hybrids under water stress. *Plant Physiol.* **1992**, *98*, 516–524. [CrossRef] [PubMed]

33. Anev, M.S.; Tzvetkova, N.P. Drought stress in four subalpine species: Gas exchange response and survivorship. *Russ. J. Ecol.* **2018**, *49*, 422–427. [CrossRef]

34. Li, Z.; Zhao, Y.J.; Song, H.; Zhang, J.; Liu, J. Effects of karst soil thickness heterogeneity on the leaf anatomical structure and photosynthetic traits of two grasses under different water treatments. *Acta Ecol. Sin.* **2018**, *38*, 721–732.

35. Dan, T.W.; John, S.A.; Lesley, T.S. Water-use efficiency of a mallee eucalypt growing naturally and in short-rotation coppice cultivation. *Plant Soil* **2004**, *262*, 111–128.

36. Volaire, F. A unified framework of plant adaptive strategies to drought: Crossing scales and disciplines. *Glob. Change Biol.* **2018**, *24*, 2929–2938. [CrossRef] [PubMed]

37. Zhu, L.; Zhang, H.L.; Gao, X.; Qi, Y.; Xu, X. Seasonal patterns in water uptake for Medicago sativa grown along an elevation gradient with shallow groundwater table in Yanchi county of Ningxia, Northwest China. *J. Arid Land* **2016**, *8*, 921–934. [CrossRef]

38. Moshelion, M.; Halperin, O.; Wallach, R.; Oren, R.; Danielle, A.W. Role of aquaporins in determining transpirationand photosynthesis in water-stressed plants: Crop water-use efficiency, growth and yield. *Plant Cell Environ.* **2015**, *38*, 1785–1793. [CrossRef]

39. Bao, S.D. *Agrochemical Soil Analysis*; Chinese Agricultural Press: Beijing, China, 2000; pp. 263–274. (In Chinese)

Article

Biogeographic Patterns of Leaf Element Stoichiometry of *Stellera chamaejasme* L. in Degraded Grasslands on Inner Mongolia Plateau and Qinghai-Tibetan Plateau

Lizhu Guo [1,2], Li Liu [3], Huizhen Meng [4], Li Zhang [4], Valdson José Silva [5], Huan Zhao [6], Kun Wang [1], Wei He [4] and Ding Huang [1,*]

[1] College of Grassland Science and Technology, China Agricultural University, Beijing 100193, China; ellenguo@sina.cn (L.G.); wangkun@cau.edu.cn (K.W.)
[2] Institute of Grassland, Flowers and Ecology, Beijing Academy of Agriculture and Forestry Sciences, Beijing 100097, China
[3] Grassland Research Institute, Chinese Academy of Agricultural Sciences, Hohhot 010010, China; liu_li530@163.com
[4] College of Life Sciences, Northwest University, Xi'an 710069, China; 17835417718@163.com (H.M.); xdlizhang@163.com (L.Z.); hewei.scu@gmail.com (W.H.)
[5] Department of Animal Science, Federal Rural University of Pernambuco, Recife 52171-900, Brazil; valdson.silva@ufrpe.br
[6] Academy of Inventory and Planning, National Forestry and Grassland Administration, Beijing 100714, China; zhaohuan333@163.com
* Correspondence: huangding@263.net

Citation: Guo, L.; Liu, L.; Meng, H.; Zhang, L.; Silva, V.J.; Zhao, H.; Wang, K.; He, W.; Huang, D. Biogeographic Patterns of Leaf Element Stoichiometry of *Stellera chamaejasme* L. in Degraded Grasslands on Inner Mongolia Plateau and Qinghai-Tibetan Plateau. *Plants* 2022, 11, 1943. https://doi.org/10.3390/plants11151943

Academic Editors: Bingcheng Xu, Zhongming Wen and Nikos Fyllas

Received: 5 May 2022
Accepted: 22 July 2022
Published: 26 July 2022

Publisher's Note: MDPI stays neutral with regard to jurisdictional claims in published maps and institutional affiliations.

Abstract: Plant leaf stoichiometry reflects its adaptation to the environment. Leaf stoichiometry variations across different environments have been extensively studied in grassland plants, but little is known about intraspecific leaf stoichiometry, especially for widely distributed species, such as *Stellera chamaejasme* L. We present the first study on the leaf stoichiometry of *S. chamaejasme* and evaluate its relationships with environmental variables. *S. chamaejasme* leaf and soil samples from 29 invaded sites in the two plateaus of distinct environments [the Inner Mongolian Plateau (IM) and Qinghai-Tibet Plateau (QT)] in Northern China were collected. Leaf C, N, P, and K and their stoichiometric ratios, and soil physicochemical properties were determined and compared with climate information from each sampling site. The results showed that mean leaf C, N, P, and K concentrations were 498.60, 19.95, 2.15, and 6.57 g kg^{-1}; the average C:N, C:P, N:P, N:K and K:P ratios were 25.20, 245.57, 9.81, 3.13, and 3.21, respectively. The N:P:K-ratios in *S. chamaejasme* leaf might imply that its growth is restricted by K- or K+N. Moreover, the soil physicochemical properties in the *S. chamaejasme*-infested areas varied remarkably, and few significant correlations between *S. chamaejasme* leaf ecological stoichiometry and soil physicochemical properties were observed. These indicate the nutrient concentrations and stoichiometry of *S. chamaejasme* tend to be insensitive to variations in the soil nutrient availability, resulting in their broad distributions in China's grasslands. Besides, different homeostasis strength of the C, N, K, and their ratios in *S. chamaejasme* leaves across all sites were observed, which means *S. chamaejasme* could be more conservative in their use of nutrients improving their adaptation to diverse conditions. Moreover, the leaf C and N contents of *S. chamaejasm* were unaffected by any climate factors. However, the correlation between leaf P content and climate factors was significant only in IM, while the leaf K happened to be significant in QT. Besides, MAP or MAT contribution was stronger in the leaf elements than soil by using mixed effects models, which illustrated once more the relatively weak effect of the soil physicochemical properties on the leaf elements. Finally, partial least squares path modeling suggested that leaf P or K contents were affected by different mechanisms in QT and IM regions, suggesting that *S. chamaejasme* can adapt to changing environments by adjusting its relationships with the climate or soil factors to improve its survival opportunities in degraded grasslands.

Keywords: biogeographic patterns; leaf stoichiometry; climatic variables; soil physicochemical properties; *Stellera chamaejasme* L.

1. Introduction

Ecological stoichiometry plays an important role in analyzing the composition, structure, and function of a concerned community and ecological system [1–3]. Over the last a few decades, one particular focus of ecological stoichiometry has been to document large-scale patterns of and the driving factors for plant carbon: nitrogen: phosphorus (C:N:P) stoichiometry [4–8]. The relationship between leaf stoichiometry, geographic patterns, and climate factors have been studied on both global and regional scales. Geographical variation in foliar ecological stoichiometry is a challenging issue to plant ecologists [5,9–11]. Meanwhile, the homeostasis (*H*) of element composition is one of the central concepts of ecological stoichiometry, and its strength is related to the ecological strategy and adaptability of species [2,12]. Stoichiometric homeostasis can help predict the strategies that are used by different plant species to cope with limited resources [2,13]. The nutrient conservatism of high *H*-species could be important mechanism contributing to their success, particularly in natural (unmodified) terrestrial ecosystems, where nutrient supply is often limited and highly variable [14,15]. Indeed, the stoichiometric homeostasis of plants varied with species, growth stages, and element types [16–19].

Stellera chamaejasme L. is a native perennial weed that has distributed abundantly in the alpine meadow on the eastern Tibetan Plateau and typical steppe on eastern Inner Mongolia Plateau of China [20,21]. It competes with forage-grass species for water, nutrition, and space, thereby decreasing the quality of the forage grass and shortening the use of grasslands [22]. The whole plant of *S. chamaejasme* is poisonous and its roots and pollens are most toxic, therefore, livestock may be poisoned by inadvertently inhaling the pollen while grazing [23]. It has become one of the most serious weeds threatening a wide range of grasslands, which were grazed heavily, posing potential hazards to the grassland ecological safety and its impact on animal husbandry sustainability [21]. Previous studies of *S. chamaejasme* focused on its nutrient uptake efficiency and water use efficiency compared to co-existing species [21], its allelochemicals and allelopathic effects on forages [24], and weed control techniques and use [25], but no similar phylogeographical study had ever been conducted on *S. chamaejasme*. Plant nutrient and stoichiometry are key foliar traits with great ecological importance, but previous publications provide limited insight into the biogeographic leaf nutrient and stoichiometry patterns for *S. chamaejasme* [21]. As habitat heterogeneity tends to increase with geographical scale, wide-ranging species can usually use a wide array of resources and tolerate broad environmental conditions or physiological stresses and flourish over a larger area [26,27]. Recent studies have assumed that wide-ranging species always have stronger homeostasis or a weak relationship with nutrient concentrations than narrow-ranging species in response to environmental factors (e.g., soil fertility) [15,26]. The widespread nature of *S. chamaejasme* may be associated with its stoichiometric homeostasis.

Several studies on a regional and global scale reported that changes of the leaf N and P stoichiometry are associated with many biotic and abiotic factors, including climate variables, soil properties, species type, and plant functional groups [4–7,10,28–30]. However, sample collection is commonly limited to a few individuals, a few populations, and averaged at the population or species level, disregarding the intraspecific variability [31]. Investigating the geographic variation within species can help uncover the mechanisms of relationships between plant tissue nutrients and environments [32] by excluding the confounding effects of taxonomic and phylogenetic structure such as those that have been found to influence the geographic patterns in leaf nutrients, and their linkages to climate and soil. Since relationships between environment and plant traits along environmental gradients could be presented as evidence of environmental control over species distribution, examining plant-environment (e.g., climate and soil nutrient availability) interactions may provide some insights into the underlying mechanisms of *S. chamaejasme* distribution in degraded grasslands. However, no studies have yet incorporated information on the geographic patterns in leaf stoichiometry of *S. chamaejasme* in relation to environmental factors.

This study aimed to assess the element stoichiometry of *S. chamaejasme* leaves in degraded grasslands across northern China. The distinct regions of the Qinghai-Tibetan Plateau (QT) and Inner Mongolia Plateau (IM) provide a unique opportunity to test whether there are significant differences in leaf stoichiometry under different environmental conditions and to examine how and to what extent soil and climate modify leaf stoichiometry of *S. chamaejasme* across degraded grasslands. In general, most researchers focused on the roles of C, N, and P stoichiometry in the ecological process from individuals to ecosystems, but potassium (K) is an essential macronutrient that has been partly overshadowed by C, N, and P [5,8,33,34]. Our study also focuses on leaf K concentrations of *S. chamaejasme*, which broadens the contents of ecological stoichiometry. We hypothesized that: (1) *S. chamaejasme*, a wide-spread weed, would exhibit small variation in leaf stoichiometry and tolerate broad environmental conditions; in other words, *S. chamaejasme* may have stoichiometric homeostasis and, (2) due to the differences in limiting factors to vegetation in QT and IM, the relationship between *S. chamaejasme* and environmental factors may be related to different factors in the two regions. To test our hypotheses, we first explored the overall biogeographic patterns of C, N, P, and K stoichiometry of *S. chamaejasme* leaves from 29 sampling sites in the two grassland ecosystems in northern China. We then disentangled the effects of soil and climate on the overall plant stoichiometry pattern and compared the difference between the two regions.

2. Results

2.1. Pattern of Leaf Ecological Stoichiometry and Soil Physicochemical Properties of S. chamaejasme

Leaf C, N, P, K, and C:N, C:P, N:P, N:K, K:P of *S. chamaejasme* varied little across all the study sites (Table 1 and Table S1). The mean leaf C, N, P, and K across all sites were 498.60 g kg^{-1}, 19.95 g kg^{-1}, 2.15 g kg^{-1}, and 6.57 g kg^{-1}, respectively, and the CV% of leaf P was the largest. Moreover, the mean leaf C:N ratio was 25.20, C:P ratio 245.57, N:P ratio 9.81, N:K ratio 3.13, and K:P ratio 3.21. Inconsistent with the pattern of leaf results, the soil physicochemical properties of *S. chamaejasme*-infested areas varied remarkably (Tables 2 and S2). The soil C, N, P, and K exhibited large variations, primarily ranging c. 5.87–84.74 g kg^{-1} for C; 0.24–7.43 g kg^{-1} for N, 0.20–0.82 g kg^{-1} for P, and 0.95–30.55 g kg^{-1} for K. The variation in the soil K content across all the study sites was about 32 times (maximum/minimum), which was the most variable element among the four total elements. The soil mean C:N, N:P, C:P, N:K, and K:P ratios were 13.54, 77.72, 6.34, 0.20, and 38.73, respectively. For the available soil nutrients, soil NN variation was considerably larger than that for the AP, AK, and AN content, as evidenced by coefficients of variation (CVs). Similarly, soil WC, pH, and Ec showed a greater variation throughout the sampling areas.

When comparing the leaf element contents and stoichiometry of *S. chamaejasme* in QT and IM, we found that only the leaf K concentrations, N:K and K:P ratio were significantly different between the two regions (Table 1). Moreover, most soil physicochemical properties were higher in QT than those in IM, except soil AN, NN, and Ec. Specifically, soil P, K, AP, WC, and pH were significantly higher in QT than IM, but soil Ec was significantly lower in QT. Similarly, soil C, N, P, and K stoichiometry showed no significant difference between QT and IM (Table 2).

Table 1. Regional *S. chamaejasme* leaf ecological stoichiometry. SD is the standard deviation and CV is the coefficient of variation. Differences between QT and IM were tested using independent *t*-test; significant differences at $p < 0.05$ are indicated by different letters.

	All (n = 29)		QT (n = 19)		IM (n = 10)	
	Mean ± SD	CV (%)	Mean ± SD	CV (%)	Mean ± SD	CV (%)
Carbon (g kg^{-1})	498.60 ± 22.07	4.43	498.97 ± 21.14	4.24	497.89 ± 24.91	5.00
Nitrogen (g kg^{-1})	19.95 ± 2.09	10.47	19.94 ± 2.02	10.15	19.97 ± 2.32	11.63
Phosphorus (g kg^{-1})	2.15 ± 0.52	24.33	2.21 ± 0.53	24.15	2.05 ± 0.52	25.20
Potassium (g kg^{-1})	6.57 ± 1.18	17.94	7.13 ± 0.99 a	13.94	5.51 ± 0.66 b	12.02
C:N	25.20 ± 2.27	8.99	25.21 ± 2.18	8.65	25.18 ± 2.543	10.09
C:P	245.57 ± 61.64	25.10	239.45 ± 61.85	25.83	257.20 ± 62.77	24.41
N:P	9.81 ± 2.60	26.54	9.63 ± 2.89	30.05	10.15 ± 2.04	20.05
N:K	3.13 ± 0.63	20.10	2.84 ± 0.45 b	15.70	3.67 ± 0.58 a	15.83
K:P	3.21 ± 0.93	28.89	3.42 ± 1.02 a	29.76	2.81 ± 0.57 b	20.44

Table 2. Regional *S. chamaejasme* soil physicochemical properties. SD is the standard deviation and CV is the coefficient of variation. Differences between QT and IM were tested using independent *t*-test; significant differences at $p < 0.05$ are indicated by different letters.

	All (n = 29)		QT (n = 19)		IM (n = 10)	
	Mean ± SD	CV (%)	Mean ± SD	CV (%)	Mean ± SD	CV (%)
Carbon (g kg^{-1})	46.11 ± 21.63	46.90	48.65 ± 18.31	37.63	41.29 ± 27.30	66.11
Nitrogen (g kg^{-1})	3.75 ± 1.70	45.24	3.93 ± 1.68	42.70	3.41 ± 1.77	51.85
Phosphorus (g kg^{-1})	0.57 ± 0.17	29.24	0.61 ± 0.13 a	22.04	0.49 ± 0.20 b	40.88
Potassium (g kg^{-1})	20.80 ± 5.86	28.17	22.23 ± 5.00 a	22.49	18.09 ± 6.66 b	36.82
C:N	13.54 ± 6.72	49.62	14.80 ± 7.89	53.30	11.16 ± 2.54	22.91
C:P	77.72 ± 26.00	33.46	79.31 ± 25.53	32.19	74.69 ± 28.01	36.06
N:P	6.34 ± 2.16	34.13	6.25 ± 2.44	39.10	6.52 ± 1.61	25.58
N:P	0.20 ± 0.14	68.81	0.18 ± 0.08	42.44	0.25 ± 0.21	85.03
K:P	38.73 ± 15.12	39.04	38.68 ± 13.27	34.31	38.83 ± 18.95	48.81
Available phosphorus (mg kg^{-1})	5.29 ± 1.96	37.07	5.84 ± 1.90 a	32.57	4.25 ± 1.70 b	40.09
Available potassium (mg kg^{-1})	175.91 ± 96.39	54.79	176.69 ± 106.27	60.14	174.43 ± 79.47	45.56
Ammonium nitrogen (mg kg^{-1})	19.17 ± 7.89	41.14	19.05 ± 7.64	40.12	19.39 ± 8.75	45.11
Nitrate nitrogen (mg kg^{-1})	14.12 ± 14.20	100.59	12.95 ± 4.46	34.42	16.35 ± 24.07	147.25
Water content	0.18 ± 0.08	44.51	0.21 ± 0.08 a	36.46	0.14 ± 0.07 b	53.50
pH	7.90 ± 0.51	6.41	8.05 ± 0.39 a	4.90	7.63 ± 0.60 b	7.89
Electrical conductivity (μs cm^{-1})	247.21 ± 221.21	89.48	193.96 ± 74.42 b	38.37	348.38 ± 351.86 a	101.00

2.2. Ecological Stoichiometry Homeostasis of S. chamaejasme in Degraded Grassland

Pearson correlations analysis indicates that there are only weak or no correlations between leaf ecological stoichiometry and soil physicochemical properties (Table S3). Furthermore, the relationships between leaf elements and stoichiometry of *S. chamaejasme* and soil by using the homeostasis model were analyzed (Table 3). For C, N (vs. soil N and nitrate N), and K (vs soil K and available K) content, and C:N, N:K, K:P ratios of *S. chamaejasme* leaves were categorized as 'strictly homeostatic' ($p > 0.1$). The leaf P content and C:P were 'weakly plastic', and the leaf N (vs. soil ammonium N) and N:P ratio were classified as 'weakly homeostatic'.

Table 3. Standardized major axis regression analysis and stoichiometric homeostasis coefficients (*H*) for leaf C, N, P, and K contents and leaf C:N:P:K ratio in *S. chamaejasme* (*n* = 29). All data have been log10-transformed before analysis. If the regression was non-significant (*p* > 0.1), 1/H was set to zero, and the organism was considered to be 'strictly homeostatic'. Species with 1/H = 1 were considered not homeostatic. All datasets with significant regressions and 0 < H < 1 were categorized as: 0 < 1/H < 0.25: 'homeostatic'; 0.25 < 1/H < 0.5: 'weakly homeostatic'; 0.5 < 1/H < 0.75: 'weakly plastic'; 1/H > 0.75 'plastic'. For 1/H > 1, 1/H close to 1 indicates weak or no stoichiometric homeostasis, and 1/H much larger than 1 indicates 'homeostatic'.

Y	X	1/H (slope)	p	r^2	Category
Leaf C	Soil C	0	0.351	0.0323	strictly homeostatic
	Soil N	0	0.829	0.0017	strictly homeostatic
Leaf N	Soil ammonium N	−0.273	0.089	0.1033	weakly homeostatic
	Soil nitrate N	0	0.291	0.0412	strictly homeostatic
Leaf P	Soil P	0.660	0.042	0.1437	weakly plastic
	Soil available P	0.622	0.002	0.2968	weakly plastic
Leaf K	Soil K	0	0.112	0.0910	strictly homeostatic
	Soil available K	0	0.154	0.0738	strictly homeostatic
Leaf C:N	Soil C:N	0	0.789	0.0028	strictly homeostatic
Leaf C:P	Soil C:P	−0.622	0.018	0.1915	weakly plastic
Leaf N:P	Soil N:P	−0.474	0.085	0.1061	weakly homeostatic
Leaf N:K	Soil N:K	0	0.956	0.0001	strictly homeostatic
Leaf K:P	Soil K:P	0	0.774	0.0031	strictly homeostatic

2.3. Spatial Variation of Leaf Elements of S. chamaejasme in Relation to Climatic Factors

No significant relationships among the leaf C and N content and two climatic factors (MAT and MAP) were found using data for all the sample sites or regions (Figure 1). For all the study sites, only the leaf K content was correlated with MAT (*p* < 0.001, Figure 1d). For the regions, it should be noted that in IM, the relationship between the leaf P and climatic factors was significant, but K was not; on the contrary, the K content of *S. chamaejasme* leaves was related to climatic factors but P was not in QT. To be specific, the leaf P concentration increased with increasing MAT and MAP in IM. Moreover, with increasing MAT, leaf K had an increasing trend, but increasing MAP showed an opposite trend in QT.

2.4. Relative Roles of Soil and Climatic Factors in Leaf Elements of S. chamaejasme

Variation in the leaf C and P heterogeneity across all the sites was mainly explained by the MAP (C: 46.46%; P: 36.49%; Table S4). MAT explained a relatively larger percentage of variation in the leaf N heterogeneity (72.62%) and leaf K heterogeneity (51.46%). The interaction of soil, MAP, and MAT explained the different percentage of the variation in four elements (C: 22.16%; N: 6.99%; P: 22.03%; K: 11.85%).

What is more, both the leaf P and K contents of *S. chamaejasme* were affected by soil and climatic factors. Thus, a more in-depth analysis using partial least squares path modeling revealed direct and indirect effects of the environmental drivers on leaf P and K content of *S. chamaejasme* in different regions (Figure 2, Table S5). Firstly, the influence of climatic factors on soil were bigger in IM than that in QT, and the effect of climatic factors on soil was significant on LP in IM. Secondly, we found that soil factors had a significant effect only on LP in QT only. Thirdly, the effect of climate factors on LP was significant in IM, but the direct effect of climate factors on LP or LK in IM and QT were greater than the indirect. These results suggest that LP or LK were affected by different mechanisms in QT and IM regions. Moreover, the goodness of fit (GOF) was 0.3205 and 0.3556 for LP and LK in QT, respectively, and 0.5490 and 0.4431 in IM. The relatively low predictive power of the model of QT suggested that most variation remained unexplained.

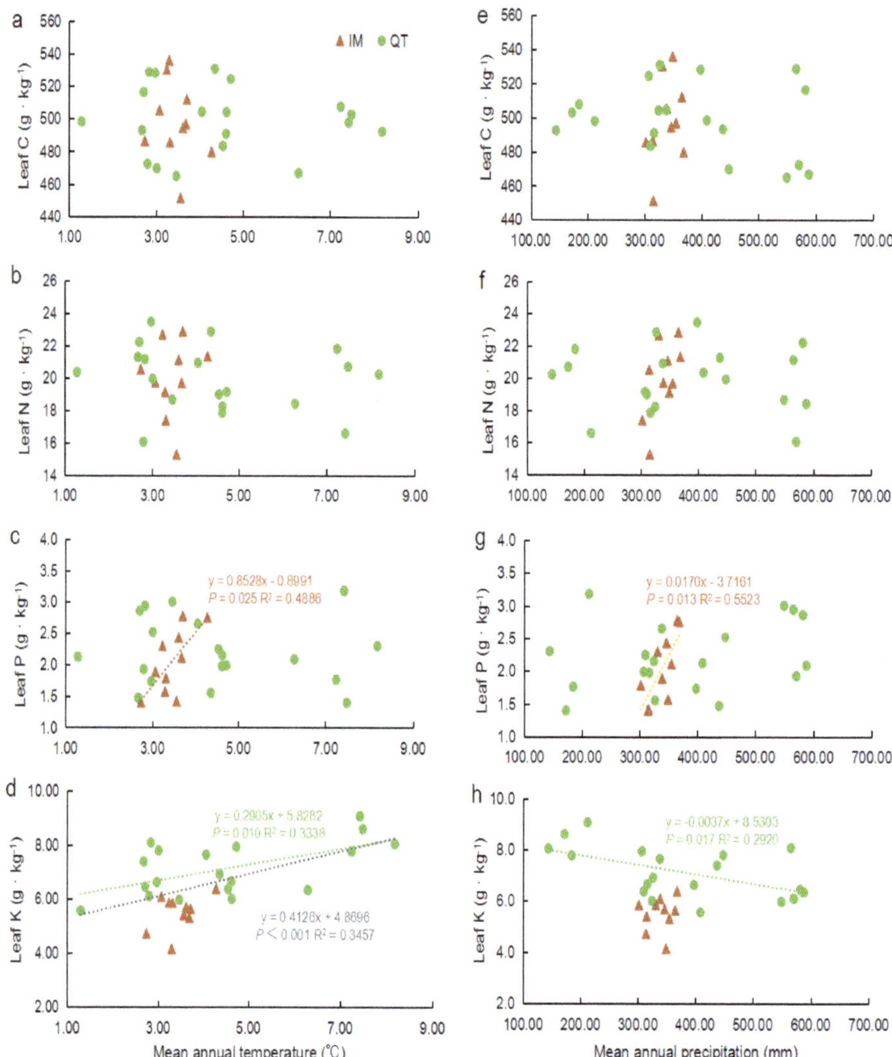

Figure 1. Relationships between the leaf C, N, P, and K content of *S. chamaejasme* with MAT & MAP in the Qinghai–Tibet Plateau (green circles, *n* = 19) and Inner Mongolia Plateau (red triangles, *n* = 10). Linear regression model analyses were utilized. Colored dotted lines represented significant relationships (*p* < 0.05) in different region (red, IM; green, QT; grey, all sampling sites). (**a**) MAT vs. leaf C; (**b**) MAT vs. leaf N; (**c**) MAT vs. leaf P; (**d**) MAT vs. leaf K; (**e**) MAP vs. leaf C; (**f**) MAP vs. leaf N; (**g**) MAP vs. leaf P; (**h**) MAP vs. leaf K.

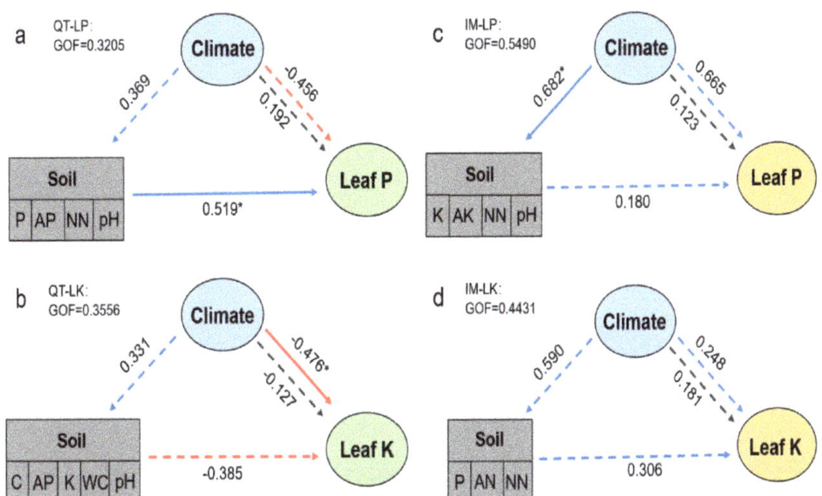

Figure 2. Effects of different soil and climatic variables on the leaf P and K of *S. chamaejasme* in the Qinghai−Tibet Plateau (QT) and Inner Mongolia Plateau (IM) based on partial least squares path modeling. The blue arrows represent positive pathways, the red arrows indicate negative pathways, both are direct effects. The grey arrows show the indirect effects. The standard path coefficients are shown on the arrow. A significant effect is indicated by an * ($p < 0.05$). GOF, goodness of fit of the statistical model. (**a,b**) PLS−PM describing the relationships in QT; (**c,d**) PLS−PM describing the relationships in IM.

3. Discussion

3.1. Leaf Ecological Stoichiometry and Soil Physicochemical Properties of S. chamaejasme

It is essential to maintain nutrient elements in sufficient amounts and relatively stable ratios for plants to survive and grow [1,2,35–38]. This study presents, to our knowledge, the first analysis of leaf element concentrations (C, N, P, K) and ratios (C:N, C:P, N:P, N:K, K:P) of *S. chamaejasme* across degraded grasslands in northern China. Our results show that the leaf C (498.60 g kg^{-1}), N (19.95 g kg^{-1}), and P (2.15 g kg^{-1}) of *S. chamaejasme* were higher than the mean value of all species average in the the China Grassland Transect [38], and that there was no obvious difference between two regions of *S. chamaejasme*. N and P are the most important limiting nutrients for primary productivity in terrestrial ecosystems [39], and a high concentration of N and P in *S. camazepams* leaves suggests its high nutrient uptake efficiency in degraded grasslands, which could facilitate its competitive advantage over other species in nutrient-poor environments. Moreover, K is one of the essential macronutrients that plays a critical role in various metabolic processes, but it has been partly overshadowed in ecological stoichiometry by nitrogen and phosphorus [40,41]. It is worth noting that K concentrations of *S. chamaejasme* were greater in QT than that in IM. The reason may be that the content of nutrients in plants are constrained by nutrient supply in the soil, and the content of soil K is significantly higher in QT, therefore generating this difference. Generally, C:N:P can be used as an effective tool to analyze coupled relationships and differences between each element in the plant-soil system [1,2]. The average leaf C:N and C:P ratio of *S. chamaejasme* were 25.20 and 245.57, respectively, which were lower than the national grassland average leaf C:N (26.86) and C:P (439.84) [38]. The results indicated that *S. chamaejasme* have higher P utilization rates and N utilization efficiency. Previous studies found that nutrient ratios in aboveground vascular plants can be used to distinguish (1) N-limited sites, (2) P- or P+N-limited sites, and (3) K- or K+N-limited sites from each [7,29,42]. The N:P < 14.5, N:K > 2.1, and K:P < 3.4 in *S. chamaejasme* leaf might imply that its growth is restricted by K- or K+N-limited. Both the leaf and soil K content were significantly different between two sampling regions and fertilizer experiments should be conducted to test the validity of this idea in the future.

We found that *S. chamaejasme* could survive in a soil environment with considerable variation, which is consistent with the fact that *S. chamaejasme* is a wide-ranging species in the grasslands of China [22]. The soil conditions for *S. chamaejasme* growth varies considerably from site to site. Soil physicochemical properties varied with a difference of more than 10 times between the maximum and the minimum included C (14.43 times), N (30.94 times), K (32.27 times), NN (26.66 times), and WC (10.60 times), Ec (21.86 times). This may provide a competitive advantage for *S. chamaejasme* against other plant species and help explain its rapid expansion in various environments, even in heavily degraded grasslands. Generally, Tibetan alpine grasslands and Inner Mongolian temperate grasslands, which have different limiting factors, are both zonal grassland types in China [43]. Alpine grasslands are mainly limited by low temperatures in the growing season, while temperate grasslands are affected by drought [38]. Accordingly, our analysis indicated that some soil physicochemical properties of *S. chamaejasme* for the regions were significantly different. Soil WC and pH for Qinghai-Tibet were significantly higher, and the Ec was lower than those for Inner Mongolia. However, apart from SP, SK and SAP, soil C and N concentrations, and other soil available nutrients (AN, NN, AK) for the regions were insignificantly different. These findings suggest that climate imposes important controls on some soil nutrients.

3.2. Relationships between Leaf Ecological Stoichiometry and Environmental Variables

Plant nutrient concentrations and their correlations with soil nutrients are considered effective tools for exploring plant adaptation and resource utilization strategies in a severe environment [28,44]. In our study, few significant correlations between leaf ecological stoichiometry of *S. chamaejasme* and grassland soil physicochemical properties were observed, implying an insensitive response to the changes in the soil nutrient supply of *S. chamaejasme*. This supports the finding of Geng et al. [26] and provides confirmation that wide-ranging species are usually able to use a wide array of resources and tolerate broad environmental conditions or physiological stresses, and hence flourish over a larger area. The poor synchronization with local edaphic conditions demonstrates a capacity of *S. chamaejasme* to maintain a high level of function at both high and low resource levels, resulting in their broad distributions in China's grasslands. Further, stoichiometric homeostasis can help the plants to maintain their element composition at a relatively stable level regardless of changes in nutrient availability via various physiological mechanisms [2,12], and the degree of stoichiometric homeostasis can be indicated by the homeostatic coefficient (H) [12,45,46]. Stoichiometric homeostasis had been reported in many dominant palatable species [15,18,47] in grasslands. However, this has not been established in poisonous species. Since unpalatable plants represent the majority of the plant species that were detected after grasslands have been degraded globally [48–50], revealing the ecophysiology characteristics of poisonous weeds will help us better understand how the communities that are dominated by poisonous weeds form. Generally, species-level stoichiometric homeostasis was positively correlated with the stability of vegetation [15,18,51]. Meanwhile, the species with the highest degree of N homeostasis consistently had the relatively highest growth rates [19] and well-developed storage systems [15,52]. Therefore, resource utilization and storage functions of these species mitigates environmental variations [53], resulting in spatiotemporal stability in abundance [54]. Our results showed that *S. chamaejasme* leaves contain different homeostasis strength of C, N, and K contents and its ratios, which means *S. chamaejasme* could be more conservative in their use of nutrients improving their adaptation to diverse conditions.

Besides, growing plants induce changes in the composition of soil communities and the physicochemical soil environment [55,56]. A previous study in an alpine meadow ecosystem has shown that *S. chamaejasme* produced more aboveground litter with higher tissue N and lower lignin:N than each of the co-occurring species, and significantly increased the surface soil organic matter [19]. Another study found that *S. chamaejasme* had different ammonia oxidizing bacterial (AOB) that were present with low ammonia oxidation rates, which means greater N availability for *S. chamaejasme* growth due to losses of N reduction [57]. *S. chamaejasme* have positive feedback on soil processes, especially soil C and N, which could be the reason that why leaf C and N contents of *S. chamaejasme* were not different among the two sites of distinct environments, and both were unaffected by local soil factors. Besides, the soil biota plays a pivotal role in modulating primary production by controlling decomposition and nutrient availability, as well as affecting root grazing and plant nutrient uptake [58,59]. Generally, roots of invasive plants enhance or reduce their mutualistic associations with different mycorrhizal fungi or N fixing bacteria [60], which potentially feedback to plant invasion by enhancing N uptake of invaders [61] or by lowering the dependence of plant invaders on arbuscular mycorrhizal fungi (AMF) compared to native species [62]. A recent study suggested that *S. chamaejasme* possessed high ratios of plant growth-promoting proteobacterial endophytes such as *Pseudomonas*, *Acinetobacter*, and *Brevundimonas* [63], and its invaded soils had a lower relative abundance of AMF, but greater pathogenic fungi [64], which in consistent with many invasive plants. Past studies have revealed that *S. chamaejasme* can cultivate the soil environment differently, facilitating it spread rapidly in degraded grassland although it is a native species.

Our results indicate that in the macro scale, leaf C and N do not directly correlate with meteorological factors (MAT and MAP), which is in agreement with previous studies that were conducted in the grassland biomes of China [7]. The weak relationships that was observed between leaf C, N, and climatic variables may result from plant growth, development, metabolism, phenological, and life-history traits rather than from the specific geographic environment. On the contrary, there were close relationships between the leaf P and K and climatic factors (Figure 1). The relationship between the leaf P and climate factors was significant only in IM, and the K content of *S. chamaejasme* leaves was significantly related to climate factors only in QT. We noticed that the correlation of leaf P and MAP ($R^2 = 0.5523$) was greater than the relationship between P and MAT ($R^2 = 0.4886$) in IM, and the relationship between K and MAT ($R^2 = 0.3338$) was greater than that with MAP ($R^2 = 0.2920$) in QT. These again reflect the different limiting factors of plant growth in different regions [38]. It is a reasonable assumption that precipitation is a more important limiting factor than the temperature for vegetation growth in arid and semi-arid regions such as Inner Mongolian Plateau temperate grasslands. However, the variation in MAP seems very small and the positive relationships among MAP, soil water content, soil P, and AP were weak (Figure S1). We suggested that altitude changes may the reason behind a positive association between MAP and leaf P in IM considering the ranges of geographical distribution and altitude (N 41.34°~44.77°, E 115.30°~118.16°, 1060 m to 1535 m). Our results for the strong relationships between leaf P, soil P and AP, and soil water content and altitude in IM supported our suggestion (Figure S1). In contrast, the temperature is more likely to have a greater effect on the leaf element concentrations than precipitation in QT alpine grasslands with high-altitude and low temperature. We also found that only leaf K was negatively correlated with MAP in QT. One possible explanation is that K leaches more easily from leaves than N and P, hence it is easy to ascertain the increase of MAP in the studies area leading to more leaf K leaching of *S. chamaejasme*. Another possible explanation may be that K plays many fundamental physiological and metabolic roles in terrestrial plants in relation to water-use efficiency [34]. Some recent studies have observed that K concentrations are related to drought resistance [65,66], therefore, the leaf K content tends to decrease with MAP increase.

Moreover, the MAP or MAT contribution was stronger in leaf elements than soil, which illustrated once more the relatively weak effect of soil physicochemical properties on leaf elements. To explore complex relationships between soil and climatic factors on the leaf P and K contents of *S. chamaejasme*, we conducted a PLS-PM analysis. We found that soil exerted a significant effect on leaf P content and climate affected leaf K directly in QT, while the leaf P content appeared to be limited mainly by climate but the leaf K content was not affected significantly in IM. This does not fit with the fact that climate factors which often affect leaf elements through their influence on soil nutrient status [67]. The arid conditions of the IM may have restricted grassland plants growth by insufficient water supply, but the results of our previous study [21] have proven that high water use efficiency plus high nutrient uptake efficiency of *S. chamaejasme* ensures its competitive advantage on degraded grasslands in IM, which makes the relationship between leaf P or K of *S. chamaejasme* and the soil factors weak in the IM region. However, the leaf P content was positively correlated with soil factors (soil P, available P, nitrate N, and pH), which was not entirely consistent with the result that was obtained in IM. The negative influence of climatic variables on leaf K was significant in QT. This may be the result of the negative relationship between MAP and leaf K, because K shows a greater loss from the plant canopy by foliar leaching than other nutrients such as N and P [34,68]. Our model suggests that the underlying mechanisms behind the leaf P or leaf K content in *S. chamaejasme* were different in the two regions that were studied, which means *S. chamaejasme* developed adjustable relationships with environmental factors to adapt to different growth conditions, thus facilitating its spread in degraded grasslands.

In addition, species' natural habitats will be subjected to more disturbances in the future due to climate change and habitat degradation that is caused by intensive anthropogenic activities [69,70]. Thus, continuing wide-scale sampling and considering the influence of human activities are required to further develop a deeper understanding of the geographic patterns of leaf stoichiometry in *S. chamaejasme*.

4. Materials and Methods

4.1. Study Area

A total of 29 sites were selected (10 sites in IM, 19 sites in QT, Figure 3), of which the longitude ranged from 99.68 to 118.16° E and latitude from 33.35 to 44.77° N. The altitudes spanned from 1060 to 3500 m (Table S6). The mean annual temperature (MAT) and mean annual precipitation (MAP) range from 1.29 to 8.19 °C and 143.84 to 587.53 mm, respectively.

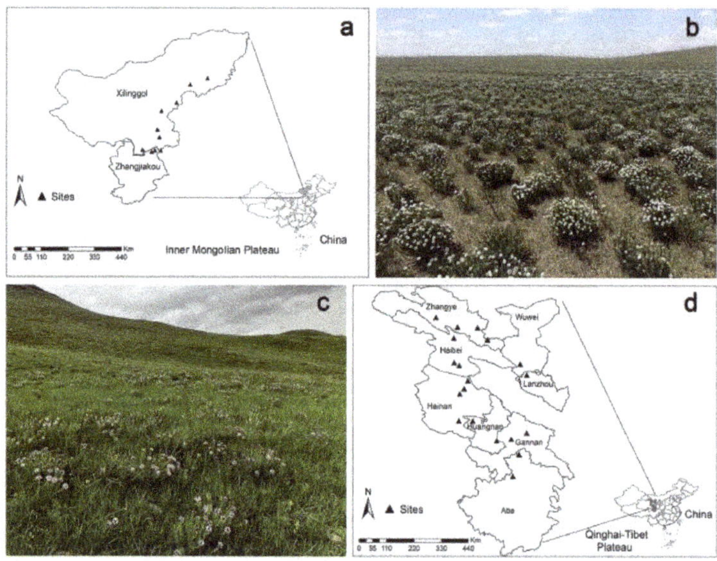

Figure 3. Location of the study and sampling sites. (**a**) Sampling sites on the Inner Mongolia Plateau; (**b**) *S. chamaejasme* coverage in Taipusi Banner in Inner Mongolia Plateau; (**c**) Sampling sites on the Qinghai-Tibetan Plateau; and (**d**) *S. chamaejasme* coverage in Qilian County in Qinghai-Tibetan Plateau.

4.2. Plant and Soil Sampling

Field measurements were conducted in June–July 2019, when it was the vigorous growth stage for *S. chamaejasme*. At least 30 *S. chamaejasme* plants were randomly collected in each sampling site, then were subdivided into three subsamples, and the leaves of the subsamples were mixed into a composite sample. The samples were ground into fine powder for measuring the content of elements (carbon, nitrogen, phosphorus, potassium). The concentrations of the total C and total N of the *S. chamaejasme* leaves were determined sequentially on a FLASH 2000 elemental analyzer (Thermo Fisher Scientific, Waltham, MA, USA). The total leaf P and K content were determined by using an AA-6300 atomic absorption spectrophotometer (Shimadzu, Japan).

A total of three soil samples (0–15 cm in depth) were collected from each sample site, and each sample was thoroughly mixed with three subsamples and air-dried. The soil samples were removed from the roots and passed through a 100-mesh sieve (30 cm in diameter). Then, the soil was analyzed for soil carbon (C), nitrogen (N), phosphorus (P), potassium (K), ammonium nitrogen (AN), nitrate nitrogen (NN), available potassium (AK), available phosphorus (AP), pH, electrical conductivity (Ec), and water content (WC). The soil physicochemical properties were measured as described by Bao [71], soil C and N by the FLASH 2000 elemental analyzer, K by NaOH fusion-flame photometry, P by NaOH fusion-Mo/Sb colorimetry, soil AN and NN by Auto Discrete analyzer, and soil AK were determined by the flame atomic absorption spectrophotometer. To measure soil AP, the air-dried and pre-weighed soil was extracted using 0.5 mol L^{-1} $NaHCO_3$ and the P concentration in the extract was determined by the ammonium molybdate method. The soil pH was measured in a 1:2.5 soil:water suspension, and the soil Ec was measured using a conductivity meter. The soil water contents were determined gravimetrically by oven-drying subsamples at 105 °C for 24 h.

4.3. Data Analysis

The means, standard deviations (SD), and coefficients of variation (CV) of the leaf element concentrations and their ratios, and soil physicochemical properties were calculated. Differences between QT and IM were evaluated by Independent-Samples *t*-test. Pearson correlations analysis was used to evaluate the relationship between *S. chamaejasme* leaf ecological stoichiometry and the soil physicochemical properties across the 29 sampling sites.

The degree of homeostasis (*H*) was calculated by plotting the log-transformed values of leaf elements and soil from 29 sites, where the *H* is the inverse of the slope [2]:

Log (leaf element concentration or stoichiometry) = α + (1/H) log (soil element concentration or stoichiometry).

To determine the degree of stoichiometric homeostasis, the method that was proposed by Persson et al. [12] was used. Standardized major axis regression analyses were conducted for C, N, P, K, and C:N:P:K ratios for leaves (The R package 'smatr 3') [72]. Since the slope was expected to be ≥ 0, one-tailed *t*-tests with $\alpha = 0.1$ were used. If the regression was nonsignificant ($p > 0.1$), 1/H was set to zero, and the organism was considered 'strictly homeostatic'. Species with 1/H = 1 were considered not homeostatic. All the datasets with significant regressions and 0 < H < 1 were categorized as: 0 < 1/H < 0.25: 'homeostatic'; 0.25 < 1/H < 0.5: 'weakly homeostatic'; 0.5 < 1/H < 0.75: 'weakly plastic'; 1/H > 0.75 'plastic'. For 1/H > 1, 1/H close to 1 indicates weak or no stoichiometric homeostasis, and 1/H much larger than 1 indicates 'homeostatic'.

In order to determine the influence of climate factors, we obtained raw daily precipitation and temperature data (2010–2019) from the China Meteorological Administration and calculated the annual precipitation and temperature using the Kriging interpolation method in ArcGIS (ESRI (Environmental Systems Research Institute), Redlands, CA, USA). Therefore, climate data for the mean annual temperature (MAT) and mean annual precipitation (MAP) for the sample sites were obtained. Regression analyses were performed to determine the correlation of the leaf element contents and climate factors (MAT, MAP).

Scatter plots were used to visualize the relationships among the leaf element contents and climate factors (MAT, MAP), and liner regression equations were developed.

We determined the relative importance of the soil, MAP, and MAT for leaf C, N, P, and K heterogeneity across all the sites, respectively, using mixed effects models. In these models, soil, MAP, MAT, and their interaction were fitted as fixed factors, and region was fitted as a random factor (The R package 'lme4', 'lmerTest', 'glmm.hp', 'readxl', 'ade4'). The soil data (soil C, N, P, K, AP, AK, AN, NN, WC, pH, Ec) were processed by principal component analysis (PCA; The R package 'Vegan', 'FactoMineR', 'factoextra'), then the number of first axis was used as the soil parameter (Tables S7–S9; Figure S2). Leaf C, N, P, K, and climatic data (MAP and MATA) heterogeneity were log-transformed to linearized the data.

Partial least squares path modeling (PLS-PM) was employed to explore the direct, indirect, and interactive effects between all the environmental variables for leaf element contents (The R package 'plspm'). The model included the following variables: Leaf elements (P, K), climate factors (MAT, MAP), and soil factors (K, AK, NN, and pH for leaf P in IM, P, AN and NN for leaf K in IM, P, AP, NN and pH for leaf P in QT, C, K, AP, WC, and pH for leaf K in QT), after testing for collinearity of soil factors with the multivariate analog of *Levene's* test using the *"betadisper"* function in the vegan package. The indirect effects are defined as multiplied path coefficients between the predictor and response variables, including all the possible paths excluding the direct effect. The final model was chosen among all constructed models based on the Goodness of Fit (GOF) statistic according to the model's overall predictive power.

5. Conclusions

To our knowledge this is the first study to comprehensively research the chemistry of multiple nutritional elements (C, N, P, K) and their ratios in *S. chamaejasme* leaves and its surrounding soil physiochemical properties and quantify the potential controls and variability at a large scale. We found that there was no obvious difference between the leaf C, N, and P content in *S. chamaejasme* from the QT and IM, but the leaf K concentration was significantly higher in QT than that in IM. Inconsistent with the variation of leaf element contents and ratios, the soil physiochemical properties of *S. chamaejasme*-infested areas varied remarkably, and most of them were greater in QT. Our results clearly showed that there was no significant correlation between *S. chamaejasme* leaf ecological stoichiometry and soil physicochemical properties, which supported the fact that the nutrient concentrations and stoichiometry of wide-ranging species tend to be insensitive to variation in soil nutrient availability.

Besides, the different homeostasis strength of C, N, K, and their ratios in *S. chamaejasme* leaves across all the sites were observed, which indicated that *S. chamaejasme* could be more conservative in their use of nutrients improving their adaptation to diverse conditions. Both the C and N content of *S. chamaejasme* leaves were unaffected by any climatic factors, but the leaf P and K were affected differently in QT and IM. Besides, the MAP or MAT contribution was stronger in the leaf elements than soil by using mixed effects models, which illustrated once more the relatively weak effect of soil physicochemical properties on leaf elements. Finally, we conducted a partial least squares path modeling analysis to examine the different effects of soil and climatic on leaf P and K of *S. chamaejasme*. These results suggest that *S. chamaejasme* adapts to changing environments by adjusting its relationships with climate or soil factors to improve their chances of survival and spread in degraded grasslands.

Supplementary Materials: The following supporting information can be downloaded at: https://www.mdpi.com/article/10.3390/plants11151943/s1, Table S1: Leaf element contents of *S. chamaejasme* in different sample sites in this study (Mean ± SD); Table S2: Soil physiochemical properties in different sample sites in this study (Mean ± SD); Table S3: Relationship between *S. chamaejasme* leaf stoichiometry and soil physicochemical properties in Northern China ; Table S4: Summary of linear mixed effects model analyzing the effects of soil, MAT, and MAP on variation of C, N, P, and K concentrations in *S. chamaejasme* leaves, using soil, MAP, MAT, and their interaction as fixed effects. Region was taken as a random factor; Table S5: Summary of the total effects on the leaf P and K of *S. chamaejasme* in Qinghai Tibet Plateau (QT) and Inner Mongolia Plateau (IM); Table S6: The geographical and climatic information associated with the sample sites in this study; Table S7: Eigenvalues of the soil PCA; Table S8: Loadings of the soil PCA; Table S9: Site scores of the soil PCA; Figure S1: Heat map of Pearson correlations among *S. chamaejasme* leaf P content, soil P and AP content, soil water content (SWC), mean annual precipitation (MAP), and altitude (ALT) in IM region; Figure S2: Biplot for the first two axes of soil physicochemical properties.

Author Contributions: Conceptualization, L.G. and D.H.; methodology, L.G. and L.L.; software, H.M.; validation, L.L., L.Z. and H.Z.; formal analysis, W.H.; investigation, H.M. and L.Z.; resources, L.L. and H.Z.; writing—original draft preparation, L.G.; writing—review and editing, W.H., V.J.S. and K.W.; visualization, K.W.; supervision, D.H.; funding acquisition, D.H. and W.H. All authors have read and agreed to the published version of the manuscript.

Funding: Please add: This research was funded by The Climate-Smart Grassland Ecosystem Management Project (CSMG-C-03), Special Aid Fund for Qinghai Province (2020-QY-210), International Collaboration Fund of Department of Science and Technology of Shaanxi Province (2020KW-030), China Agriculture Research System (CARS-34).

Institutional Review Board Statement: Not applicable.

Informed Consent Statement: Not applicable.

Data Availability Statement: Not applicable.

Conflicts of Interest: The authors declare no conflict of interest.

References

1. Elser, J.J.; Sterner, R.W.; Gorokhova, E.; Fagan, W.F.; Markow, T.A.; Cotner, J.B.; Harrison, J.F.; Hobbie, S.E.; Odell, G.M.; Weider, L.W. Biological stoichiometry from genes to ecosystems. *Ecol. Lett.* **2000**, *3*, 540–550. [CrossRef]
2. Sterner, R.W.; Elser, J.J. *Ecological Stoichiometry: The Biology of Elements from Molecules to the Biosphere*; Princeton University Press: Princeton, NJ, USA, 2002.
3. Sardans, J.; Janssens, I.A.; Ciais, P.; Obersteiner, M.; Peñuelas, J. Recent advances and future research in ecological stoichiometry. *Perspect. Plant Ecol. Evol. Syst.* **2021**, *50*, 125611. [CrossRef]
4. Güsewell, S. N:P ratios in terrestrial plants: Variation and functional significance. *New Phytol.* **2004**, *164*, 243–266. [CrossRef]
5. Reich, P.B.; Oleksyn, J. Global patterns of plant leaf N and P in relation to temperature and latitude. *Proc. Natl. Acad. Sci. USA* **2004**, *101*, 11001–11006. [CrossRef]
6. He, J.-S.; Wang, L.; Flynn, D.; Wang, X.; Ma, W.; Fang, J. Leaf nitrogen:phosphorus stoichiometry across Chinese grassland biomes. *Oecologia* **2008**, *155*, 301–310. [CrossRef]
7. Wu, P.; Zhou, H.; Cui, Y.C.; Zhao, W.J.; Hou, Y.J.; Zhu, J.; Ding, F.J. Stoichiometric characteristics of leaf nutrients in karst plant species during natural restoration in Maolan national nature reserve, Guizhou, China. *J. Sustain. For.* **2021**, 1–25. [CrossRef]
8. Sardans, J.; Rivas-Ubach, A.; Peñuelas, J. The elemental stoichiometry of aquatic and terrestrial ecosystems and its relationships with organismic lifestyle and ecosystem structure and function: A review and perspectives. *Biogeochemistry* **2012**, *111*, 1–39. [CrossRef]
9. Chown, S.L.; Gaston, K.J.; Robinson, D. Macrophysiology: Large-scale patterns in physiological traits and their ecological implications. *Funct. Ecol.* **2004**, *18*, 159–167. [CrossRef]
10. Yang, X.J.; Huang, Z.Y.; Zhang, K.L.; Cornelissen, J.H.C. C:N:P stoichiometry of Artemisia species and close relatives across northern China: Unravelling effects of climate, soil and taxonomy. *J. Ecol.* **2015**, *103*, 1020–1031. [CrossRef]
11. Wang, X.G.; Lü, X.T.; Dijkstra, F.A.; Zhang, H.Y.; Wu, Y.N.; Wang, Z.W.; Feng, J.; Han, X.G. Changes of plant N:P stoichiometry across a 3000-km aridity transect in grasslands of northern China. *Plant Soil* **2019**, *443*, 107–119. [CrossRef]
12. Persson, J.; Fink, P.; Goto, A.; Hood, J.M.; Jonas, J.; Kato, S. To be or not to be what you eat: Regulation of stoichiometric homeostasis among autotrophs and heterotrophs. *Oikos* **2010**, *119*, 741–751. [CrossRef]
13. Hessen, D.O.; Agren, G.I.; Anderson, T.R.; Elser, J.J.; De Ruiter, P.C. Carbon sequestration in ecosystems: The role of stoi-chiometry. *Ecology* **2004**, *85*, 1179–1192. [CrossRef]

14. Elser, J.J.; Fagan, W.F.; Kerkhoff, A.J.; Swenson, N.G.; Enquist, B.J. Biological stoichiometry of plant production: Metabolism, scaling and ecological response to global change. *New Phytol.* **2010**, *186*, 593–608. [CrossRef] [PubMed]

15. Yu, Q.; Chen, Q.S.; Elser, J.J.; He, N.P.; Wu, H.H.; Zhang, G.M.; Wu, J.G.; Bai, Y.F.; Han, X.G. Linking stoichiometric homoeostasis with ecosystem structure, functioning and stability. *Ecol. Lett.* **2010**, *13*, 1390–1399. [CrossRef] [PubMed]

16. Peng, H.Y.; Chen, Y.H.; Yan, Z.B.; Han, W.X. Stage-dependent stoichiometric homeostasis and responses of nutrient resorption in *Amaranthus mangostanus* to nitrogen and phosphorus addition. *Sci. Rep.* **2016**, *6*, 37219. [CrossRef] [PubMed]

17. Yan, Z.B.; Guan, H.Y.; Han, W.X.; Han, T.S.; Guo, Y.L.; Fang, J.Y. Reproductive organ and young tissues show constrained elemental composition in *Arabidopsis thaliana*. *Ann. Bot.* **2016**, *117*, 431–439. [CrossRef]

18. Yu, Q.; Elser, J.J.; He, N.P.; Wu, H.H.; Chen, Q.S.; Zhang, G.M.; Han, X.G. Stoichiometric homeostasis of vascular plants in the Inner Mongolia grassland. *Oecologia* **2011**, *166*, 1–10. [CrossRef] [PubMed]

19. Yu, Q.; Wilcox, K.; La Pierre, K.; Knapp, A.K.; Han, X.G.; Smith, M.D. Stoichiometric homeostasis predicts plant species dominance, temporal stability, and responses to global change. *Ecology* **2015**, *96*, 2328–2335. [CrossRef] [PubMed]

20. Sun, G.; Luo, P.; Wu, N.; Qiu, P.F.; Gao, Y.H.; Chen, H.; Shi, F.S. *Stellera chamaejasme* L. increases soil N availability, turnover rates and microbial biomass in an alpine meadow ecosystem on the eastern Tibetan Plateau of China. *Soil Biol. Biochem.* **2009**, *41*, 86–91. [CrossRef]

21. Guo, L.Z.; Li, J.H.; He, W.; Liu, L.; Huang, D.; Wang, K. High nutrient uptake efficiency and high water use efficiency facilitate the spread of *Stellera chamaejasme* L. in degraded grasslands. *BMC Ecol.* **2019**, *19*, 1–12. [CrossRef] [PubMed]

22. Zhang, Y.-H.; Zhang, J.-W.; Li, Z.-M.; Sun, H. Genetic diversity of the weed species, *Stellera chamaejasme* in China inferred from amplified fragment length polymorphism analysis. *Weed Biol. Manag.* **2015**, *15*, 165–174. [CrossRef]

23. Zhao, M.L.; Gao, X.L.; Wang, J.; He, X.L.; Hsn, B. A review of the most economically important poisonous plants to the live-stock industry on temperate grasslands of China. *J. Appl. Toxicol.* **2013**, *33*, 9–17. [CrossRef] [PubMed]

24. Jin, H.; Guo, H.R.; Yang, X.Y.; Xin, A.Y.; Liu, H.Y.; Qin, B. Effect of allelochemicals, soil enzyme activity and environmental factors from *Stellera chamaejasme* L. on rhizosphere bacterial communities in the northern Tibetan Plateau. *Arch. Agron. Soil Sci.* **2020**, *68*, 547–560. [CrossRef]

25. Guo, L.Z.; Zhao, H.; Zhai, X.J.; Wang, K.L.; Liu, L.; Wang, K.; Huang, D. Study on life history traits of *Stellera chamaejasme* provide insights into its control on degraded typical steppe. *J. Environ. Manag.* **2021**, *291*, 112716. [CrossRef]

26. Geng, Y.; Wang, Z.H.; Liang, C.Z.; Fang, J.Y.; Baumann, F.; Kuhn, P.; Scholten, T.; He, J.S. Effect of geographical range size on plant functional traits and the relationships between plant, soil and climate in Chinese grasslands. *Glob. Ecol. Biogeogr.* **2011**, *21*, 416–427. [CrossRef]

27. Zhang, H.Y.; Wu, H.H.; Yu, Q.; Wang, Z.W.; Wei, C.Z.; Long, M.; Kattge, J.; Smith, M.; Han, X.G. Sampling Date, Leaf age and root size: Implications for the study of plant C:N:P stoichiometry. *PLoS ONE* **2013**, *8*, e60360. [CrossRef] [PubMed]

28. Ordoñez, J.C.; van Bodegom, P.M.; Witte, J.-P.M.; Wright, I.J.; Reich, P.B.; Aerts, R. A global study of relationships between leaf traits, climate and soil measures of nutrient fertility. *Glob. Ecol. Biogeogr.* **2009**, *18*, 137–149. [CrossRef]

29. Luo, Y.; Peng, Q.W.; He, M.S.; Zhang, M.X.; Liu, Y.Y.; Gong, Y.M.; Eziz, A.; Liu, K.H.; Han, W.X. N, P and K stoichiometry and resorption efficiency of nine dominant shrub species in the deserts of Xinjiang, China. *Ecol. Res.* **2020**, *35*, 625–637. [CrossRef]

30. Sun, L.K.; Zhang, B.G.; Wang, B.; Zhang, G.S.; Zhang, W.; Zhang, B.L.; Chang, S.J.; Chen, T.; Liu, G.X. Leaf elemental stoichiometry of *Tamarix* Lour. species in relation to geographic, climatic, soil, and genetic components in China. *Ecol. Eng.* **2019**, *106*, 448–457. [CrossRef]

31. Albert, C.H.; Thuiller, W.; Yoccoz, N.G.; Soudant, A.; Boucher, F.; Saccone, P.; Lavorel, S. Intraspecific functional variability: Extent, structure and sources of variation. *J. Ecol.* **2010**, *98*, 604–613. [CrossRef]

32. Hu, Y.-K.; Zhang, Y.-L.; Liu, G.-F.; Pan, X.J.; Yang, X.; Li, W.B.; Dai, W.-H.; Tang, S.-L.; Xiao, T.; Chen, L.-Y.; et al. Intraspecific N and P stoichiometry of *Phragmites australis*: Geographic patterns and variation among climatic regions. *Sci. Rep.* **2017**, *7*, 43018. [CrossRef]

33. Han, W.X.; Fang, J.Y.; Reich, P.B.; Woodward, F.I.; Wang, Z.H. Biogeography and variability of eleven mineral elements in plant leaves across gradients of climate, soil and plant functional type in China. *Ecol. Lett.* **2011**, *14*, 788–796. [CrossRef] [PubMed]

34. Sardans, J.; Peñuelas, J. Potassium: A neglected nutrient in global change. *Glob. Ecol. Biogeogr.* **2015**, *24*, 261–275. [CrossRef]

35. Koerselman, W.; Meuleman, A.F.M. The Vegetation N:P Ratio: A New Tool to Detect the Nature of Nutrient Limitation. *J. Appl. Ecol.* **1996**, *33*, 1441–1450. [CrossRef]

36. Han, W.X.; Fang, J.Y.; Guo, D.L.; Zhang, Y. Leaf nitrogen and phosphorus stoichiometry across 753 terrestrial plant species in China. *New Phytol.* **2005**, *168*, 377–385. [CrossRef]

37. Song, Z.L.; Liu, H.Y.; Zhao, F.J.; Xu, C.Y. Ecological stoichiometry of N:P:Si in China's grasslands. *Plant Soil* **2014**, *380*, 165–179. [CrossRef]

38. Fan, J.W.; Harris, W.; Zhong, H.P. Stoichiometry of leaf nitrogen and phosphorus of grasslands of the Inner Mongolian and Qinghai-Tibet Plateaus in relation to climatic variables and vegetation organization levels. *Ecol. Res.* **2016**, *31*, 821–829. [CrossRef]

39. Elser, J.J.; Bracken, M.E.S.; Cleland, E.E.; Gruner, D.S.; Harpole, W.S.; Hillebrand, H.; Ngai, J.T.; Seabloom, E.W.; Shurin, J.B.; Smith, J.E. Global analysis of nitrogen and phosphorus limitation of primary producers in freshwater, marine and terrestrial ecosystems. *Ecol. Lett.* **2007**, *10*, 1135–1142. [CrossRef]

40. Shin, R. Strategies for improving potassium use efficiency in plants. *Mol. Cells* **2014**, *37*, 575–584. [CrossRef]

41. Adams, E.; Shin, R. Transport, signaling, and homeostasis of potassium and sodium in plants. *J. Integr. Plant Biol.* **2014**, *56*, 231–249. [CrossRef] [PubMed]
42. Venterink, H.O.; Wassen, M.J.; Verkroost, A.W.M.; De Ruiter, P.C. Species richness- productivity patterns differ between N-, P-, and K-limited wetlands. *Ecology* **2003**, *84*, 2191–2199. [CrossRef]
43. Li, L.H.; Chen, J.Q.; Han, X.G.; Zhang, W.H.; Shao, C.L. *Grassland Ecosystems of China: A Synthesis and Resume*; Springer Nature: Singapore, 2020.
44. Hong, J.T.; Wang, X.D.; Wu, J.B. Effects of soil fertility on the N:P stoichiometry of herbaceous plants on a nutrient-limited alpine steppe on the northern Tibetan Plateau. *Plant Soil* **2015**, *391*, 179–194. [CrossRef]
45. Giordano, M. Homeostasis: An underestimated focal point of ecology and evolution. *Plant Sci.* **2013**, *211*, 92–101. [CrossRef] [PubMed]
46. Wang, J.N.; Wang, J.Y.; Wang, L.; Zhang, H.; Guo, Z.W.; Wang, G.G.; Smith, W.K.; Wu, T.G. Does stoichiometric homeostasis differ among tree organs and with tree age? *For. Ecol. Manag.* **2019**, *453*, 117637. [CrossRef]
47. Li, Y.F.; Li, Q.Y.; Guo, D.Y.; Liang, S.; Wang, Y.J. Ecological stoichiometry homeostasis of *Leymus chinensis* in degraded grassland in western Jilin Province, NE China. *Ecol. Eng.* **2016**, *90*, 387–391. [CrossRef]
48. Ralphs, M.H. Ecological relationships between poisonous plants and rangeland condition: A review. *Rangel. Ecol. Manag.* **2002**, *55*, 285. [CrossRef]
49. Kleijn, D.; Müller-Schärer, H. The Relation Between Unpalatable Species, Nutrients and Plant Species Richness in Swiss Montane Pastures. *Biodivers. Conserv.* **2006**, *15*, 3971–3982. [CrossRef]
50. Zhang, Z.; Sun, J.; Liu, M.; Xu, M.; Wang, Y.; Wu, G.L.; Zhou, H.; Ye, C.; Tsechoe, D.; Wei, T. Don't judge toxic weeds on whether they are native but on their ecological effects. *Ecol. Evol.* **2020**, *10*, 9014–9025. [CrossRef]
51. Bai, X.J.; Wang, B.R.; An, S.S.; Zeng, Q.C.; Zhang, H.X. Response of forest species to C:N:P in the plant-litter-soil system and stoichiometric homeostasis of plant tissues during afforestation on the Loess Plateau, China. *CATENA* **2019**, *183*, 104186. [CrossRef]
52. Johnson, N.C.; Rowland, D.L.; Corkidi, L.; Allen, E.B. Plant winners and losers during grassland N-eutrophication differ in biomass allocation and mycorrhizas. *Ecology* **2008**, *89*, 2868–2878. [CrossRef]
53. Aerts, R.; Chapin, F.S., III. The mineral nutrition of wild plants revisited: A reevaluation of processes and patterns. *Adv. Ecol. Res.* **2000**, *30*, 1–67.
54. Tilman, D.; Reich, P.B.; Knops, J.M.H. Biodiversity and ecosystem stability in a decade-long grassland experiment. *Nature* **2006**, *441*, 629–632. [CrossRef] [PubMed]
55. Berg, G.; Smalla, K. Plant species and soil type cooperatively shape the structure and function of microbial communities in the rhizosphere. *FEMS Microbiol. Ecol.* **2009**, *68*, 1–13. [CrossRef] [PubMed]
56. Manrubia, M.; van der Putten, W.; Weser, C.; Veen, C. Rhizosphere and litter feedbacks to range-expanding plant species and related natives. *J. Ecol.* **2019**, *108*, 353–365. [CrossRef] [PubMed]
57. Ma, J.; Bowatte, S.; Wang, Y.; Newton, P.; Hou, F. Differences in soil ammonia oxidizing bacterial communities under unpalatable (*Stellera chamaejasme* L.) and palatable (*Elymus nutans* Griseb.) plants growing on the Qinghai Tibetan Plateau. *Soil Biol. Biochem.* **2020**, *144*, 107779. [CrossRef]
58. Wardle, D.A.; Bardgett, R.D.; Klironomos, J.N.; Setälä, H.; van der Putten, W.H.; Wall, D.H. Ecological Linkages between Aboveground and Belowground Biota. *Science* **2004**, *304*, 1629–1633. [CrossRef]
59. Bardgett, R.D.; Wardle, D.A. *Aboveground-Belowground Linkages: Biotic Interactions, Ecosystem Processes, and Global Change*; Oxford University Press: Oxford, UK, 2010.
60. Zhang, P.; Li, B.; Wu, J.H.; Hu, S.J. Invasive plants differentially affect soil biota through litter and rhizosphere pathways: A meta-analysis. *Ecol. Lett.* **2018**, *22*, 200–210. [CrossRef]
61. Pringle, A.; Bever, J.D.; Gardes, M.; Parrent, J.L.; Rillig, M.C.; Klironomos, J.N. Mycorrhizal symbioses and plant invasions. *Annu. Rev. Ecol. Evol. Syst.* **2009**, *40*, 699–715. [CrossRef]
62. Vogelsang, K.M.; Bever, J.D. Mycorrhizal densities decline in association with nonnative plants and contribute to plant in-vasion. *Ecology* **2009**, *90*, 399–407. [CrossRef]
63. Jiang, Y.J.; Li, Q.H.; Mao, W.Q.; Tang, W.T.; White, J.F., Jr.; Li, H.Y. Endophytic bacterial community of *Stellera chamaejasme* L. and its role in improving host plants' competitiveness in grasslands. *Environ. Microbiol.* **2022**. *peer-review*. [CrossRef]
64. He, W.; Detheridge, A.; Liu, Y.M.; Wang, L.; Wei, H.C.; Griffith, G.W.; Scullion, J.; Wei, Y.H. Variation in soil fungal composition associated with the invasion of *Stellera chamaejasme* L. in Qinghai-Tibet Plateau grassland. *Microorganisms* **2019**, *7*, 587. [CrossRef] [PubMed]
65. Rivas-Ubach, A.; Sardans, J.; Pérez-Trujillo, M.; Estiarte, M.; Peñuelas, J. Strong relationship between elemental stoichiometry and metabolome in plants. *Proc. Natl. Acad. Sci. USA* **2012**, *109*, 4181–4186. [CrossRef] [PubMed]
66. Sardans, J.; Peñuelas, J.; Coll, M.; Vayreda, J.; Rivas-Ubach, A. Stoichiometry of potassium is largely determined by water availability and growth in Catalonian forests. *Funct. Ecol.* **2012**, *26*, 1077–1089. [CrossRef]
67. Luo, Y.; Peng, Q.W.; Li, K.H.; Gong, Y.M.; Liu, Y.Y.; Han, W.X. Patterns of nitrogen and phosphorus stoichiometry among leaf, stem and root of desert plants and responses to climate and soil factors in Xinjiang, China. *CATENA* **2021**, *199*, 105100. [CrossRef]
68. Kopáček, J.; Turek, J.; Hejzlar, J.; Šantrůčková, H. Canopy leaching of nutrients and metals in a mountain spruce forest. *Atmospheric Environ.* **2009**, *43*, 5443–5453. [CrossRef]

69. Chen, J.H.; Yan, F.; Lu, Q. Spatiotem-poral variation of vegetation on the Qinghai-Tibet plateau and the influence of climatic factors and human activities on vegetation trend (2000–2019). *Remote Sens.* **2020**, *12*, 3150. [CrossRef]
70. Wei, Y.Q.; Lu, H.Y.; Wang, J.N.; Sun, J.; Wang, X.F. Responses of vegetation zones, in the Qinghai-Tibetan Plateau, to climate change and anthropogenic influences over the last 35 years. *Pratacultural Sci.* **2019**, *36*, 250–263. (In Chinese)
71. Bao, S.D. *Soil Agricultural Chemical Analysis*, 3rd ed.; China Agricultural Press: Beijing, China, 2000. (In Chinese)
72. Warton, D.I.; Duursma, R.A.; Falster, D.S.; Taskinen, S. SMATR 3—An R package for estimation and inference about allometric lines. *Methods Ecol. Evol.* **2011**, *3*, 257–259. [CrossRef]

Article

Leaf Photosynthetic and Functional Traits of Grassland Dominant Species in Response to Nutrient Addition on the Chinese Loess Plateau

Yuan Jin [1,†], Shuaibin Lai [1,†], Zhifei Chen [2], Chunxia Jian [1], Junjie Zhou [1], Furong Niu [3] and Bingcheng Xu [1,4,*]

1. State Key Laboratory of Soil Erosion and Dryland Farming on the Loess Plateau, Northwest A&F University, Xianyang 712100, China
2. College of Life Sciences, Guizhou University, Guiyang 550025, China
3. College of Forestry, Gansu Agricultural University, Lanzhou 730070, China
4. Institute of Soil and Water Conservation, Chinese Academy of Sciences and Ministry of Water Resources, Xianyang 712100, China
* Correspondence: bcxu@ms.iswc.ac.cn
† These authors contributed equally to this work.

Citation: Jin, Y.; Lai, S.; Chen, Z.; Jian, C.; Zhou, J.; Niu, F.; Xu, B. Leaf Photosynthetic and Functional Traits of Grassland Dominant Species in Response to Nutrient Addition on the Chinese Loess Plateau. *Plants* 2022, 11, 2921. https://doi.org/10.3390/plants11212921

Academic Editors: Jianhui Huang and Giuseppe Fenu

Received: 11 June 2022
Accepted: 27 October 2022
Published: 30 October 2022

Publisher's Note: MDPI stays neutral with regard to jurisdictional claims in published maps and institutional affiliations.

Abstract: Leaf photosynthetic and functional traits of dominant species are important for understanding grassland community dynamics under imbalanced nitrogen (N) and phosphorus (P) inputs. Here, the effects of N (N0, N50, and N100, corresponding to 0, 50, and 100 kg ha^{-1} yr^{-1}, respectively) or/and P additions (P0, P40, and P80, corresponding to 0, 40, and 80 kg ha^{-1} yr^{-1}) on photosynthetic characteristics and leaf economic traits of three dominant species (two grasses: *Bothriochloa ischaemum* and *Stipa bungeana*; a leguminous subshrub: *Lespedeza davurica*) were investigated in a semiarid grassland community on the Loess Plateau of China. Results showed that, after a three-year N addition, all three species had higher specific leaf area (SLA), leaf chlorophyll content (SPAD value), maximum net photosynthetic rate (P_{Nmax}), and leaf instantaneous water use efficiency (WUE), while also having a lower leaf dry matter content (LDMC). The two grasses, *B. ischaemum* and *S. bungeana*, showed greater increases in P_{Nmax} and SLA than the subshrub *L. davurica*. P addition alone had no noticeable effect on the P_{Nmax} of the two grasses while it significantly increased the P_{Nmax} of *L. davurica*. There was an evident synergetic effect of the addition of N and P combined on photosynthetic traits and most leaf economic traits in the three species. All species had relatively high P_{Nmax} and SLA under the addition of N50 combined with P40. Overall, this study suggests that N and P addition shifted leaf economic traits towards a greater light harvesting ability and, thus, elevated photosynthesis in the three dominant species of a semiarid grassland community, and this was achieved by species–specific responses in leaf functional traits. These results may provide insights into grassland restoration and the assessment of community development in the context of atmospheric N deposition and intensive agricultural fertilization.

Keywords: semiarid grassland; fertilization; leaf functional trait; leaf photosynthesis; atmospheric nitrogen deposition

1. Introduction

The semiarid Loess Plateau region in China is characterized by severe water scarcity, soil erosion, and nutrient-poor soils, which all greatly limit vegetation growth [1]. Grassland, with a size of ~2.7 × 10^5 km^2, accounts for ca. 43% of the regional total land area and is the dominant vegetation type on the Plateau [2]. It provides essential ecosystem functions and services such as carbon sequestration, soil and water conservation, and biodiversity [3,4]. Grassland management and restoration are of great ecological and economic significance in the region. Fertilization, as an effective management practice to increase grassland productivity and promote grassland restoration, is adopted in many natural/semi-natural grasslands, e.g., in the Inner Mongolian steppe [5], the alpine meadow on the Qinghai–Tibet Plateau [6] in China, and in the European alpine grasslands [7] and the North American Great Plains [8]. However, fertilization may also lead to undesirable consequences such as exacerbated eutrophication [9], reduced ecosystem resilience (e.g., increased grassland drought sensitivity [8,10]), and biodiversity loss [11–13], which subsequently alter community composition and structure, as well as decrease the effects of diversity stability on maintaining grassland productivity [14]. Soils in the semiarid Loess Plateau are commonly deficient in both nitrogen (N) and phosphorus (P) [15]. Previous studies in the region have shown that appropriate N and P fertilization could improve soil N and P availability and play a positive role in increasing grassland productivity and recovery of degraded grasslands [16–18]. On the other hand, the effects of the N and P addition on photosynthetic and leaf functional traits of dominant species, which are important for understanding underlying mechanisms of grassland community dynamics under N and P inputs, have only received limited attention in the regional grassland.

Plant photosynthesis is the basis of plant growth, and its diurnal patterns reflect the sustained carbon assimilation ability of plants and have been extensively studied across a wide range of arid and semiarid grassland species, particularly in North America (e.g., [19–22]). Previous studies have explored the photosynthetic diurnal dynamics—under elevated CO_2 conditions—in dry and wet years [22,23], the diurnal photosynthetic performance of grasses with contrasting functional types (e.g., C_3 vs. C_4, invasive vs. native) [20,21], and biotic and abiotic controls of photosynthetic diurnal courses [19,24]. However, there is limited information on the impacts of nutrient addition on photosynthetic diurnal courses in dryland grassland species. As an essential element of all proteins in plants (e.g., nucleic acid, enzymes, and chlorophyll), N primarily determines plant photosynthetic performance [25,26]. Extensive studies have documented that N addition increases plant photosynthetic rate and promotes plant growth in grasslands [27,28], and this promotion may be mediated by soil moisture [10]. However, when N addition exceeds a threshold, it will not continue to increase plant photosynthesis or, even, inhibit plant growth [26,29]. P, as another essential macronutrient, is also vital for plant photosynthesis, and it is the main component of ATP, NADPH, and phospholipids, which all play important roles in regulating photosynthesis machinery and electron transport activities [30,31]. Apart from the direct regulation of N/P on plant photosynthesis, N and P addition could indirectly or interactively affect plant photosynthesis. For instance, P addition could improve photosynthesis by increasing leaf area and stomatal aperture, particularly under soil water deficit conditions [32,33]. P addition could increase the activity of N-fixing bacteria, nodule biomass, and nitrogenase activity in legumes, which subsequently increases leaf N and P content and photosynthetic rate [34]. The combined fertilization of N and P is, thereby, often considered an effective management tool for sustaining productivity in many grassland communities, while such effects need to be evaluated in the regional semiarid grassland on the Loess Plateau.

Besides photosynthetic characteristics, other important leaf functional traits, such as specific leaf area (SLA), leaf dry matter content (LDMC), leaf nitrogen mass (N_{mass}), and leaf phosphorus mass (P_{mass}), are also the intuitive representation of strategies adopted by plants to cope with environmental changes [35,36]. According to the leaf economic spectrum (LES) theory [37], angiosperm plants could generally be divided into a rapid/slow growing strategy according to a set of leaf functional traits, with the rapid-growing ones having low LDMC and high P_n, SLA, N_{mass}, and P_{mass}; contrarily, the slow-growing ones have the opposite leaf traits. This kind of functional trait-based theory, from leaf to plant levels, provides great insights into understanding species resource utilization and species distribution [37]. Nevertheless, species-specific responses in these leaf economic traits exist under varied environmental (e.g., under different soil water and nutrient availability) conditions, which should be systematically assessed [38]. Efforts have been made to quantify the variation of LES of grassland species under differed soil nutrient availability, which confirmed species-specific patterns [39], while the assessment of species-specific responses to varying N and P fertilization conditions is seldom conducted on the semiarid Loess Plateau, which is needed to better understand grassland community development.

The dominant species occupy important ecological niches and play vital roles in maintaining community structure and function [40,41]. Biomass increases after N/P additions tend to be achieved by decreasing species diversity or increasing the biomass of dominant species [42]. Quantifying physiological and growth characteristics of dominant species under N and/or P addition could, thereby, be important for the evaluation of community productivity and dynamics, as well as provide valuable information for grassland management and restoration. *Bothriochloa ischaemum* (L.) Keng (a C_4 perennial grass), *Stipa bungeana* Trin. (a C_3 perennial grass), and *Lespedeza davurica* (Laxm.) Schindl. (a C_3 N-fixing subshrub) are co-dominant species in the natural/restored grasslands on the semiarid Loess Plateau of China [43]. The previous study on a regional grassland community, targeting these species, has shown that the addition of N and P combined improved grassland productivity and decreased species diversity, primarily via effects on tall clonal and annual species [44], which, once again, suggested species or functional-type-specific responses within a community, while the variation of leaf functional traits in these dominant species, after N and P fertilization, have not been fully assessed. Thus, we examined photosynthetic diurnal change, SPAD value, and leaf economic traits, including N_{mass}, P_{mass}, N_{mass}/P_{mass} ratio, SLA, and LDMC of the three dominant species, following a three-year N and P addition experiment in a typical semiarid grassland community on the Loess Plateau. We tested the hypotheses that: (1) N/P addition would increase photosynthetic rates and alter the photosynthetic diurnal dynamics of the three dominant species in the peak growing season, and these photosynthetic responses would be related to species-specific shifts in leaf functional traits; (2) addition of N and P combined would further promote photosynthesis compared with N/P additions alone.

2. Results

2.1. Environmental Factors

Photosynthetically active radiation (PAR) and air temperature (T_a) showed a single-peaked diurnal curve during the experimental period, and the maximum values appeared at 12:00 h and 14:00 h, with the values of 1854 $\mu mol·m^{-2}·s^{-1}$ and 30.2 °C, respectively (Figure 1). The relative humidity (RH) remained stable during the daytime (~13%) (Figure 1).

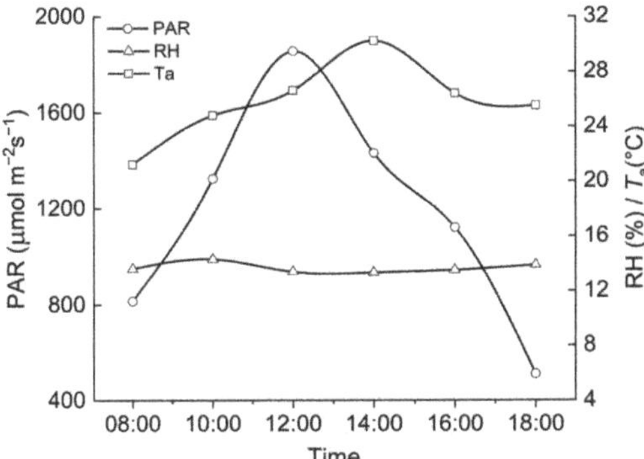

Figure 1. Diurnal changes of photosynthetically active radiation (PAR), air relative humidity (RH), and air temperature (T_a) during the measurement period (20–22 July 2019).

2.2. Diurnal Changes in Photosynthesis

The diurnal changes of net photosynthetic rate (P_n) and leaf instantaneous water use efficiency (WUE) of the three dominant species showed a double-peak curve under different N and P addition treatments. The first peak appeared at 10:00 h, the second at 14:00 h, and the midday depression of the photosynthesis (so-called "noon break") appeared at around 12:00 h (Figure 2). The leaf transpiration rate (T_r) of *B. ischaemum* mostly showed a double-peak diurnal course. While diurnal changes of T_r in *S. bungeana* and *L. davurica* showed a single peak.

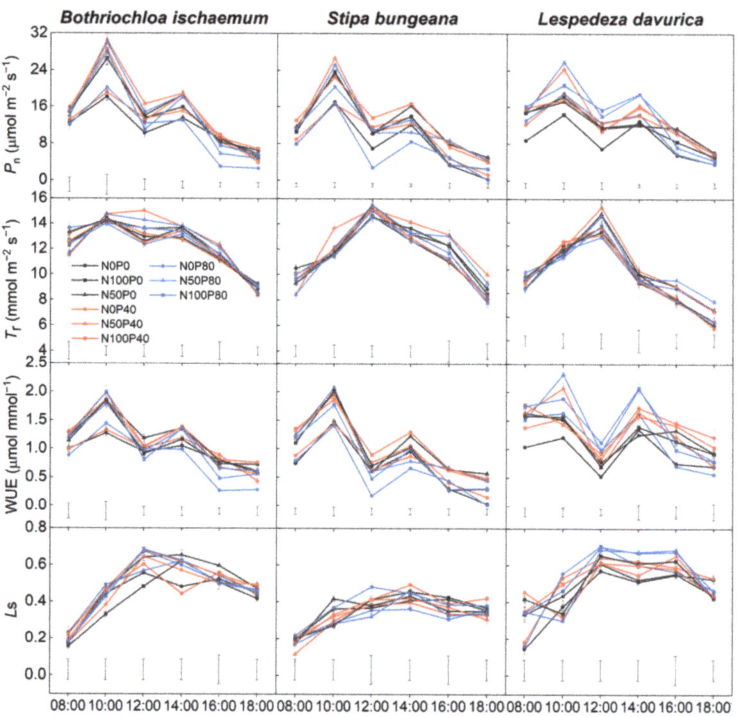

Figure 2. Diurnal changes of net photosynthetic rate (P_n), transpiration rate (T_r), instantaneous water use efficiency (WUE), and stomatal limitation value (L_s) of the three species under different N and P addition treatments. Vertical bars indicate LSD values.

Compared with CK (i.e., N0P0), N addition alone and addition of N and P combined significantly increased the P_n values at 10:00 h and 14:00 h in the three species (except under N50P40 and N50P80 treatments in *S. bungeana*). The greatest P_n values appeared at 10:00 h under N and P combined treatments (i.e., N50P40 and N50P80) for the three species. The WUE of the three species significantly increased by N addition alone compared with CK (Table 1). The WUE of *L. davurica* increased significantly under all levels of P alone additions, while the WUE of the two grasses only significantly increased under N0P40 (Table 1). Under N and P combined addition, the maximum WUE values of *B. ischaemum*, *S. bungeana*, and *L. davurica* were 1.17, 1.09, and 1.47 μmol mmol^{-1}, respectively (Table 1).

N addition, alone, significantly increased the L_s values of *B. ischaemum* and *L. davurica* (Table 1). P addition, alone, increased ($p < 0.05$) and decreased ($p < 0.05$) the L_s of *L. davurica* and *S. bungeana*, respectively, while only significantly increasing the L_s of *B. ischaemum* under N0P80 treatment. N and P interaction significantly affected the L_s values of the three species (Table 1). Under the addition of N50 combined with P, the L_s of *L. davurica* increased, and those of *B. ischaemum* decreased (both $p < 0.05$). Under the addition of N100 combined with P, the L_s of *B. ischaemum* significantly increased, while those of *S. bungeana* decreased significantly ($p < 0.05$; Table 1).

Table 1. Instantaneous water use efficiency (WUE) and stomatal limitation value (L_s) of the three species under different N and P additions (mean ± SD, n = 3).

Species	Treatment	WUE (µmol mmol^{-1})			L_s		
		P0	P40	P80	P0	P40	P80
B. ischaemum	N0	0.95 ± 0.02 [B,b]	0.97 ± 0.02 [A,b]	0.81 ± 0.01 [C,b]	0.44 ± 0.00 [B,c]	0.45 ± 0.01 [B,b]	0.48 ± 0.00 [A,a]
	N50	1.15 ± 0.02 [A,a]	1.17 ± 0.01 [A,a]	1.10 ± 0.01 [B,a]	0.50 ± 0.01 [A,a]	0.46 ± 0.01 [B,b]	0.48 ± 0.01 [B,a]
	N100	1.10 ± 0.05 [A,a]	1.21 ± 0.02 [A,a]	1.11 ± 0.04 [A,a]	0.42 ± 0.00 [B,b]	0.50 ± 0.01 [A,a]	0.48 ± 0.01 [A,a]
S. bungeana	N0	0.69 ± 0.01 [B,b]	0.78 ± 0.01 [A,b]	0.59 ± 0.01 [C,b]	0.68 ± 0.00 [A,a]	0.35 ± 0.01 [B,a]	0.33 ± 0.01 [B,b]
	N50	1.05 ± 0.01 [A,a]	0.97 ± 0.02 [B,a]	0.97 ± 0.01 [B,a]	0.36 ± 0.02 [A,c]	0.35 ± 0.01 [A,a]	0.32 ± 0.01 [B,b]
	N100	0.91 ± 0.02 [A,a]	1.09 ± 0.02 [A,a]	0.84 ± 0.02 [B,a]	0.57 ± 0.03 [A,b]	0.36 ± 0.00 [B,a]	0.37 ± 0.00 [B,a]
L. davurica	N0	0.93 ± 0.01 [C,b]	1.36 ± 0.03 [A,a]	1.18 ± 0.01 [B,b]	0.45 ± 0.01 [B,b]	0.52 ± 0.01 [A,a]	0.51 ± 0.01 [A,b]
	N50	1.24 ± 0.00 [C,a]	1.33 ± 0.02 [B,b]	1.47 ± 0.06 [A,a]	0.48 ± 0.01 [B,a]	0.50 ± 0.00 [A,a]	0.52 ± 0.01 [A,b]
	N100	1.22 ± 0.02 [B,a]	1.36 ± 0.01 [A,a]	1.43 ± 0.03 [A,a]	0.51 ± 0.00 [B,a]	0.51 ± 0.02 [B,a]	0.55 ± 0.01 [A,a]
B. ischaemum		N ***(269); P ***(47.9); N × P ***(8.4)			N ***(23.3); P ***(36.6); N × P ***(63.2)		
S. bungeana		N ***(1084); P ***(223); N × P ***(75.5)			N ***(66.6); P **(8.62); N × P ***(14.1)		
L. davurica		N ***(156); P ***(223); N × P ***(46.8)			N ***(26.4); P ***(45.8); N × P ***(8.93)		

Data with different capital letters indicate significant differences among P additions under each N addition rate, while different small letters indicate significant differences among N additions under each P addition rate. Numbers in parentheses are *F*-values, while '**' and '***' indicate $p \leq 0.01$, and $p \leq 0.001$, respectively.

N addition, alone, increased the P_{Nmax} values of the three species ($p < 0.05$), while there was no difference between them under N50 and N100 (Figure 3). The increase in P_{Nmax} in the two grasses was about two times larger than those of *L. davurica* under N addition alone (Figure 3). P addition, alone, increased the P_{Nmax} values of *L. davurica* and *S. bungeana* ($p < 0.05$; Figure 3). N and P combined addition only significantly affected the P_{Nmax} values of *S. bungeana* and *L. davurica* ($p < 0.05$; Figure 3).

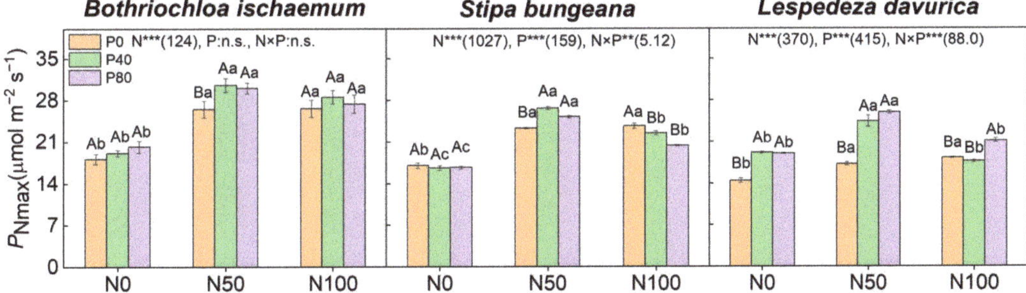

Figure 3. Leaf maximum net photosynthetic rate (P_{Nmax}) of the three species under different N and P addition treatments. Values are mean ± SD. Different capital letters above the column indicate significant differences among P additions under each N addition rate, while different small letters indicate significant differences among N additions under each P addition rate. Numbers in parentheses are *F*-values, while '**', and '***' indicate $p \leq 0.01$, and $p \leq 0.001$, respectively. 'n.s.' indicates no significant difference.

2.3. Leaf SPAD Value

N addition, alone, significantly increased the SPAD values of the three species (Figure 4). P addition, alone, significantly increased the SPAD values of *B. ischaemum* and *L. davurica* ($p < 0.05$), while SPAD values only increased under low levels of P addition alone (i.e., N0P40) in *S. bungeana* ($p < 0.05$). Addition of N and P combined significantly increased the SPAD values of both *B. ischaemum* and *L. davurica* (Figure 4).

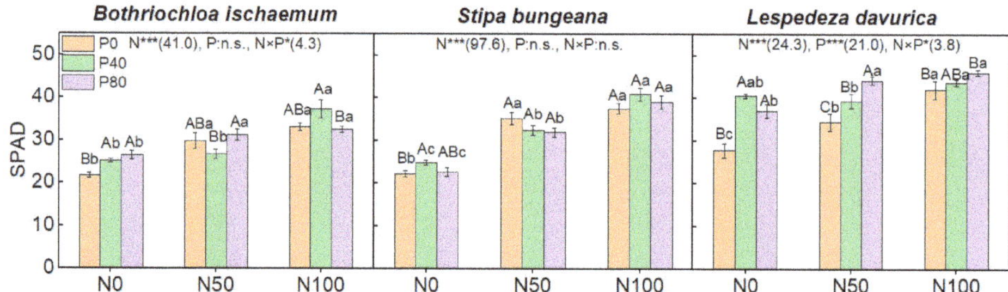

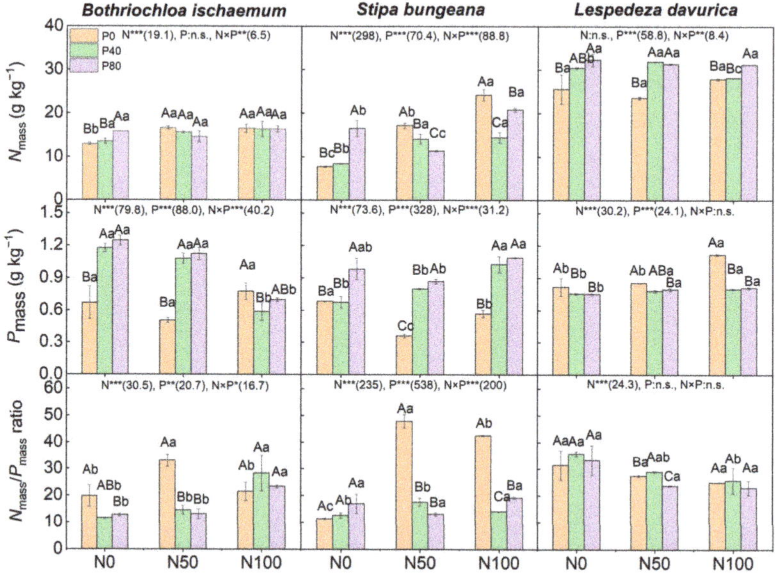

Figure 4. Leaf SPAD values of the three species under different N and P addition treatments. Different capital letters above the column indicate significant differences among P additions under each N addition rate, while different small letters indicate significant differences among N additions under each P addition rate. Numbers in parentheses are F-values, while '*' and '***' indicate $p \leq 0.05$ and $p \leq 0.001$, respectively. 'n.s.' indicates no significant difference.

2.4. Leaf N and P Concentration (N_{mass} and P_{mass}) and N_{mass}/P_{mass} Ratio

N addition, alone, significantly increased the N_{mass} values of the two grasses, but it had no effects on N_{mass} of *L. davurica*. The high level of P addition, alone (N0P80), significantly increased P_{mass} of all species except *L. davurica* (Figure 5). N and P interaction significantly affected N_{mass} of the three species, while it only significantly affected P_{mass} of the two grasses. Under the addition of N50 combined with P, the P_{mass} of *B. ischaemum* and *S. bungeana*, as well as the N_{mass} of *L. davurica*, increased significantly. Under the addition of N100 combined with P, the P_{mass} of *S. bungeana* increased significantly, while those of *L. davurica* decreased significantly (Figure 5).

Figure 5. Leaf nitrogen (N) and phosphorus (P) content (N_{mass} and P_{mass}), as well as the N_{mass}/P_{mass} ratio, of the three species under different N and P addition treatments. Values are mean $\pm$ SD. Different capital letters above the column indicate significant differences among P additions under each N addition rate, while different small letters indicate significant differences among N additions under each P addition rate. Numbers in parentheses are F-values, while '*', '**', and '***' indicate $p \leq 0.05$, $p \leq 0.01$, and $p \leq 0.001$, respectively. 'n.s.' indicates no significant difference.

N addition, alone, significantly affected the N_{mass}/P_{mass} ratios of all three species. The low level of N addition alone (N50P0) significantly increased the N_{mass}/P_{mass} of the two grasses; the high level of N addition, alone (N100P0), only increased the N_{mass}/P_{mass} of *S. bungeana* grass ($p < 0.05$, Figure 5). N and P addition interaction significantly affects the N_{mass}/P_{mass} of the two grasses but has no effect on the subshrub. Under the addition of N50 combined with P, the N_{mass}/P_{mass} of *B. ischaemum* and *S. bungeana* decreased significantly, while the N_{mass}/P_{mass} of *L. davurica* increased under N50P40 treatment ($p < 0.05$). Under the addition of N100 combined with P, the N_{mass}/P_{mass} of *S. bungeana* decreased significantly, while the N_{mass}/P_{mass} of *B. ischaemum* and *L. davurica* had no significant changes (Figure 5).

2.5. Specific Leaf Area (SLA) and Leaf Dry Matter Content (LDMC)

N addition, alone, significantly increased the SLA values of the three species, and the low level of N addition alone (N50P0) decreased the LDMC of the two grasses ($p < 0.05$, Figure 6). P addition, alone, significantly increased SLA of *L. davurica*, while it had limited effects on the two grasses (Figure 6). N and P addition interaction significantly affected both the SLA and LDMC values of all three species. Under the addition of N50 combined with P, the SLA of *B. ischaemum* and *L. davurica* increased significantly ($p < 0.05$), and the SLA of *S. bungeana* increased under N50P40, while it decreased significantly under N50P80 ($p < 0.05$); LDMC was comparable between different levels of P additions in the two grasses, but it decreased in *L. davurica*. Under the addition of N100 combined with P, the SLA values of *B. ischaemum* and *L. davurica* increased significantly, and the SLA of *S. bungeana* increased ($p < 0.05$) only under the high level of N addition (N100P80); LDMC, among different levels of P addition (i.e., N100P40 and N100P80), was comparable in *B. ischaemum* and *L. davurica*, while it significantly decreased under N100P80 in *S. bungeana* (Figure 6).

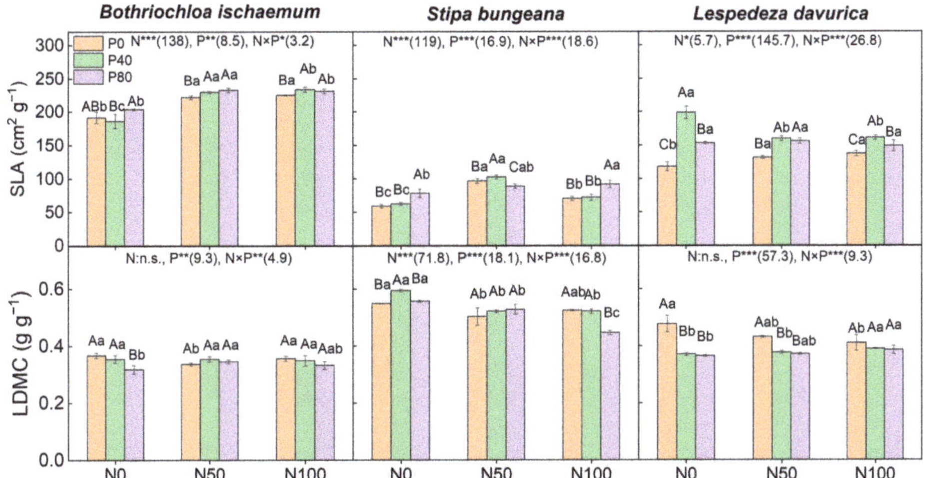

Figure 6. Specific leaf area (SLA) and leaf dry matter content (LDMC) of the three species under different N and P addition treatments. Different capital letters above the column indicate significant differences among P additions under each N addition rate, while different small letters indicate significant differences among N additions under each P addition rate. Numbers in parentheses are *F*-values, while '*', '**', and '***' indicate $p \leq 0.05$, $p \leq 0.01$, and $p \leq 0.001$, respectively. 'n.s.' indicates no significant difference.

P_{Nmax} was positively correlated with WUE, SPAD, N_{mass}, and SLA in all three species, while it was negatively correlated with LDMC in *S. bungeana* and *L. davurica* (Figure 7). PCA analysis showed that the variance explained by the first and second principal components was 37.5% and 23.5%, respectively, with a total value of 61% (Figure 8). The first principal component had a high correlation with P_{Nmax}, WUE, P_{mass}, and SLA; the second principal component had high correlation with LDMC. SPAD, N_{mass}, and $N_{\text{mass}}/P_{\text{mass}}$ ratio are correlated with both principal components (Figure 8). In a score plot of PCA analysis, under N addition alone, *B. ischaemum* and *L. davurica* gradually moved to the right with N addition level, and *S. bungeana* gradually moved to the upper right. Under P addition, alone, *L. davurica* gradually moved to the right with P addition level, and *B. ischaemum* slightly moved to the upper right. Under the addition of N combined with P, with the increase in fertilizer application, all three species moved towards higher P_{Nmax}, WUE, SPAD, and SLA values (Figure 9).

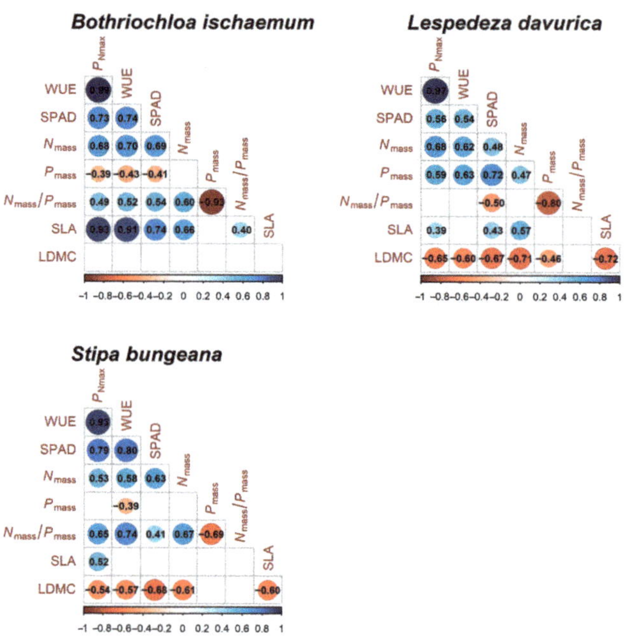

Figure 7. Heatmap of Pearson correlation testing relationships between P_{Nmax}, WUE, and leaf functional traits (N_{mass}, P_{mass}, SLA, and LDMC) in the three species. Numbers in circles indicate the Pearson correlation coefficients.

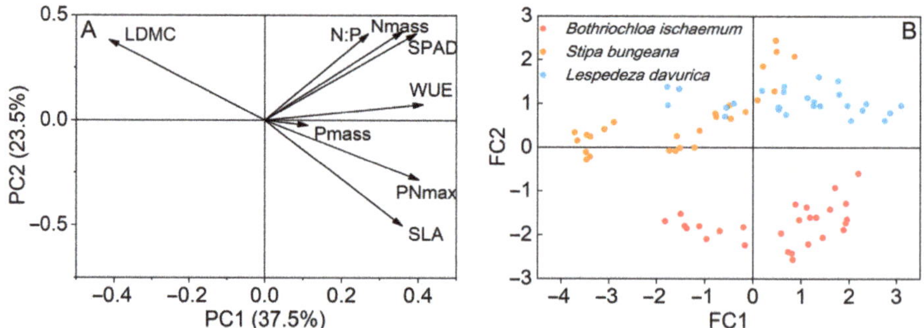

Figure 8. Principal component analysis (PCA) of photosynthetic characteristics (P_{Nmax} and WUE) and leaf functional traits (SPAD, N_{mass}, P_{mass}, SLA, and LDMC) in the three species under N and P additions. (**A**) Loadings for each leaf trait; (**B**) Factor scores for each species.

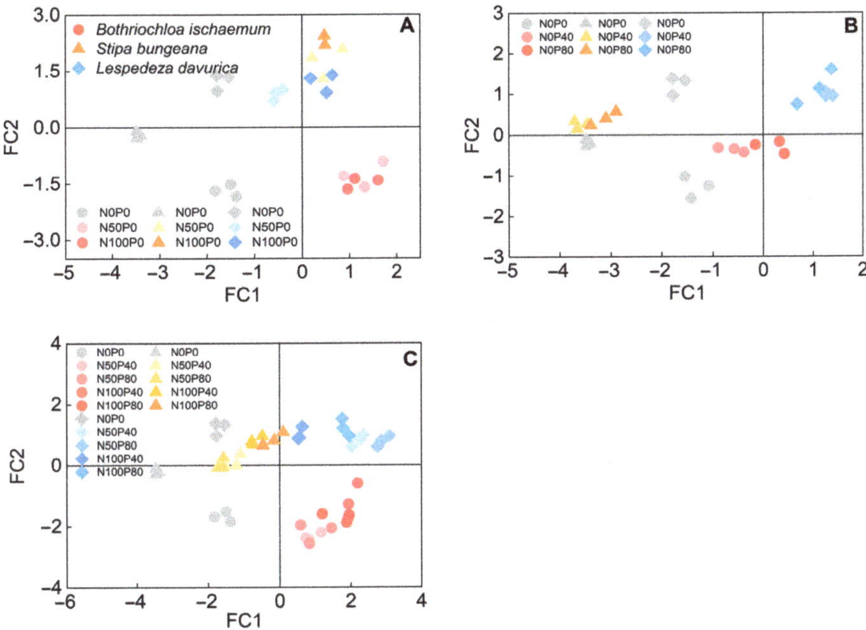

Figure 9. Score plots of PCA analysis of photosynthetic characteristics (P_{Nmax} and WUE) and leaf economic traits (SPAD, N_{mass}, P_{mass}, SLA, and LDMC) in the three species under N and P additions. (**A**) N addition alone; (**B**) P addition alone; (**C**) N and P combined addition.

3. Discussion

The diurnal dynamics of photosynthesis reflect plants' sustained ability to carry out physiological metabolism and biomass accumulation throughout the daytime [45,46], which have been extensively studied in numerous dryland species (e.g., [19–22]). Our results, corroborated with others, showed that the diurnal course of photosynthetic rate showed a double-peaked curve in the three grassland dominant species, and they showed an evident "noon break" of photosynthesis. The noon break is a mechanism to avoid stresses such as excess light, high temperature, and water deficit during the midday [47], and it is a result of stomatal or non-stomatal restriction [48]. Our measurements showed that, during the period of 10:00–12:00 h, the P_n values gradually decreased while the L_s values increased (Figure 2 and Table 1), suggesting the "noon break" was likely caused by stomatal limitation [46,48]. In line with our first hypothesis and consistent with others (e.g., [49,50]), the P_n values of the three species considerably increased by nutrient addition, particularly during the peak photosynthetic period (~10:00 h). This is expected since environmental conditions (e.g., light and temperature) are relatively optimal for photosynthesis during this period, hence the promotion of nutrient addition would be most effective. In addition, during the noon time with high air temperature and light radiation, the P_n was slightly increased after fertilization, which may be due to increased stomatal conductance due to N and P addition, and this ostensibly alleviated the "noon break" [51]. Our study indicated that N and P fertilization could improve the photosynthetic ability of the three species at the diurnal scale and increase the daily accumulative carbon assimilation. However, we only focused on short-term responses during the peak growing season (i.e., July), so future studies should be taken to further assess intra- or interannual patterns of their photosynthesis to better understand the long-term effects of fertilization.

Both N and P are essential elements of key compounds involved in the photosynthetic process, and appropriate N and P additions would increase the content of these compounds and, subsequently, the photosynthetic rate [17,26,30]. This was observed in our study: the

P_{Nmax}, SPAD, and N_{mass} of the three species increased significantly after N addition alone, and there were strong positive correlations between N_{mass}, SPAD, and P_{Nmax} ($p < 0.05$; Figures 3–5 and 7). The increase in P_{Nmax} and WUE with the N addition level was much greater (larger regression slopes) in the two grass species than in the legume *L. davurica* (Table 2), with the greatest increase (the largest slope) of P_{Nmax} and WUE, along with the N addition level, in C_4 grass *B. ischaemum* (Table 2). Together, this suggested that the two grasses were more sensitive to N addition alone than the legume. This is consistent with our previous study quantifying the plant biomass of *B. ischaemum* and *L. davurica* mixtures under varying soil moisture and nutrient supplies [52]. We suspect that the subshrub *L. davurica* may not be N-limited due to its N fixation ability and is, thereby, insensitive to exogenous N fertilization. On the other hand, the photosynthetic rate does not continuously increase with N addition amounts after passing a threshold [29,53], which was also observed, here, as the P_{Nmax} values of the three species were not significantly different between N50 and N100 (Figures 3 and 6; Table 1). Fossil fuel combustion and extensive fertilization have greatly increased atmospheric N deposition globally in recent decades [54]. Chronic N input by long-term N deposition may, hence, alleviate N limitation and promote plant photosynthesis and growth of regional grassland species, but it may, meanwhile, intensify plant P limitation by increasing P demand [55,56].

Table 2. Regression slopes (SE) derived from the multiple linear regression analysis between photosynthetic characteristics (P_{Nmax} and WUE) and N addition, P addition, and N and P addition interaction in the three species.

Species	Variable	P_{Nmax}	WUE
	N	**0.0894 (0.0161) *****	**0.0060 (0.0010) *****
B. ischaemum	P	0.0182 (0.0201) n.s.	0.0012 (0.0013) n.s.
	N × P	−0.0001 (0.0003) n.s.	0.0000 (0.0000) n.s.
	N	**0.0640 (0.0131) *****	**0.0049 (0.0009) *****
S. bungeana	P	−0.0250 (0.0164) n.s.	**−0.0029 (0.0012) ***
	N × P	−0.0002 (0.0003) n.s.	−0.0001 (0.0001) n.s.
	N	0.0238 (0.0205) n.s.	0.0022 (0.0019) n.s.
L. davurica	P	**0.0782 (0.0257) ****	**0.0071 (0.0024) ****
	N × P	−0.0002 (0.0004) n.s.	0.0000 (0.0000) n.s.

'n.s.', '*', '**', and '***' indicate $p > 0.05$, $p \leq 0.05$, $p \leq 0.01$, and $p \leq 0.001$, respectively. Significant slopes are in bold.

Here, the P addition, alone, had greater effects on P_{Nmax}, SPAD, and SLA of the leguminous *L. davurica* among the three species (Figures 3, 4 and 6; Table 2). This may be ascribed to P, as it could promote the activity of nitrogenase in the root nodules of legumes and enhance their photosynthesis [34,57], and elevated leaf P content can also directly improve photosynthetic capacity by promoting ATP and NADPH synthesis, as well as regeneration of RuBP [33,58]. Compared with N or P addition alone, the three species had higher P_{Nmax}, WUE, and SLA values under N and P combined additions, suggesting a synergetic effect of N and P on plant photosynthesis (Figures 5 and 6). This confirms our second hypothesis and suggests that appropriate N and P combined fertilization should be considered to maintain regional grassland productivity. A myriad of studies have documented this synergetic effect in grasslands worldwide (e.g., [59,60]). Previous studies in the Loess Plateau grasslands also reported the N and P combination had synergetic effects on community productivity [44]. A recent long-term (over 66 years) nutrient addition study in a mesic grassland in South Africa also highlighted that N and P combined addition promoted plant P acquisition and uptake (e.g., increased organic P storage, P recycling, and plant P utilization), which may contribute to the synergetic effect of N and P combined addition [59].

Drylands (e.g., the semiarid Loess Plateau) are often co-limited by water and nutrients [8], as well as characterized by frequent drought events, which greatly impact plant N and P uptake [61]. Nutrient addition, such as N, at an appropriate rate could improve post-drought recovery of grassland and increase the aboveground biomass production [62].

Contrarily, some studies reported that nutrient addition increased grassland drought sensitivity and constrained its recovery from drought events [10]. Besides, grass species with different photosynthetic pathways (C_3 vs. C_4) may respond differentially to drought and rewatering under nutrient addition conditions [63]. For the regional grassland, previous studies have quantified the photosynthetic responses of dominant species following rainfall events and reported species-specific patterns [64]. Nevertheless, the interaction of soil moisture (especially drought) and fertilization on dominant species performance remains less understood in the regional grassland and should be assessed, considering recurrent drought events, under future climate scenarios [65].

Leaf functional traits, particularly those so-called economic traits, are invoked to explain plant resource acquisition and utilization [36]. Among them, the leaf N_{mass}/P_{mass} ratio indicates environmental N and P availability where the plant grows [56]. In general, N_{mass}/P_{mass} ratio less than 10 indicates the N limitation, and greater than 20 indicates the P limitation [56]. The leaf N_{mass}/P_{mass} ratio of the three species, averaged across treatments, was 18.8 (*B. ischaemum*), 11.3 (*S. bungeana*), and 24.3 (*L. davurica*), respectively (Figure 5), suggesting species-specific N and P limitations. The N_{mass}/P_{mass} ratio of the two grasses increased significantly with N addition, while no noticeable change was found in *L. davurica* (Figure 5). This indicates that N addition may lead to P limitation in the two grass species. Meanwhile, increased soil N and P availability would release plants from nutrient competition to other resource competition, such as light and water [66]. Grassland dominant species may accordingly alter their leaf functional traits to maximize light harvesting to maintain dominance. According to the LES theory, plants with higher light capture, resource acquisition, and turnover capacity show higher SLA, N_{mass}, and P_{mass} in contrast to the slow-growth ones with higher LDMC and conservative nutrient resource use [36]. Similar to other studies (e.g., [66]), the three species studied here shifted to a fast-growth strategy after N addition with larger, thinner, and N-rich leaves (higher SLA, N_{mass}, and SPAD), as well as higher assimilation rate per unit leaf area (higher P_{Nmax}). Though score plots from PCA analysis indicated that, under N addition, three species adopted different strategies to improve their light harvesting: C_4 grass *B. ischaemum* mainly by increasing SLA and P_{Nmax}, while C_3 grass *S. bungeana* and C_3 subshrub *L. davurica* primarily increased leaf N content and SPAD (Figure 9), and only *L. davurica* had notable shifts in photosynthetic and leaf functional traits under P additions (Figure 9B), which suggests that the three species had different trade-off strategies in photosynthetic performance and leaf economic traits in response to N and/or P addition [17,66], and these should be considered when assessing N and P fertilization effects on community structure and functions. The P_{Nmax} values of the three species were mostly highest under the 'N50P40' treatment among all treatments, indicating that it could be considered an optimal fertilization measure for improving grassland production.

4. Materials and Methods

4.1. Site Description

This work was conducted at the Zhifanggou watershed (109°13′46″–109°16′03″ E, 36°42′42″–36°46′28″ N), located in the Ansai District, Yan'an City, Shaanxi Province, China. It has a semiarid continental monsoon climate. The mean annual temperature is 8.8 °C, with the lowest temperature being -6.9 °C in January and the highest being 22.6 °C in July. The mean annual rainfall is 507 mm. The soil is classified as Calcaric Cambisol. Rainfall shows a highly seasonal variability with ca. 82% occurring from May to September (the growing season). The soil available N, P, and K were 20.9–71.3 mg kg^{-1}, 1.6–2.8 mg kg^{-1}, and 10.07–30.97 g kg^{-1}, respectively, and soil pH was 8.4–8.8 [67]. The targeted grassland is dominated by xerophytic plants, e.g., *B. ischaemum*, *S. bungeana*, *L. davurica*, *Artemisia sacrorum*, and *Artemisia giraldii*.

4.2. N and P addition

A grassland community (20 × 30 m) was fenced to exclude grazing since May 2017. A randomized split-plot design with three N addition rates at the main plot level and three P addition rates at the subplot level was carried out. The main plot was 4 × 4 m, and N addition rates were N0 (0 kg N), N50 (50 kg N ha^{-1} yr^{-1}), and N100 (100 kg N ha^{-1} yr^{-1}). The N50 and N100 treatments were about 2 and 4 times the annual average N deposition rate in the loess hilly area [~21.76 kg (N) ha^{-1} yr^{-1}] [68]. N was applied as calcium ammonium nitrate [5Ca(NO$_3$)$_2$ NH$_4$NO$_3$ 10H$_2$O] (15.5% of N). Each main plot was divided into four subplots (2 × 2 m). P was applied as triple superphosphate [Ca(H$_2$PO$_4$)$_2$·H$_2$O] (45% of P), and the addition rates were set to 0, 1, and 2 times the local fertilization rate, corresponding to P0 (0 kg P$_2$O$_5$), P40 (40 kg P$_2$O$_5$ ha^{-1} yr^{-1}), and P80 (80 kg P$_2$O$_5$ ha^{-1} yr^{-1}) [44].

Totally, there were 9 treatments, including a control (N0P0), two N addition alone treatments (N50P0, N100P0), two P addition alone treatments (N0P40, N0P80), four N and P combined addition treatments (N50P40, N50P80, N100P40, N100P80), and three replicates per treatment. N and P additions were conducted once a year, on rainy days, from 2017–2019 (4 June 2017, 21 May 2018, and 13 June 2019).

4.3. Ecophysiological Measurements

4.3.1. Diurnal Variations of Photosynthesis

The portable photosynthesis system (CIRAS-2, PP Systems, Amesbury, MA, USA) was used to measure the diurnal changes of photosynthesis of *B. ischaemum*, *S. bungeana*, and *L. davurica*, successively, and all measurements were conducted on three consecutive sunny days from 20–22 July 2019 (one species per day). The measurement was taken on one newly fully-expanded healthy leaf per species per treatment from 8:00–18:00 h with 2 h intervals. The measured parameters include net photosynthetic rate (P_n, μmol·m^{-2}·s^{-1}), transpiration rate (T_r, mmol·m^{-2}·s^{-1}), intercellular CO$_2$ concentration (C_i, μmol·mol^{-1}), and environmental factors, including photosynthetically active radiation (PAR, μmol·m^{-2}·s^{-1}), air temperature (T_a, °C), and relative humidity (RH, %). The photosynthetic rate at 10:00 h was taken as the maximum net photosynthetic rate (P_{Nmax}, μmol·m^{-2}·s^{-1}). Instantaneous water use efficiency (WUE, μmol mmol^{-1}) was calculated as P_n/T_r. Stomatal limitation value (L_s) was derived by $1 - C_i/C_a$ [48].

4.3.2. Leaf SPAD Value

Leaf SPAD value (a measure of leaf relative chlorophyll content) was measured on three newly fully-expanded healthy leaves per species per treatment using a chlorophyll meter (SPAD-502 model, Konica-Minolta, Osaka, Japan) on 20–22 July 2019.

4.3.3. Leaf Functional Traits

The 10–20 newly-fully expanded healthy leaves were randomly sampled per species per treatment, stored in zipped plastic bags, and quickly taken back to the laboratory, in an insulated box with ice packs, for leaf functional traits measurements. Leaves were weighed with an analytical balance (d = 0.0001 g). The fresh leaves were scanned (Epson duplex scanner, Epson, Tokyo, Japan), and the leaf area was derived using ImageJ (National Institutes of Health, Bethesda, MD, USA). Then, leaves were oven-dried at 75 °C for 24 h and ground with a high-throughput tissue grinder (MM-400, Retsch, Haan, Germany). Specific leaf area (SLA, m^2 g^{-1}) was calculated as leaf area divided by leaf dry mass. Leaf dry matter content (LDMC, g g^{-1}) was calculated as leaf dry mass divided by fresh mass. After digestion with H$_2$SO$_4$-HClO$_4$, the mass-based leaf N concentration (N_{mass}) was obtained using a Kjeldahl N analyzer (FOSS-8400, Foss, Höganäs, Denmark). The mass-based leaf P concentration (P_{mass}) was determined by a molybdenum blue colorimetry (UV-2600 ultraviolet-visible spectrophotometer, Shimadzu, Kyoto, Japan). N_{mass}/P_{mass} ratio was then calculated.

4.4. Statistical Analysis

All statistical analyses were performed with SPSS 20.0. One-way analysis of variance (ANOVA) was used to compare the differences in leaf photosynthetic characteristics (WUE, L_s, and P_{Nmax}) and leaf functional traits (SPAD, N_{mass}, P_{mass}, N_{mass}/P_{mass}, SLA, and LDMC) of the three species under different N and P addition treatments. Tukey's HSD test was used for multiple comparisons. Two-way ANOVA was used to test the effects of N addition, P addition, and their interaction on P_{Nmax}, SPAD value, N_{mass}, P_{mass}, SLA, and LDMC. Pearson correlation was used to explore the relationship between leaf photosynthetic characteristics (P_{Nmax}, WUE) and leaf functional traits (SPAD value, N_{mass}, P_{mass}, N_{mass}/P_{mass}, SLA, and LDMC). Multiple linear regression was used to explore the relationship between N addition, P addition, and their interaction, as well as P_{Nmax} and WUE. Principal component analysis (PCA) was conducted on photosynthetic characteristics and leaf functional traits. Graphing was performed with Origin 2021 (Origin Lab Software, Chicago, IL, USA).

5. Conclusions

Our three-year field fertilization study suggested that N addition—alone or combined with P—improved the photosynthesis of the three grassland dominant species on the semiarid Loess Plateau of China. All three species shifted to a fast-growth strategy with increased P_{Nmax}, SLA, and N_{mass}, as well as reduced LDMC under N and/or P addition. Furthermore, species-specific shifts in leaf functional traits were observed among the three species following N and/or P addition, of which C_4 grass *B. ischaemum* increased SLA and P_{Nmax}, and C_3 grass *S. bungeana* and subshrub *L. davurica* mainly increased leaf N and SPAD. P addition seems to only effectively impact the P_n of *L. davurica*. Evident N and P synergetic effects on the photosynthetic performance in all three species were observed, and a combination of 50 kg ha^{-1} yr^{-1} N and 40 kg ha^{-1} yr^{-1} P addition could be considered optimal fertilization for improving grassland productivity locally.

Author Contributions: Conceptualization: B.X.; Methodology: Y.J., S.L. and Z.C.; data collection and curation: Y.J. and S.L.; data analysis: Y.J. and S.L.; field investigation: Y.J., S.L., C.J. and J.Z.; writing-review and editing: Y.J., F.N. and B.X. All authors have read and agreed to the published version of the manuscript.

Funding: This study was funded by National Key Research and Development Program of China (2016YFC0501703).

Data Availability Statement: The data presented in this study are available on reasonable request from the corresponding author.

Conflicts of Interest: The authors declare no conflict of interest.

References

1. Huang, L.; Shao, M. Advances and perspectives on soil water research in China's Loess Plateau. *Earth Sci. Rev.* **2019**, *199*, 102962. [CrossRef]
2. Gang, C.C.; Zhao, W.; Zhao, T.; Zhang, Y.; Gao, X.R.; Wen, Z.G. The impacts of land conversion and management measures on the grassland net primary productivity over the Loess Plateau, Northern China. *Sci. Total Environ.* **2018**, *645*, 827–836. [CrossRef] [PubMed]
3. Sun, C.J.; Hou, H.X.; Chen, W. Effects of vegetation cover and slope on soil erosion in the Eastern Chinese Loess Plateau under different rainfall regimes. *Peer J.* **2021**, *9*, e11226. [CrossRef] [PubMed]
4. Yang, Y.; Liu, H.; Yang, X.; Yao, H.J.; Deng, X.Q.; Wang, Y.Q.; An, S.S.; Kuzyakov, Y.; Chang, S.X. Plant and soil elemental C:N:P ratios are linked to soil microbial diversity during grassland restoration on the Loess Plateau, China. *Sci. Total Environ.* **2022**, *806*, 150557. [CrossRef] [PubMed]
5. Bai, Y.; Wu, J.; Clark, C.M.; Naeem, S.; Pan, Q.; Huang, J.; Zhang, L.; Han, X. Tradeoffs and thresholds in the effects of nitrogen addition on biodiversity and ecosystem functioning: Evidence from inner Mongolia Grasslands. *Glob. Change Biol.* **2010**, *16*, 358–372. [CrossRef]
6. Wang, Y.; Sun, Y.; Chang, S.; Wang, Z.; Fu, H.; Zhang, W.; Hou, F. Restoration practices affect alpine meadow ecosystem coupling and functions. *Rangel. Ecol. Manag.* **2020**, *73*, 441–451. [CrossRef]

7. Botter, M.; Zeeman, M.; Burlando, P.; Fatichi, S. Impacts of fertilization on grassland productivity and water quality across the European Alps under current and warming climate: Insights from a mechanistic model. *Biogeosciences* **2021**, *18*, 1917–1939. [CrossRef]

8. Bharath, S.; Borer, E.T.; Biederman, L.A.; Blumenthal, D.M.; Fay, P.A.; Gherardi, L.A.; Knops, J.M.H.; Leakey, A.D.B.; Yahdjian, L.; Seabloom, E.W. Nutrient addition increases grassland sensitivity to droughts. *Ecology* **2020**, *101*, e02981. [CrossRef]

9. Smith, V.H.; Schindler, D.W. Eutrophication science: Where do we go from here? *Trends Ecol. Evol.* **2009**, *24*, 201–207. [CrossRef]

10. Meng, B.; Li, J.; Maurer, G.E.; Zhong, S.; Yao, Y.; Yang, X.; Collins, S.L.; Sun, W. Nitrogen addition amplifies the nonlinear drought response of grassland productivity to extended growing-season droughts. *Ecology* **2021**, *102*, e03483. [CrossRef]

11. Suding, K.N.; Collins, S.L.; Gough, L.; Clark, C.; Cleland, E.E.; Gross, K.L.; Milchunas, D.G.; Pennings, S. Functional-and abundance-based mechanisms explain diversity loss due to N fertilization. *Proc. Natl. Acad. Sci. USA* **2005**, *102*, 4387–4392. [CrossRef] [PubMed]

12. Socher, S.A.; Prati, D.; Boch, S.; Müller, J.; Klaus, V.H.; Hölzel, N.; Fischer, M. Direct and productivity-mediated indirect effects of fertilization, mowing and grazing on grassland species richness. *J. Ecol.* **2012**, *100*, 1391–1399. [CrossRef]

13. Harpole, W.S.; Sullivan, L.L.; Lind, E.M.; Firn, J.; Adler, P.B.; Borer, E.T.; Chase, J.; Fay, P.A.; Hautier, Y.; Hillebrand, H.; et al. Addition of multiple limiting resources reduces grassland diversity. *Nature* **2016**, *537*, 93–96. [CrossRef] [PubMed]

14. Hautier, Y.; Zhang, P.F.; Loreau, M.; Wilcox, K.R.; Seabloom, E.W.; Borer, E.T.; Wang, S. General destabilizing effects of eutrophication on grassland productivity at multiple spatial scales. *Nat. Commun.* **2020**, *11*, 5375. [CrossRef]

15. Liu, Z.P.; Shao, M.A.; Wang, Y.Q. Spatial patterns of soil total nitrogen and soil total phosphorus across the entire Loess Plateau region of China. *Geoderma* **2013**, *197*, 67–68. [CrossRef]

16. Zhang, L.; Wei, X.; Hao, M.; Zhang, M. Changes in aggregate-associated organic carbon and nitrogen after 27 years of fertilization in a dryland alfalfa grassland on the loess plateau of china. *J. Arid Land* **2015**, *7*, 429–437. [CrossRef]

17. Chen, Z.F.; Xiong, P.F.; Zhou, J.J.; Lai, S.B.; Jian, C.X.; Wang, Z.; Xu, B.C. Photosynthesis and nutrient-use efficiency in response to N and P addition in three dominant grassland species on the semiarid Loess Plateau. *Photosynthetica* **2020**, *58*, 1028–1039. [CrossRef]

18. Qu, Q.; Wang, M.; Xu, H.; Yan, Z.; Liu, G.; Xue, S. Role of soil biotic and abiotic properties in plant community composition in response to nitrogen addition. *Land Degrad. Dev.* **2022**, *33*, 904–915. [CrossRef]

19. Senock, R.S.; Devine, D.L.; Sisson, W.B.; Donart, G.B. Ecophysiology of three C_4 perennial grasses in the northern Chihuahuan Desert. *Southwest. Nat.* **1994**, *39*, 122–127. [CrossRef]

20. Nippert, J.B.; Fay, P.A.; Knapp, A.K. Photosynthetic traits in C_3 and C_4 grassland species in mesocosm and field environments. *Environ. Exp. Bot.* **2007**, *60*, 412–420. [CrossRef]

21. Huxman, T.E.; Smith, S.D. Photosynthesis in an invasive grass and native forb at elevated CO_2 during an El Nino year in the Mojave Desert. *Oecologia* **2001**, *128*, 193–201. [CrossRef] [PubMed]

22. Adam, N.R.; Owensby, C.E.; Ham, J.M. The effect of CO_2 enrichment on leaf photosynthetic rates and instantaneous water use efficiency of Andropogon gerardii in the tallgrass prairie. *Photosynth. Res.* **2000**, *65*, 121–129. [CrossRef] [PubMed]

23. Nijs, I.; Impens, I.; Van Hecke, P. Diurnal changes in the response of canopy photosynthetic rate to elevated CO2 in a coupled temperature-light environment. *Photosynth. Res.* **1992**, *32*, 121–130. [CrossRef] [PubMed]

24. Haase, P.; Pugnaire, F.I.; Clark, S.C.; Incoll, L.D. Environmental control of canopy dynamics and photosynthetic rate in the evergreen tussock grass *Stipa tenacissima*. *Plant Ecol.* **1999**, *145*, 327–339. [CrossRef]

25. Evans, J.R.; Terashima, I. Effects of nitrogen nutrition on electron transport components and photosynthesis in spinach. *Funct. Plant Biol.* **1987**, *14*, 59–68. [CrossRef]

26. Sinha, D.; Tandon, P.K. An Overview of Nitrogen, Phosphorus and Potassium: Key Players of Nutrition Process in Plants. *Sustain. Solut. Elem. Defic. Excess Crop Plants* **2020**, *5*, 85–117. [CrossRef]

27. Wang, D.; Ling, T.Q.; Wang, P.P.; Fan, J.Z.; Wang, H.; Zhang, Y.Q. Effects of 8-year nitrogen and phosphorus treatments on the ecophysiological traits of two key species on tibetan plateau. *Front. Plant Sci.* **2018**, *9*, 1290. [CrossRef]

28. Lai, S.B.; Xu, S.; Jian, C.X.; Chen, Z.F.; Zhou, J.J.; Yang, Q.; Wang, Z.; Xu, B.C. Leaf photosynthetic responses to nitrogen and phosphorus additions of dominant species in farm-withdrawn grassland in the loess hilly-gully region. *Acta Ecol. Sin.* **2021**, *41*, 5454–5464. (In Chinese) [CrossRef]

29. Xing, H.; Zhou, W.; Wang, C.; Li, L.; Li, X.; Cui, N.; Hao, W.; Liu, F.; Wang, Y. Excessive nitrogen application under moderate soil water deficit decreases photosynthesis, respiration, carbon gain and water use efficiency of maize. *Plant Physiol. Biochem.* **2021**, *166*, 1065–1075. [CrossRef]

30. Holford, I.C.R. Soil phosphorus: Its measurement, and its uptake by plants. *Aust. J. Soil Res.* **1997**, *35*, 227–240. [CrossRef]

31. Carstensen, A.; Herdean, A.; Schmidt, S.B.; Sharma, A.; Spetea, C.; Pribil, M.; Husted, S. The impacts of phosphorus deficiency on the photosynthetic electron transport chain. *Plant Physiol.* **2018**, *177*, 271–284. [CrossRef] [PubMed]

32. Chaudhary, M.I.; Adu-Gyamfi, J.J.; Saneoka, H.; Nguyen, N.T.; Suwa, R.; Kanai, S.; El-Shemy, H.A.; Lightfoot, D.A.; Fujita, K. The effect of phosphorus deficiency on nutrient uptake, nitrogen fixation and photosynthetic rate in mashbean, mungbean and soybean. *Acta Physio. Plant* **2008**, *30*, 537–544. [CrossRef]

33. Liu, C.; Wang, Y.; Pan, K. Effects of phosphorus application on photosynthetic carbon and nitrogen metabolism, water use efficiency and growth of dwarf bamboo (*Fargesia rufa*) subjected to water deficit. *Plant Physiol. Bioch.* **2015**, *96*, 20–28. [CrossRef] [PubMed]

34. Suriyagoda, L.; Lambers, H.; Ryan, M.H.; Renton, M. Effects of leaf development and phosphorus supply on the photosynthetic characteristics of perennial legume species with pasture potential: Modelling photosynthesis with leaf development. *Funct. Plant Biol.* **2010**, *37*, 713–725. [CrossRef]
35. Walker, A.P.; Beckerman, A.P.; Gu, L.H.; Kattge, J.; Cernusak, L.A.; Domingues, T.F.; Scales, J.C.; Wohlfahrt, G.; Wullschlrger, S.D.; Woodward, F.L. The relationship of leaf photosynthetic traits–V_{cmax} and J_{max}–to leaf nitrogen, leaf phosphorus, and specific leaf area: A meta-analysis and modeling study. *Ecol. Evol.* **2015**, *4*, 3218–3235. [CrossRef]
36. Wright, I.J.; Reich, P.B.; Westoby, M.; Ackerly, D.D.; Baruch, Z.; Bongers, F.; Cavender-Bares, J.; Chapin, T.; Cornelissen, J.H.C.; Diemer, M.; et al. The worldwide leaf economics spectrum. *Nature* **2004**, *428*, 821–827. [CrossRef]
37. Reich, P.B. The world-wide 'fast-slow' plant economics spectrum: A traits manifesto. *J. Ecol.* **2014**, *102*, 275–301. [CrossRef]
38. Wright, J.P.; Sutton-Grier, A. Does the leaf economic spectrum hold within local species pools across varying environmental conditions? *Funct. Ecol.* **2012**, *26*, 1390–1398. [CrossRef]
39. Mao, W.; Li, Y.L.; Zhao, X.Y.; Zhang, T.H.; Liu, X.P. Variations of leaf economic spectrum of eight dominant plant species in two successional stages under contrasting nutrient supply. *Pol. J. Ecol.* **2016**, *64*, 14–24. [CrossRef]
40. Sasaki, T.; Lauenroth, W.K. Dominant species, rather than diversity, regulates temporal stability of plant communities. *Oecologia* **2011**, *166*, 761–768. [CrossRef]
41. Avolio, M.L.; Forrestel, E.J.; Chang, C.C.; La Pierre, K.J.; Burghardt, K.T.; Smith, M.D. Demystifying dominant species. *New Phytol.* **2019**, *223*, 1106–1126. [CrossRef] [PubMed]
42. Avolio, M.L.; Koerner, S.E.; La Pierre, K.J.; Wilcox, K.R.; Wilson, G.W.T.; Smith, M.D.; Collins, S.L. Changes in plant community composition, not diversity, during a decade of nitrogen and phosphorus additions drive above-ground productivity in a tallgrass prairie. *J. Ecol.* **2014**, *102*, 649–1660. [CrossRef]
43. Yang, Y.; Dou, Y.; An, S.S. Environmental driving factors affecting plant biomass in natural grassland in the Loess Plateau, China. *Ecol. Indic.* **2017**, *82*, 250–259. [CrossRef]
44. Chen, Z.F.; Xiong, P.F.; Zhou, J.J.; Yang, Q.; Wang, Z.; Xu, B.C. Grassland productivity and diversity changes in responses to N and P addition depend primarily on tall clonal and annual species in semiarid Loess Plateau. *Ecol. Eng.* **2020**, *145*, 105727. [CrossRef]
45. Geiger, D.R.; Servaites, J.C. Diurnal regulation of photosynthetic carbon metabolism in C3 plants. *Annu. Rev. Plant Biol.* **1994**, *45*, 235–256. [CrossRef]
46. Du, Y.C.; Nose, A.; Kondo, A. Diurnal changes in photosynthesis in sugarcane leaves. I. carbon dioxide exchange rate, photosynthesic enzyme activities and metabolite levels relating to the C_4 pathway and the calvin cycle. *Plant Prod. Sci.* **2008**, *3*, 3–8. [CrossRef]
47. Muraoka, H.; Tang, Y.; Terashima, I.; Koizumi, H.; Washitani, I. Contributions of diffusional limitation, photoinhibition and photorespiration to midday depression of photosynthesis in *Arisaema heterophyllum* in natural high light. *Plant Cell Environ.* **2000**, *23*, 235–250. [CrossRef]
48. Farquhar, G.D.; Sharkey, T.D. Stomatal conductance and photosynthesis. *Annu. Rev. Plant Physiol.* **1982**, *33*, 317–345. [CrossRef]
49. Wu, F.Z.; Bao, W.K.; Li, F.L.; Wu, N. Effects of water stress and nitrogen supply on leaf gas exchange and fluorescence parameters of *Sophora davidii* seedlings. *Photosynthetica* **2008**, *46*, 40–48. [CrossRef]
50. Peng, Y.; Li, C.; Fritschi, F.B. Diurnal dynamics of maize leaf photosynthesis and carbohydrate concentrations in response to differential N availability. *Environ. Exp. Bot.* **2014**, *99*, 18–27. [CrossRef]
51. Zhao, H.B.; Qi, L.; Liu, Y.G. Effects of combined application of nitrogen and phosphorus on diurnal variation of photosynthesis at grain-filling stage and grain yield of super high-yielding wheat. *Chin. J. Appl. Ecol.* **2010**, *21*, 2545–2550. (In Chinese)
52. Xu, B.C.; Xu, W.Z.; Wang, Z.; Chen, Z.F.; Palta, J.A.; Chen, Y.L. Accumulation of N and P in the legume *Lespedeza davurica* in controlled mixtures with the grass *Bothriochloa ischaemum* under varying water and fertilization conditions. *Front. Plant Sci.* **2018**, *9*, 165. [CrossRef] [PubMed]
53. Tian, Z.W.; Liu, X.X.; Gu, S.L.; Yu, J.H.; Zhang, L.; Zhang, W.W.; Jiang, D.; Cao, W.X.; Dai, T.B. Postponed and reduced basal nitrogen application improves nitrogen use efficiency and plant growth of winter wheat. *J. Integr. Agric.* **2018**, *17*, 2648–2661. [CrossRef]
54. Ackerman, D.; Millet, D.B.; Chen, X. Global estimates of inorganic nitrogen deposition across four decades. *Glob. Biogeochem. Cycles* **2019**, *33*, 100–107. [CrossRef]
55. Li, Y.; Niu, S.; Yu, G. Aggravated phosphorus limitation on biomass production under increasing nitrogen loading: A meta-analysis. *Global Change Biol.* **2016**, *22*, 934–943. [CrossRef]
56. Güsewell, S. N: P ratios in terrestrial plants: Variation and functional significance. *New Phytol.* **2004**, *164*, 243–266. [CrossRef]
57. Naeem, M.; Khan, M.M.A.; Moinuddin; Idrees, M.; Aftab, T. Phosphorus ameliorates crop productivity, photosynthetic efficiency, nitrogen-fixation, activities of the enzymes and content of nutraceuticals of *Lablab purpureus* L. *Sci. Hortic.* **2010**, *126*, 205–214. [CrossRef]
58. Reich, P.B.; Oleksyn, J.; Wright, I.J. Leaf phosphorus influences the photosynthesis–nitrogen relation: A cross-biome analysis of 314 species. *Oecologia* **2009**, *160*, 207–212. [CrossRef]
59. Schleuss, P.M.; Widdig, M.; Heintz-Buschart, A.; Kirkman, K.; Spohn, M. Interactions of nitrogen and phosphorus cycling promote P acquisition and explain synergistic plant-growth responses. *Ecology* **2020**, *101*, e03003. [CrossRef]

60. Elser, J.J.; Bracken, M.E.; Cleland, E.E.; Gruner, D.S.; Harpole, W.S.; Hillebrand, H.; Ngai, J.T.; Seabloom, E.W.; Shurin, B.J.; Smith, J.E. Global analysis of nitrogen and phosphorus limitation of primary producers in freshwater, marine and terrestrial ecosystems. *Ecol. Lett.* **2007**, *10*, 1135–1142. [CrossRef]

61. Mariotte, P.; Cresswell, T.; Johansen, M.P.; Harrison, J.J.; Keitel, C.; Dijkstra, F.A. Plant uptake of nitrogen and phosphorus among grassland species affected by drought along a soil available phosphorus gradient. *Plant Soil* **2020**, *448*, 121–132. [CrossRef]

62. Kinugasa, T.; Tsunekawa, A.; Shinoda, M. Increasing nitrogen deposition enhances post-drought recovery of grassland productivity in the Mongolian steppe. *Oecologia* **2012**, *170*, 857–865. [CrossRef] [PubMed]

63. Zhong, S.Z.; Xu, Y.; Meng, B.; Loik, M.E.; Ma, J.Y.; Sun, W. Nitrogen addition increases the sensitivity of photosynthesis to drought and re-watering differentially in C_3 versus C_4 grass species. *Front. Plant Sci.* **2019**, *10*, 815. [CrossRef]

64. Niu, F.R.; Duan, D.P.; Chen, J.; Xiong, P.F.; Zhang, H.; Wang, Z.; Xu, B.C. Eco-physiological responses of dominant species to watering in a natural grassland community on the semi-arid Loess Plateau of China. *Front. Plant Sci.* **2016**, *7*, 663. [CrossRef]

65. Sun, C.X.; Huang, G.H.; Fan, Y.; Zhou, X.; Lu, C.; Wang, X.Q. Drought occurring with hot extremes: Changes under future climate change on Loess Plateau, China. *Earth's Future* **2019**, *7*, 587–604. [CrossRef]

66. Wan, H.W.; Yang, Y.; Bai, S.Q.; Xu, Y.H.; Bai, Y.F. Variations in leaf functional traits of six species along a nitrogen addition gradient in *Leymus chinensis* steppe in Inner Mongolia. *J. Plant Ecol.* **2008**, *32*, 611–621. [CrossRef]

67. Zhao, W.; Zhang, R.; Huang, C.Q.; Wang, B.Q.; Cao, H.; Koopal, L.K.; Tan, W.F. Effect of different vegetation cover on the vertical distribution of soil organic and inorganic carbon in the Zhifanggou watershed on the Loess Plateau. *Catena* **2016**, *139*, 191–198. [CrossRef]

68. Liang, T.; Tong, Y.A.; Xu, W.; Wei, Y.; Lin, W.; Pang, Y.; Liu, F.; Liu, X.J. Atmospheric nitrogen deposition in the Loess area of China. *Atmos. Pollut. Res.* **2016**, *7*, 447–453. [CrossRef]

Article

Using Trait-Based Methods to Study the Response of Grassland to Fertilization in the Grassland in Semiarid Areas in the Loess Plateau of China

Yuting Yang [1], Zhifei Chen [2], Bingcheng Xu [3], Jiaqi Wei [1], Xiaoxu Zhu [1], Hongbin Yao [1] and Zhongming Wen [1,3,*]

[1] College of Grassland Agriculture, Northwest A&F University, Yangling 712100, China
[2] College of Life Sciences, Guizhou University, Guiyang 550025, China
[3] Institute of Soil and Water Conservation, Northwest A&F University, Yangling 712100, China
* Correspondence: zmwen@ms.iswc.ac.cn

Abstract: Grassland is the dominant vegetation type in the Loess Plateau, and grassland productivity and processes are limited by nitrogen (N) and phosphorus (P). Studies have shown that productivity would change following fertilization in the grassland. The response of productivity to fertilization mainly depends on the dominant species traits. Trait-based methods provide a useful tool for explaining the variations in grassland productivity following fertilization. However, the relative contribution of plant functional traits to grassland productivity under N and P addition in the Loess Plateau is not clear. We measured aboveground biomass (AGB) and leaf N content (LN), leaf P content (LP), leaf N/P ratio (LN/P), specific leaf area (SLA), leaf tissue density (LTD), leaf dry matter content (LDMC), and maximum plant height (H_{max}) to study how these plant functional traits regulate the relative biomass of different species and grassland productivity following fertilization. Our results showed, that under different nutrient addition levels, the linkages between plant functional traits and the relative biomass of different species were different. Community AGB was positively related to community−weighted mean LN (CWM_LN), CWM_LN/P, CWM_SLA, and CWM_H_{max}, but negatively related to CWM_LTD and CWM_LDMC. Dominant species traits largely determined grassland productivity, in line with the mass ratio hypothesis. These findings further highlight the close linkages between community-level functional traits and grassland productivity. Our study contributes to the mechanisms underlying biodiversity−ecosystem function relationships and has significance for guiding semiarid grassland management.

Keywords: functional traits; productivity; grassland; fertilization; Loess Plateau

Citation: Yang, Y.; Chen, Z.; Xu, B.; Wei, J.; Zhu, X.; Yao, H.; Wen, Z. Using Trait-Based Methods to Study the Response of Grassland to Fertilization in the Grassland in Semiarid Areas in the Loess Plateau of China. *Plants* **2022**, *11*, 2045. https://doi.org/10.3390/plants11152045

Academic Editor: Dimitris L. Bouranis

Received: 10 July 2022
Accepted: 1 August 2022
Published: 4 August 2022

Publisher's Note: MDPI stays neutral with regard to jurisdictional claims in published maps and institutional affiliations.

1. Introduction

Anthropogenic drivers of environmental changes (e.g., fertilization) will cause changes in biodiversity and ecosystem function [1]. Recent studies have also shown that the decline in biodiversity may change ecosystem functioning [2,3]. In the context of global change, studies on biodiversity and ecosystem function are increasing [4,5]. Grassland is one of the largest ecosystems in the world [6]. Thus, it is necessary to study the relationship between biodiversity and grassland ecosystem productivity under environmental change. In most previous field experiments, the effect of fertilization on productivity was mainly dependent on the species, and species diversity was often used to explain the changes in productivity following fertilization [1,7,8]. However, it is being recognized more and more that functional diversity rather than species diversity finally drives the biodiversity−ecosystem function relationship [9]. Furthermore, plant functional traits can better help to explain and understand ecosystem function than species-based metrics (e.g., species richness and species abundance) [10,11].

Plant functional traits are defined as plant physiological and morphological characteristics that influence plant growth, survival, and so on [12]. They are closely related to the community productivity or plant adaptation [13,14]. Trait-based methods have become a useful method to understand ecosystem function [15,16]. The literature has shown that plant functional traits can modulate grassland productivity along nutrient gradients [17]. The leaf is the main organ of photosynthesis in plants, and has a significant impact on ecosystem function [18,19], such as specific leaf area (SLA), which represents changes in the leaf economics spectrum, indicating the ability of species to respond to rapid growth [18]. In view of this, understanding the linkages between plant functional traits and productivity under nitrogen (N) and phosphorus (P) addition in the Loess Plateau is of increasing importance for guiding grassland management and restoration.

The mass ratio hypothesis suggests that the ecosystem process is determined to a great extent by the dominant species traits, and community productivity is mainly determined by the mean functional traits of plant species [20,21]. An increasing body of studies suggested that the traits of dominant species related to competition play key roles in the response of the community to environment [22]. The mass ratio hypothesis used community−weighted mean (CWM) traits to characterize plant functional traits [21], and the CWM traits of these species can describe the community property accurately [23]. For example, rapid growing plant species from nutrient-abundant environments usually have high SLA and leaf N content (LN) and low leaf dry matter content (LDMC), but the opposite traits are found in species from nutrient-poor environments [24,25]. Studies have shown that SLA affects community productivity by affecting the maximum photosynthetic rate and increasing light interception [26]. Some studies also showed that plant nutrient characteristics would influence productivity [17]. Thus, the challenge is to identify the key plant functional traits of dominant species that have important effects on community aboveground biomass (community AGB) under N and P addition.

N addition would influence community productivity, but it will also aggravate P limitation [27]. As in the case of the Loess Plateau, grassland is the main vegetation type in the region [28,29]. However, the grassland productivity here is usually limited by both N and P in the Loess Plateau [30,31]. Thus, if we only add N, it will not be beneficial to the restoration of the grassland in the Loess Plateau. Accordingly, it is necessary to apply N and P together to promote grassland restoration. A better understanding of mechanisms affecting ecosystem function under N and P addition is essential for guiding grassland management. However, to date, few studies have investigated the linkages between plant functional traits and grassland productivity under both N and P addition in the Loess Plateau.

Accordingly, in this study, we designed a four-year N and P addition experiment to assess which functional traits have an important influence on community AGB and how dominant species traits affect ecosystem function in a fertilized grassland in the Loess Plateau. We focused on one important grassland ecosystem function—productivity. Specifically, we studied the following: (1) the relationship between the functional traits and relative biomass of different species under fertilization. (2) The relationship between community-level functional traits and community AGB. (3) The effects of CWM traits on grassland productivity.

2. Results

2.1. Principal Component Analysis of Functional Traits and Relative Biomass

The relative biomass was used to represent the dominance degree of the six main species. We found that the addition of N and P resulted in various relative biomasses of different species (Table S1). The effect of N addition was significant for relative biomass of *Lespedeza davurica*, *Artemisia sacrorum*, and *Artemisia scoparia*. P addition significantly affected relative biomass of all species except *A. sacrorum* and *Potentilla tanacetifolia*. There was significant interaction between N and P for relative biomass of *Bothriochloa ischaemum*, *L. davurica*, and *A. sacrorum* (Table S1).

Different relationships between functional traits and relative biomass were recorded among species following the addition of different levels of N and P (Figure 1). Increased relative biomass of *B. ischaemum* was positively correlated with LN, H_{max}, and LN/P under N addition alone and N100 combined with P addition. Increased relative biomass of *L. davurica* was positively correlated with LP under P addition alone. Increased relative biomass of *Stipa bungeana* was positively correlated with LN and LN/P under N addition alone. Increased relative biomass of *A. sacrorum* was positively correlated with H_{max} under N100 combined with P addition. Increased relative biomass of *P. tanacetifolia* was positively correlated with LTD and LDMC under N25 combined with P addition. Increased relative biomass of *A. scoparia* was positively correlated with LN, H_{max}, and SLA under N100 combined with P addition (Figure 1; Tables S1 and S2).

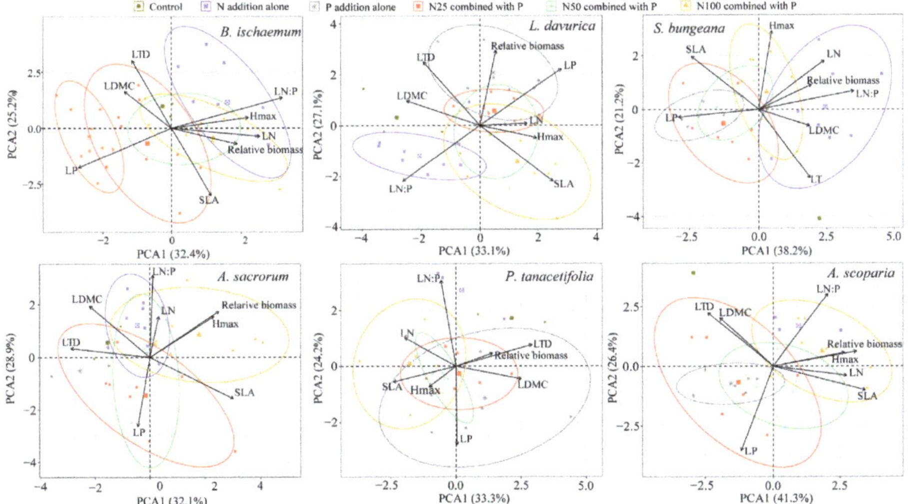

Figure 1. Principal components analysis (PCA) of the relationship between the functional traits and relative biomass of each species under different N and P levels. LN, leaf N content; LP, leaf P content; LN/P, leaf N/P ratio; SLA, specific leaf area; LTD, leaf tissue density; LDMC, leaf dry matter content; H_{max}, maximum plant height; RB, relative biomass.

2.2. The Relationship between Community–Weighted Mean Traits and Community Aboveground Biomass

The community AGB was significantly positively correlated with CWM_H_{max} ($r = 0.80$, $p < 0.001$), CWM_SLA ($r = 0.54$, $p < 0.001$), and CWM_LN ($r = 0.49$, $p < 0.001$), while there was a significantly negative correlation with CWM_LDMC ($r = -0.54$, $p < 0.001$) and CWM_LTD ($r = -0.53$, $p < 0.001$; Figure 2). The CWM_H_{max} was significantly positively correlated with CWM_SLA ($r = 0.64$, $p < 0.01$), whereas it was significantly negatively correlated with CWM_LTD ($r = -0.62$, $p < 0.001$) and CWM_LDMC ($r = -0.60$, $p < 0.001$). The CWM_LDMC had a significantly positive correlation with CWM_LTD ($r = 0.67$, $p < 0.001$), while a significantly negative correlation with CWM_SLA ($r = -0.68$, $p < 0.001$). The CWM_SLA had a significantly positive correlation with CWM_LN ($r = 0.31$, $p < 0.05$) and CWM_LP ($r = 0.32$, $p < 0.01$; Figure 2).

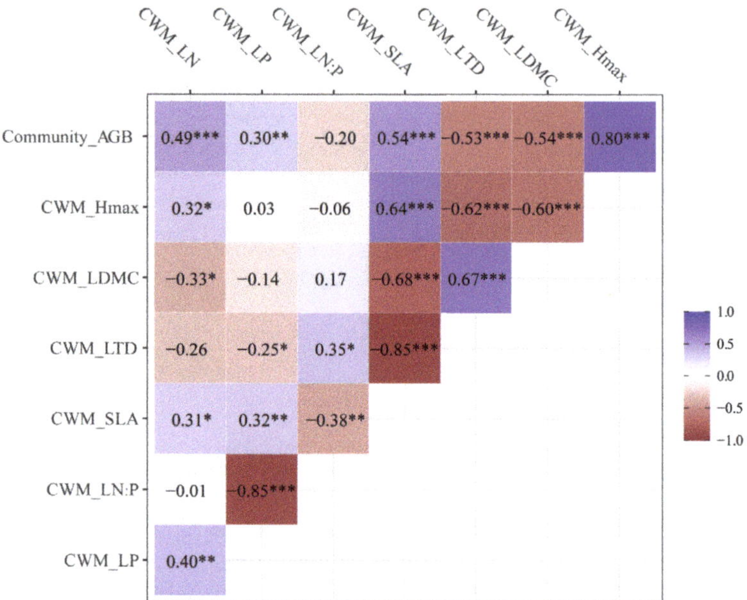

Figure 2. Correlation coefficients of community AGB and community level functional traits. * $p < 0.05$; ** $p < 0.01$; *** $p < 0.001$; LN, leaf N content; LP, leaf P content; LN/P, leaf N/P ratio; SLA, specific leaf area; LTD, leaf tissue density; LDMC, leaf dry matter content; H_{max}, maximum plant height.

2.3. Principal Component Analysis of Community—Weighted Mean Traits

Principal component analysis (PCA) compressed the variation of CWM traits into two principal components, representing 73.4% of the total variation (Figure 3A). The first principal component (CWM1) captured 51.1% of the total variation, exhibiting positive correlations with CWM_SLA and CWM_H_{max}, but negative correlations with CWM_LTD and CWM_LDMC. The contribution of CWM_SLA and CWM_H_{max} to CWM1 variables was 23.5% and 15.0%, respectively. The contribution of CWM_LTD and CWM_LDMC to CWM1 variables was 21.9% and 17.9%, respectively (Figure 3B).

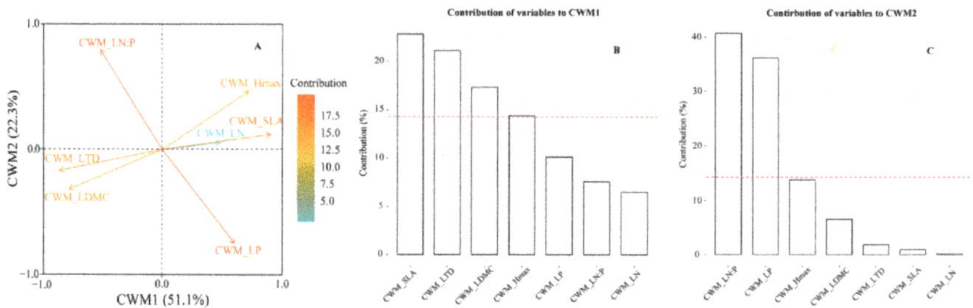

Figure 3. The principal component analysis (PCA) for community-weighted mean (CWM) traits (**A**). The contribution of each community-level trait to CWM1 (**B**) and CWM2 variables (**C**). The red dotted line on bar charts indicates the mean contribution of all seven community-level traits to CWM1 and CWM2 variables. LN, leaf N content; LP, leaf P content; LN/P, leaf N/P ratio; SLA, specific leaf area; LTD, leaf tissue density; LDMC, leaf dry matter content; H_{max}, maximum plant height.

The second principal component (CWM2), which accounted for 22.3% of the total information, exhibited positive correlations with CWM_LN/P and negative correlations with CWM_LP (Figure 3A). The contribution of CWM_LN/P to CWM2 variables was 37.7%. The contribution of CWM_LP to CWM2 variables was 41.2%. (Figure 3C).

2.4. Influence Factors of Grassland Productivity

The structural equation modeling (SEM) explained significant variation in productivity ($p = 0.996$) (Figure 4). The explanatory variables explained 75% of the variation in productivity. Productivity was positively influenced by CWM1 and CWM2 (the standardized path coefficients: $r = 0.34$, $p < 0.001$ and $r = 0.18$, $p < 0.05$, respectively). Additionally, N and P addition increased CWM1 ($r = 0.45$, $p < 0.001$ and $r = 0.41$, $p < 0.05$, respectively), and N addition increased CWM2 ($r = 0.57$, $p < 0.001$), while P addition decreased CWM2 ($r = -0.65$, $p < 0.001$; Figure 4).

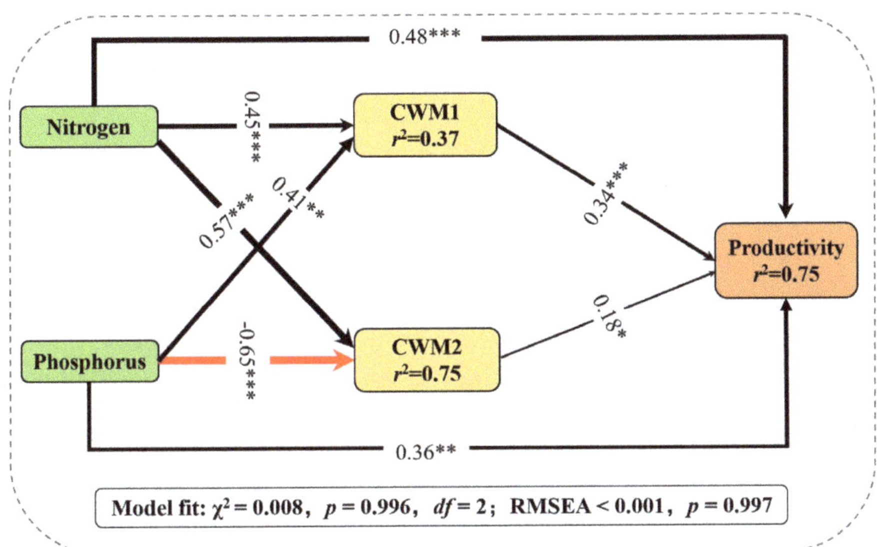

Figure 4. Results of the structural equation model. The arrows represent the hypothesized causal relationships between the variables. The solid lines represent significant relationships ($p < 0.05$). Black arrows represent positive effects, and the red arrows represent negative effects. The values next to the arrows are the standardized path coefficients. The line thickness is proportional to the standardized path coefficient. * $p < 0.05$; ** $p < 0.01$; *** $p < 0.001$. CWM1 and CWM2: the first two PCA axes of CWM traits.

3. Discussion

Grasslands in the Loess Plateau play a key role in offering ecosystem service and function [29]. Our study highlights the important role of CWM traits for predicting grassland productivity in the Loess Plateau. In this study, we explained the change in grassland productivity from the perspective of plant functional traits, suggesting that species with specific traits such as a high competitive ability or high nutrient use efficiency may be the important drivers in productivity. Studying the relationship between nutrient-induced variations in grassland productivity and plant functional traits was important for grassland restoration and management.

For these dominant species, we evaluated the linkages between plant functional traits and relative biomass of each species under different nutrient levels. Our results showed that different species exhibit different relationships between functional traits and relative biomass under different nutrient addition levels (Figure 1). Different linkages between functional traits and relative biomass may reflect different adaptation strategies to nutrient addition. For *L.davurica*, aboveground biomass increased under P addition alone, and was positively correlated with LP. This may because *L.davurica* is a legume plant with N-fixing function, and P limiting lifted as a result of the P addition. Moreover, a high level N addition may break symbiotic N fixation [32], thus leading to aboveground biomass increasing under only P addition. Accordingly, P fertilization was supposed to be considered when using fertilization to promote grassland restoration in the Loess Plateau.

The mass ratio hypothesis describes the advantage of traits in the community [20]. We found that community AGB was closely associated to CWM traits. This result indicates that the mass ratio hypothesis plays a role in the grassland of the Loess Plateau. The dominant effect of the leaf nutrient-use strategies on productivity may be related to the leaf economics spectrum [33]. Explorative plant species generally have higher SLA and LN and lower LDMC than conservative plant species [34]. In this study, the positive association of CWM_LN, CWM_SLA, and CWM_H_{max}, and the negative association of CWM_LTD and CWM_LDMC with community AGB implies acquisitive and conservative strategies across the studied species within communities. Studies have shown that the competition of plants for nutrient and light increased following fertilization [25,35]. The dominant species could maintain high grassland aboveground biomass through the traits related to competition ability. LN is closely correlated with leaf growth and defense strategies [36]. A high SLA tends to lead to high community AGB, and SLA is closely related to the relative growth rate and is a good predictor of plant responses to resource availability [37]. The positive relationship between CWM_H_{max} and community AGB indicates that high community AGB might be associated with tall or fast growing species (Figure 2). Previous studies have also shown that SLA, LDMC, and LN were correlated with aboveground biomass [38]. We also found that community-level traits showed stronger relationships with aboveground biomass than species-level traits. This may because community-level traits could better reflect the community structure [39].

Moreover, we used two PCA axes to obtain important information of community-level traits. CWM1 is positively correlated with CWM_SLA and CWM_H_{max}, but negatively correlated with CWM_LTD and CWM_LDMC. CWM2 is positively correlated with CWM_LN/P, but negatively correlated with CWM_LP (Figure 3; Table S3). In view of this, CWM1 and CWM2 represented the gradients of functional trait composition from species with slow growth and conservative resource use strategies to species with fast growth and acquisitive resource use strategies [39–41]. We found that grassland productivity increased with the increase in CWM1 and CWM2 (Figure 4; Table S3), which indicated that the mass-ratio effect plays a significant role. This is consistent with previous studies [42,43]. Species with acquisitive characteristics would have larger individual aboveground biomass, and thus increased community productivity [21]. For example, high SLA and fast nutrient acquisition would be conducive to fast growth and high productivity [34]. Plant functional traits are considered to be key drivers of ecosystem function [44]. Our results verified this relationship. These findings further prove the close relationship between plant functional traits and productivity, and emphasize the importance of plant functional traits.

4. Materials and Methods

4.1. Study Area and Experimental Design

The study area was located in Zhifanggou watershed in the Chinese Loess Plateau (36°42′–36°46′ N, 109°13′–109°16′ E). The altitude ranges from 1010 m to 1431 m. The mean annual rainfall and mean temperature in the study location was 528.8 mm and 8.8 °C, respectively. Monthly and daily precipitation and temperature in 2020 are shown in Figure 5. *Bothriochloa ischaemum, Lespedeza davurica, Stipa bungeana, Artemisia sacrorum, Potentilla tanacetifolia,* and *Artemisia scoparia* were the dominant species in this grassland.

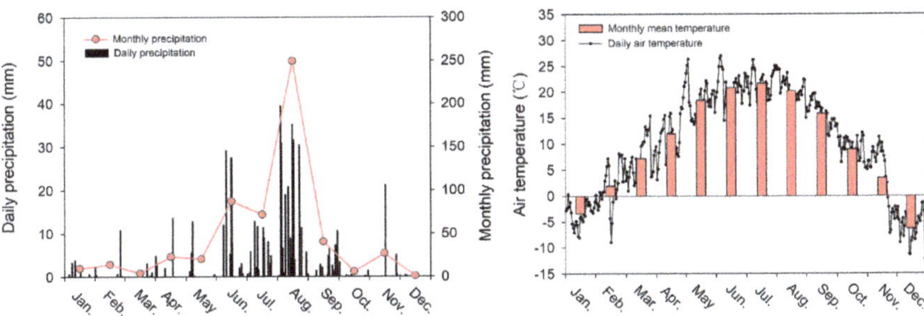

Figure 5. Monthly and daily precipitation and temperature in 2020.

In August 2017, we conducted the N and P addition experiment in this grassland. We set up twelve main-plots (4 × 4 m) in a randomized block design with three replicated blocks, and each main-plot was divided into four subplots (2 × 2 m). In each block, four N levels (0, 25, 50, and 100 kg N ha^{-1} yr^{-1}) were randomly assigned to four main-plots and four levels of P addition rate (0, 20, 40, and 80 kg P$_2$O$_5$ ha^{-1} yr^{-1}) were randomly assigned to four subplots. For more experimental details, see [35]. We used one 1 × 1 m quadrat for the community survey and one 1 × 1 m quadrat for functional trait measurement in each subplot. The soil properties of this grassland under different N and P additions are shown in Table 1.

Table 1. Soil total nitrogen (STN) and soil total phosphorus (STP) of the grassland under different N and P additions (mean ± s.e.; $n = 3$). * $p < 0.05$; ** $p < 0.01$; *** $p < 0.001$.

Soil Properties	Treatments	P0	P20	40	P80
STN (g kg^{-1})	N0	0.54 ± 0.02	0.57 ± 0.02	0.60 ± 0.02	0.53 ± 0.05
	N25	0.65 ± 0.01	0.61 ± 0.01	0.64 ± 0.03	0.53 ± 0.02
	N50	0.70 ± 0.02	0.65 ± 0.02	0.64 ± 0.04	0.64 ± 0.06
	N100	0.74 ± 0.04	0.63 ± 0.03	0.76 ± 0.03	0.67 ± 0.06
STP (g kg^{-1})	N0	0.52 ± 0.01	0.56 ± 0.01	0.60 ± 0.01	0.66 ± 0.02
	N25	0.52 ± 0.01	0.58 ± 0.01	0.61 ± 0.02	0.65 ± 0.01
	N50	0.52 ± 0.02	0.57 ± 0.01	0.57 ± 0.00	0.67 ± 0.02
	N100	0.52 ± 0.01	0.56 ± 0.01	0.58 ± 0.03	0.64 ± 0.02
STN		N ***(0.05)		P *(0.05)	N × P: ns
STP		N: ns		P ***(0.03)	N × P: ns

4.2. Community Survey and Trait Measurements

In the summer of 2020, we investigated species abundance, coverage, and maximum plant height (H$_{max}$) of each species in each 1 × 1 m quadrat [45]. Then, the aboveground parts were cut and brought to the laboratory, and oven-dried for 48 h (80 °C) to obtain the AGB of each species. We used the sum of the AGB of all species within this quadrat as a surrogate for productivity in this study.

We measured the plant functional traits of the most dominant species. All functional traits were measured from two to three individuals of each species. Leaf functional traits were measured according to standard methods [46]. SLA (cm^2 g^{-1}) = leaf area/leaf dry mass. Leaf tissue density (LTD; g cm^{-3}) = dry mass/leaf volume. LDMC (mg g^{-1}) = dry mass/fresh mass. LN (g kg^{-1}) was digested with sulfuric acid and determined using a Kjeldahl instrument (Kjektec System 2300 Distilling Unit, Foss, Tecator AB, Hoganas, Sweden), and leaf P content (LP; g kg^{-1}) was measured with a molybdenum antimony anti-colorimetric spectrophotometer (UV-2600 spectrophotometer, Shimadzu, Kyoto, Japan). The leaf N/P ratio (LN/P) was calculated as LN divided by LP [46]. CWM traits were calculated as the average of trait values weighted by the AGB of each species within a community [47].

4.3. Data Calculation and Analysis

Principal component analysis (PCA) was used to analyze the relationship between the relative biomass and functional traits of the six species under different nutrient levels and to compress the variation in CWM traits into a few principal components using R 4.1.3 (R Development Core Team) using the corrplot package, retaining the first two components (which describe most of the total variance) in the ensuing analysis. The relationship heat map between CWM traits and community AGB was produced in R 4.1.3 (R Development Core Team) using the corrplot package. Structural equation modeling (SEM) was used to explore how CWM traits influence productivity using IBM AMOS version 24.0 (Amos Development Co., Armonk, NY, USA).

5. Conclusions

Assessing how plant functional traits influence ecosystem function is important for understanding ecological processes. Our results indicate that dominant species traits could predict ecosystem functioning (productivity) in the Loess Plateau grassland. Thus, the mass ratio hypothesis is proved. Communities dominated by species with fast-growing acquisitive strategies have high productivity in the Loess Plateau grassland. The main novelty of this study is investigating the effect of functional traits after fertilization by adding N and P instead of just N. Overall, our study has strengthened the understanding of mechanisms affecting productivity in the Loess Plateau, and could help predict semiarid grassland responses to future environment change. Future studies need to further clarify the relationship between more species traits and ecosystem function to better understand the effect of fertilization on grassland.

Supplementary Materials: The following supporting information can be downloaded at: https://www.mdpi.com/article/10.3390/plants11152045/s1, Table S1: Relative biomass (%) of the six species under different N and P addition additions (mean $\pm$ s.e.; n = 3); Table S2: Analysis of variance results (F values) for the effects of N addition (N), P addition (P), species and their interactions on leaf trait and maximum plant height (H_{max}). ns, *, ** and *** indicated non-significant, significant at $p < 0.05, 0.01$ and 0.001, respectively; Table S3: Community weighted leaf traits, maximum plant height (Hmax) and community aboveground biomass under different N and P addition additions (mean $\pm$ s.e.; n = 3).

Author Contributions: Conceptualization: Z.W.; Methodology: Y.Y., B.X. and Z.C.; data collection and curation: Y.Y. and J.W.; data analysis: Y.Y. and Z.C.; field investigation: Y.Y., Z.C., X.Z. and H.Y.; writing—review and editing: Y.Y., Z.W. and B.X. All authors have read and agreed to the published version of the manuscript.

Funding: This study was funded by the National Natural Science Foundation of China (Nos. 41977077 and 41671289).

Institutional Review Board Statement: Not applicable.

Informed Consent Statement: Not applicable.

Data Availability Statement: All available data can be obtained by contacting the corresponding author.

Conflicts of Interest: The authors declare no conflict of interest.

References

1. Isbell, F.; Reich, P.B.; Tilman, D.; Hobbie, S.E.; Polasky, S.; Binder, S. Nutrient enrichment, biodiversity loss, and consequent declines in ecosystem productivity. *Proc. Natl. Acad. Sci. USA* **2013**, *110*, 11911–11916. [CrossRef] [PubMed]
2. Duffy, J.E. Why biodiversity is important to the functioning of real-world ecosystems. *Front. Ecol. Environ.* **2009**, *7*, 437–444. [CrossRef]
3. Cardinale, B.J.; Duffy, J.E.; Gonzalez, A.; Hooper, D.U.; Perrings, C.; Venail, P.; Narwani, A.; Mace, G.M.; Tilman, D.; Wardle, D.A.; et al. Biodiversity loss and its impact on humanity. *Nature* **2012**, *486*, 59–67. [CrossRef] [PubMed]
4. Hector, A.; Schmid, B.; Beierkuhnlein, C.; Caldeira, M.C.; Diemer, M.; Dimitrakopoulos, P.G.; Finn, J.A.; Freitas, H.; Giller, P.S.; Good, J.; et al. Plant diversity and productivity experiments in European grasslands. *Science* **1999**, *286*, 1123–1127. [CrossRef] [PubMed]

5. Hooper, D.U.; Adair, E.C.; Cardinale, B.J.; Byrnes, J.E.K.; Hungate, B.A.; Matulich, K.L.; Gonzalez, A.; Duffy, J.E.; Gamfeldt, L.; O'Connor, M.I. A global synthesis reveals biodiversity loss as a major driver of ecosystem change. *Nature* **2012**, *486*, U105–U129. [CrossRef]
6. Suttie, J.V.; Reynolds, S.G.; Batello, C. *Grasslands of the World*; FAO: Rome, Italy, 2005.
7. Naeem, S.; Hakansson, K.; Lawton, J.H.; Crawley, M.J.; Thompson, L.J. Biodiversity and plant productivity in a model assemblage of plant species. *Oikos* **1996**, *76*, 259–264. [CrossRef]
8. Stevens, C.J.; Dise, N.B.; Mountford, J.O.; Gowing, D.J. Impact of nitrogen deposition on the species richness of grasslands. *Science* **2004**, *303*, 1876–1879. [CrossRef]
9. Milcu, A.; Allan, E.; Roscher, C.; Jenkins, T.; Meyer, S.T.; Flynn, D.; Bessler, H.; Buscot, F.; Engels, C.; Gubsch, M.; et al. Functionally and phylogenetically diverse plant communities key to soil biota. *Ecology* **2013**, *94*, 1878–1885. [CrossRef]
10. Midolo, G.; Alkemade, R.; Schipper, A.M.; Benitez-Lopez, A.; Perring, M.P.; De Vries, W. Impacts of nitrogen addition on plant species richness and abundance: A global meta-analysis. *Glob. Ecol. Biogeogr.* **2019**, *28*, 398–413. [CrossRef]
11. Rosenfield, M.F.; Muller, S.C. Plant traits rather than species richness explain ecological processes in subtropical forests. *Ecosystems* **2020**, *23*, 52–66. [CrossRef]
12. Tilman, D.; Wedin, D.; Knops, J. Productivity and sustainability influenced by biodiversity in grassland ecosystems. *Nature* **1996**, *379*, 718–720. [CrossRef]
13. Reich, P.B.; Wright, I.J.; Cavender-Bares, J.; Craine, J.M.; Oleksyn, J.; Westoby, M.; Walters, M.B. The evolution of plant functional variation: Traits, spectra, and strategies. *Int. J. Plant Sci.* **2003**, *164*, S143–S164. [CrossRef]
14. Westoby, M.; Wright, I.J. Land-plant ecology on the basis of functional traits. *Trends Ecol. Evol.* **2006**, *21*, 261–268. [CrossRef]
15. Lavorel, S.; Garnier, E. Predicting changes in community composition and ecosystem functioning from plant traits: Revisiting the Holy Grail. *Funct. Ecol.* **2002**, *16*, 545–556. [CrossRef]
16. Li, Y.; Reich, P.B.; Schmid, B.; Shrestha, N.; Feng, X.; Lyu, T.; Maitner, B.S.; Xu, X.; Li, Y.; Zou, D.; et al. Leaf size of woody dicots predicts ecosystem primary productivity. *Ecol. Lett.* **2020**, *23*, 1003–1013. [CrossRef]
17. Zhang, D.; Peng, Y.; Li, F.; Yang, G.; Wang, J.; Yu, J.; Zhou, G.; Yang, Y. Trait identity and functional diversity co-drive response of ecosystem productivity to nitrogen enrichment. *J. Ecol.* **2019**, *107*, 2402–2414. [CrossRef]
18. Reich, P.B.; Ellsworth, D.S.; Walters, M.B.; Vose, J.M.; Gresham, C.; Volin, J.C.; Bowman, W.D. Generality of leaf trait relationships: A test across six biomes. *Ecology* **1999**, *80*, 1955–1969. [CrossRef]
19. He, N.; Liu, C.; Tian, M.; Li, M.; Yang, H.; Yu, G.; Guo, D.; Smith, M.D.; Yu, Q.; Hou, J. Variation in leaf anatomical traits from tropical to cold-temperate forests and linkage to ecosystem functions. *Funct. Ecol.* **2018**, *32*, 10–19. [CrossRef]
20. Grime, J.P. Benefits of plant diversity to ecosystems: Immediate, filter and founder effects. *J. Ecol.* **1998**, *86*, 902–910. [CrossRef]
21. Garnier, E.; Cortez, J.; Billes, G.; Navas, M.L.; Roumet, C.; Debussche, M.; Laurent, G.; Blanchard, A.; Aubry, D.; Bellmann, A.; et al. Plant functional markers capture ecosystem properties during secondary succession. *Ecology* **2004**, *85*, 2630–2637. [CrossRef]
22. McGill, B.J.; Enquist, B.J.; Weiher, E.; Westoby, M. Rebuilding community ecology from functional traits. *Trends Ecol. Evol.* **2006**, *21*, 178–185. [CrossRef] [PubMed]
23. Cadotte, M.W. Functional traits explain ecosystem function through opposing mechanisms. *Ecol. Lett.* **2017**, *20*, 989–996. [CrossRef] [PubMed]
24. Díaz, S.; Hodgson, J.G.; Thompson, K.; Cabido, M.; Cornelissen, J.H.C.; Jalili, A.; Montserrat-Marti, G.; Grime, J.P.; Zarrinkamar, F.; Asri, Y.; et al. The plant traits that drive ecosystems: Evidence from three continents. *J. Veg. Sci.* **2004**, *15*, 295–304. [CrossRef]
25. Zheng, S.; Chi, Y.; Yang, X.; Li, W.; Lan, Z.; Bai, Y. Direct and indirect effects of nitrogen enrichment and grazing on grassland productivity through intraspecific trait variability. *J. Appl. Ecol.* **2022**, *59*, 598–610. [CrossRef]
26. Li, Y.; Li, Q.; Xu, L.; Li, M.; Chen, Z.; Song, Z.; Hou, J.; He, N. Plant community traits can explain variation in productivity of selective logging forests after different restoration times. *Ecol. Indic.* **2021**, *131*, 108181. [CrossRef]
27. Crowley, K.F.; McNeil, B.E.; Lovett, G.M.; Canham, C.D.; Driscoll, C.T.; Rustad, L.E.; Denny, E.; Hallett, R.A.; Arthur, M.A.; Boggs, J.L.; et al. Do Nutrient Limitation Patterns Shift from Nitrogen Toward Phosphorus with Increasing Nitrogen Deposition Across the Northeastern United States? *Ecosystems* **2012**, *15*, 940–957. [CrossRef]
28. White, R.; Murry, S.; Rohweder, M. *Pilot Analysis of Global Ecosystems: Grassland Ecosystems*; World Resources Institute: Washington, DC, USA, 2000.
29. Gang, C.; Zhao, W.; Zhao, T.; Zhang, Y.; Gao, X.; Wen, Z. The impacts of land conversion and management measures on the grassland net primary productivity over the Loess Plateau, Northern China. *Sci. Total Environ.* **2018**, *645*, 827–836. [CrossRef]
30. Liu, Z.P.; Shao, M.A.; Wang, Y.Q. Spatial patterns of soil total nitrogen and soil total phosphorus across the entire Loess Plateau region of China. *Geoderma* **2013**, *197*, 67–78. [CrossRef]
31. Chen, Z.; Xiong, P.; Zhou, J.; Yang, Q.; Wang, Z.; Xu, B. Grassland productivity and diversity changes in responses to N and P addition depend primarily on tall clonal and annual species in semiarid Loess Plateau. *Ecol. Eng.* **2020**, *145*, 105727. [CrossRef]
32. Regus, J.U.; Wendlandt, C.E.; Bantay, R.M.; Gano-Cohen, K.A.; Gleason, N.J.; Hollowell, A.C.; O'Neill, M.R.; Shahin, K.K.; Sachs, J.L. Nitrogen deposition decreases the benefits of symbiosis in a native legume. *Plant Soil.* **2017**, *414*, 159–170. [CrossRef]
33. Reich, P.B. The world-wide 'fast-slow' plant economics spectrum: A traits manifesto. *J. Ecol.* **2014**, *102*, 275–301. [CrossRef]
34. Wright, I.J.; Reich, P.B.; Westoby, M.; Ackerly, D.D.; Baruch, Z.; Bongers, F.; Cavender-Bares, J.; Chapin, T.; Cornelissen, J.H.C.; Diemer, M.; et al. The worldwide leaf economics spectrum. *Nature* **2004**, *428*, 821–827. [CrossRef]

35. Chen, Z.; Xiong, P.; Zhou, J.; Lai, S.; Jian, C.; Xu, W.; Xu, B. Effects of plant diversity on semiarid grassland stability depends on functional group composition and dynamics under N and P addition. *Sci. Total Environ.* **2021**, *799*, 149482. [CrossRef]

36. Reich, P.B.; Walters, M.B.; Ellsworth, D.S. Leaf life-span in relation to leaf, plant, and stand characteristics among diverse ecosystems. *Ecol. Monogr.* **1992**, *62*, 365–392. [CrossRef]

37. Wright, I.J.; Reich, P.B.; Westoby, M. Strategy shifts in leaf physiology, structure and nutrient content between species of high- and low-rainfall and high- and low-nutrient habitats. *Funct. Ecol.* **2001**, *15*, 423–434. [CrossRef]

38. Reich, P.B.; Wright, I.J.; Lusk, C.H. Predicting leaf physiology from simple plant and climate attributes: A global GLOPNET analysis. *Ecol. Appl.* **2007**, *17*, 1982–1988. [CrossRef]

39. Bruelheide, H.; Dengler, J.; Purschke, O.; Lenoir, J.; Jimenez-Alfaro, B.; Hennekens, S.M.; Botta-Dukat, Z.; Chytry, M.; Field, R.; Jansen, F.; et al. Global trait-environment relationships of plant communities. *Nat. Ecol. Evol.* **2018**, *2*, 1906–1917. [CrossRef]

40. Zheng, Z.; Ma, P. Changes in above and belowground traits of a rhizome clonal plant explain its predominance under nitrogen addition. *Plant Soil.* **2018**, *432*, 415–424. [CrossRef]

41. Fang, Z.; Li, D.-D.; Jiao, F.; Yao, J.; Du, H.-T. The Latitudinal Patterns of Leaf and Soil C:N:P Stoichiometry in the Loess Plateau of China. *Front. Plant Sci.* **2019**, *10*, 85. [CrossRef]

42. Zhou, X.L.; Guo, Z.; Zhang, P.F.; Du, G.Z. Shift in community functional composition following nitrogen fertilization in an alpine meadow through intraspecific trait variation and community composition change. *Plant Soil.* **2018**, *431*, 289–302. [CrossRef]

43. Griffin-Nolan, R.J.; Blumenthal, D.M.; Collins, S.L.; Farkas, T.E.; Hoffman, A.M.; Mueller, K.E.; Ocheltree, T.W.; Smith, M.D.; Whitney, K.D.; Knapp, A.K. Shifts in plant functional composition following long-term drought in grasslands. *J. Ecol.* **2019**, *107*, 2133–2148. [CrossRef]

44. Reichstein, M.; Bahn, M.; Mahecha, M.D.; Kattge, J.; Baldocchi, D.D. Linking plant and ecosystem functional biogeography. *Proc. Natl. Acad. Sci. USA* **2014**, *111*, 13697–13702. [CrossRef]

45. Jing, Z.B.; Cheng, J.M.; Su, J.S.; Bai, Y.; Jin, J.W. Changes in plant community composition and soil properties under 3-decade grazing exclusion in semiarid grassland. *Ecol. Eng.* **2014**, *64*, 171–178. [CrossRef]

46. Pérez-Harguindeguy, N.; Díaz, S.; Garnier, E.; Lavorel, S.; Poorter, H.; Jaureguiberry, P.; Bret-Harte, M.S.; Cornwell, W.K.; Craine, J.M.; Gurvich, D.E.; et al. New handbook for standardised measurement of plant functional traits worldwide. *Aust. J. Bot.* **2013**, *61*, 167–234. [CrossRef]

47. Violle, C.; Navas, M.-L.; Vile, D.; Kazakou, E.; Fortunel, C.; Hummel, I.; Garnier, E. Let the concept of trait be functional. *Oikos* **2007**, *116*, 882–892. [CrossRef]

Article

Selected Indices to Identify Water-Stress-Tolerant Tropical Forage Grasses

Alan Mario Zuffo [1,*], Fábio Steiner [2], Jorge González Aguilera [3], Rafael Felippe Ratke [3], Leandra Matos Barrozo [1], Ricardo Mezzomo [1], Adaniel Sousa dos Santos [4], Hebert Hernán Soto Gonzales [5], Pedro Arias Cubillas [6] and Sheda Méndez Ancca [7]

[1] Department of Agronomy, State University of Maranhão, Balsas, MA 65800-000, Brazil
[2] Department of Crop Science, State University of Mato Grosso do Sul, Cassilândia, MS 79540-000, Brazil
[3] Department of Agronomy, Federal University of Mato Grosso do Sul, Chapadão do Sul, MS 79560-000, Brazil
[4] Department of Plant Sciences, Federal University of Piauí, Bom Jesus, PI 64900-000, Brazil
[5] Escuela Profesional de Ingeniería Ambiental, Universidad Nacional de Moquegua (UNAM), Ilo 18601, Peru
[6] Escuela de Posgrado-Doctorado en Ciencias Ambientales, Universidad Nacional Jorge Basadre Grohmann (UNJBG), Tacna 23001, Peru
[7] Escuela Profesional de Ingeniería Pesquera, Universidad Nacional de Moquegua (UNAM), Ilo 18601, Peru
* Correspondence: alan_zuffo@hotmail.com or alanzuffo@professor.uema.br

Citation: Zuffo, A.M.; Steiner, F.; Aguilera, J.G.; Ratke, R.F.; Barrozo, L.M.; Mezzomo, R.; Santos, A.S.d.; Gonzales, H.H.S.; Cubillas, P.A.; Ancca, S.M. Selected Indices to Identify Water-Stress-Tolerant Tropical Forage Grasses. *Plants* **2022**, *11*, 2444. https://doi.org/10.3390/plants11182444

Academic Editor: Bingcheng Xu

Received: 14 August 2022
Accepted: 13 September 2022
Published: 19 September 2022

Publisher's Note: MDPI stays neutral with regard to jurisdictional claims in published maps and institutional affiliations.

Abstract: Periods of soil water stress have been recurrent in the Cerrado region and have become a growing concern for Brazilian tropical pasture areas. Thus, the search for forage grasses more tolerant to water stress has intensified recently in order to promote more sustainable livestock. In a greenhouse experiment, the degree of water stress tolerance of nine tropical forage grass cultivars was studied under different soil water regimes. The investigation followed a 9 × 3 factorial design in four randomized blocks. Nine cultivars from five species of perennial forage grasses were tested: *Urochloa brizantha* ('BRS Piatã', 'Marandu', and 'Xaraés'), *Panicum maximum* ('Aruana', 'Mombaça', and 'Tanzânia'), *Pennisetum glaucum* ('ADR 300'), *Urochloa ruziziensis* ('Comum'), and *Paspalum atratum* ('Pojuca'). These cultivars were grown in pots under three soil water regimes (high soil water regime—HSW (non-stressful condition), middle soil water regime—MSW (moderate water stress), and low soil water regime—LSW (severe water stress)). Plants were exposed to soil water stress for 25 days during the tillering and stalk elongation phases. Twelve tolerance indices, including tolerance index (TOL), mean production (MP), yield stability index (YSI), drought resistance index (DI), stress tolerance index (STI), geometric mean production (GMP), yield index (YI), modified stress tolerance (k_1STI and k_2STI), stress susceptibility percentage index (SSPI), abiotic tolerance index (ATI), and harmonic mean (HM), were calculated based on shoot biomass production under non-stressful (Y_P) and stressful (Y_S) conditions. Soil water stress decreased leaf area, plant height, tillering capacity, root volume, and shoot and root dry matter production in most cultivars, with varying degrees of reduction among tropical forage grasses. Based on shoot biomass production under controlled greenhouse conditions, the most water-stress-tolerant cultivars were *P. maximum* cv. Mombaça and cv. Tanzânia under the MSW regime and *P. maximum* cv. Aruana and cv. Mombaça under the LSW regime. *P. maximum* cv. Mombaça has greater adaptability and stability of shoot biomass production when grown under greenhouse conditions and subjected to soil water stress. Therefore, this forage grass should be tested under field conditions to confirm its forage production potential for cultivation in tropical regions with the occurrence of water stress. The MP, DI, STI, GMP, YI, k_2STI, and HM tolerance indices were the most suitable for identifying forage grass cultivars with greater water stress tolerance and a high potential for shoot biomass production under LSW regime.

Keywords: soil water regime; stress tolerance indices; forage yield; *Panicum maximum*; *Urochloa* sp.

1. Introduction

The large territorial extension and the edaphoclimatic conditions in Brazil are fundamental elements for the country to have an expressive development of its livestock and agriculture activities. Brazil is one of the world's largest producers and exporters of animal food, as well as the origin of a great number of plant species. The country has the world's second-largest cattle herd, with 252 million head, and its production is based on grass pastures [1]. Brazil has a pasture area of approximately 172 million hectares, of which 102 million hectares are cultivated with forage plants and 70 million hectares are native pastures [2].

Because national meat and milk production are highly dependent on the natural feeding of grass and/or legume pastures, the quality of pastures is essential for Brazilian livestock activity [3]. In addition, many Brazilian producers are diversifying their agricultural production systems by cultivating tropical forage grasses in the off-season of cash crops to produce forage for cattle in the autumn/winter and straw in the spring for the agricultural production system [4]. This production system, called the Integrated Crop-Livestock System (ICLS), is an important strategy for producing quality pasture at a time of low rainfall in the Brazilian Cerrado region. However, this dry season during southern winter poses many challenges to Brazilian livestock activity, especially concerning the supply of quality pasture. Therefore, studies that evaluate and identify genotypes of forage grasses with greater water stress tolerance are essential for boosting animal production systems.

The most important tropical forage grasses used in livestock and agriculture systems in the Cerrado region are species of the genera *Urochloa, Cynodon, Panicum, Paspalum*, and *Pennisetum*. These forage grasses are currently the basis of food for meat and milk cattle, mainly due to their excellent nutritional quality and adequate adaptation to Brazilian production systems [5]. However, each forage grass species or cultivar has a distinct biomass production potential that depends on morphological and genetic characteristics. Pearl millet (*Pennisetum glaucum* (L.) R. Br.), palisade grass (*Urochloa brizantha* (Hochst. Ex A. Rich.) R.D. Webster), and ruzigrass (*U. ruziziensis* (R. Germ. & C.M. Evrard) Crins) have been cultivated due to their high biomass production capacity [6]. Improvement of forage grasses' productivity in quantity, as well as its quality, would have a significant impact on livestock production. Furthermore, palisade grass and ruzigrass have also been described as water-stress-tolerant forage grasses [7]. Therefore, looking for species and/or cultivars tolerant to water stress is of fundamental importance to mitigate the negative impacts of low soil water availability and increase forage production in the dry off-season [3,5,8], a period of the year with a deficient supply of pasture for cattle in Brazil.

Soil water stress results in a dramatic decline in leaf expansion rate and photosynthesis rate, which inhibits plant development and reduces the biomass production of forage grasses [8,9], especially by causing changes in root growth, leaf initiation rate, nutrient uptake, and carbohydrate metabolism [10–13]. However, the response of forage grasses to water stress depends on genotype, plant developmental stage, severity, and duration of the water stress period [6–8]. Generally, forage grasses are more susceptible to soil water stress during the tillering and stalk elongation phases, and stalk and leaf growth are more affected than other plant organs [9,12]. Typical effects of water stress on forage grasses include leaf rolling, stomatal closure, stalk and leaf growth inhibition, early leaf senescence, reduced leaf area, and reduced biomass production [8,11]. Some of these responses are part of the plant's strategies that aim to mitigate the adverse effects of low soil water availability and, therefore, constitute water stress tolerance mechanisms.

Water stress tolerance refers to the degree to which a plant is adapted to low soil water availability or drought conditions [14]. Forage plants are subject to periods of water stress during their growth and development phase and must adapt to these adverse conditions. Thus, plants in adverse environments have the ability to endure water stress through certain biochemical or morphological adaptations and avoidance of cell injury [9,12,15,16].

Understanding forage grass responses to water stress periods is essential to pasture biomass productivity mainly because the periodic and repeated water shortage have increasingly concerned Brazilian tropical grasslands [5]. Thus, identifying and understanding water stress tolerance mechanisms are fundamental factors for developing tolerant forage grass cultivars. The relative performance of forage grass biomass production under non-stressful and stressful conditions (i.e., high and low soil water availability) seems to be the starting point for identifying species with greater water stress tolerance [17]. Therefore, the main conditions that must be considered when determining water-stress-tolerant and water-stress-sensitive cultivars are cropping under non-stressful and stressful conditions with high and low soil water availability, respectively [18,19]. However, identifying water-stress-tolerant genotypes is not an easy task because the productive performance of forage grasses is the result of a genotypic expression modulated by continuous interaction with the growing environment [8].

Some studies have proposed using different selection indices to evaluate and identify water-stress-tolerant genotypes. Some of these selection indices were used to assess genetic differences in genotypes of sorghum [18], soybeans [19], maize [20], wheat [21,22], sunflower [23], and common beans [24]. However, these studies for tropical forage grasses are still unknown. Therefore, our research constitutes the first report on selection indices to assess the degree of water stress tolerance of the main forage grasses used in the Brazilian Cerrado region. This information will help Brazilian farmers choose the best forage cultivars to be planted in areas subject to soil water stress.

This study aimed to determine the degree of water stress tolerance of nine cultivars of tropical forage grasses grown under different soil water regimes.

2. Results and Discussion

2.1. Analysis of Variance

Analysis of variance revealed that the effect of soil water regime was significant ($p < 0.01$) on all forage grass growth traits. The interaction between soil water regime and cultivars showed a significant effect ($p < 0.05$) on all plant growth traits except for the number of tillers and shoot dry matter (Table 1). The significant interaction between the main effects of cultivars and soil water regimes on most morphological traits indicates that the forage grasses have distinct responses when exposed to soil water availability levels.

Table 1. Summary of analysis of variance for morphological traits of tropical forage grass cultivars under the effect of soil water regimes.

Causes of Variation	Probability > F							
	PH	**NT**	**NL**	**LA**	**SDM**	**RDM**	**TDM**	**RV**
Forage cultivar (C)	<0.01	<0.01	<0.01	<0.01	<0.01	<0.01	<0.01	<0.01
Soil water regime (W)	<0.01	<0.01	<0.01	<0.01	<0.01	<0.01	<0.01	<0.01
C × W	<0.01	0.914	<0.01	0.045	0.518	0.020	0.029	<0.01
CV (%)	13.65	16.41	18.61	17.82	15.99	20.66	15.70	21.92

PH: plant height; NT: number of tillers; NL: number of leaves; LA: leaf area; SDM: shoot dry matter; RDM: root dry matter; TDM: total dry matter; RV: root volume. CV: coefficient of variation.

2.2. Morphological Responses of Forage Grasses to Water Stress

Soil water stress resulted in a lower growth rate in plant height of *U. brizantha* cv. BRS Piatã, *P. glaucum* cv. ADR 300 and *P. maximum* cv. Mombaça compared to high soil water regime (Table 2). The plant height of *P. maximum* cv. Mombaça, *U. brizantha* cv. BRS Piatã and *P. glaucum* cv. ADR 300 was 22%, 28%, and 44% lower in plants grown under a low soil water (LSW) regime when compared to plants under a high soil water (HSW) regime. However, soil water regimes did not inhibit the height growth rate of other forage grasses.

Table 2. Plant height, number of green leaves, and leaf area of nine tropical forage grass cultivars grown under different soil water regimes.

Forage Grass Cultivar	Soil Water Regime		
	High	Middle	Low
	Plant Height (cm)		
U. brizantha cv. BRS Piatã	74.0 ± 5.0 bA	63.7 ± 8.1 bB	53.3 ± 4.7 bC
U. brizantha cv. Marandu	62.7 ± 10.1 cA	51.3 ± 4.8 cA	50.0 ± 2.5 bA
U. brizantha cv. Xaraés	62.0 ± 9.9 cA	56.0 ± 2.0 cA	50.7 ± 2.4 bA
U. ruziziensis cv. Comum	47.7 ± 6.4 cA	47.3 ± 2.4 cA	43.3 ± 2.9 bA
P. glaucum cv. ADR 300	154.1 ± 5.3 aA	124.6 ± 5.1 aB	86.2 ± 3.3 aC
P. maximum cv. Aruana	72.7 ± 3.7 bA	71.0 ± 7.5 bA	71.3 ± 1.8 aA
P. maximum cv. Mombaça	79.7 ± 6.4 bA	62.7 ± 4.1 bB	62.0 ± 2.0 bB
P. maximum cv. Tanzânia	68.7 ± 5.8 bA	63.7 ± 5.3 bA	63.7 ± 1.3 bA
P. atratum cv. Pojuca	62.0 ± 9.9 cA	52.3 ± 1.4 cA	52.7 ± 3.7 bA
	Number of leaves per plant		
U. brizantha cv. BRS Piatã	40 ± 1 dA	33 ± 3 bA	20 ± 1 bA
U. brizantha cv. Marandu	41 ± 4 dA	32 ± 3 bA	24 ± 1 bA
U. brizantha cv. Xaraés	36 ± 2 dA	26 ± 1 bA	18 ± 1 bA
U. ruziziensis cv. Comum	119 ± 6 bA	88 ± 2 aB	51 ± 3 aC
P. glaucum cv. ADR 300	26 ± 2 dA	21 ± 1 bA	15 ± 2 bA
P. maximum cv. Aruana	75 ± 5 cA	47 ± 5 bB	38 ± 3 bB
P. maximum cv. Mombaça	52 ± 6 dA	30 ± 2 bB	23 ± 3 bC
P. maximum cv. Tanzânia	53 ± 2 dA	32 ± 1 bA	32 ± 5 bA
P. atratum cv. Pojuca	195 ± 20 aA	94 ± 19 aB	70 ± 10 aC
	Leaf area (dm^2/plant)		
U. brizantha cv. BRS Piatã	16.9 ± 1.2 bA	10.2 ± 0.3 bB	7.6 ± 0.5 aB
U. brizantha cv. Marandu	20.8 ± 2.2 bA	17.5 ± 5.1 aA	9.9 ± 0.9 aB
U. brizantha cv. Xaraés	21.0 ± 1.2 bA	14.8 ± 1.5 aB	8.8 ± 1.5 aC
U. ruziziensis cv. Comum	31.6 ± 6.8 aA	18.9 ± 2.9 aB	11.9 ± 1.6 aC
P. glaucum cv. ADR 300	1.6 ± 0.2 dA	1.4 ± 0.1 bA	1.3 ± 0.2 bA
P. maximum cv. Aruana	28.9 ± 1.6 aA	17.0 ± 2.0 aB	11.9 ± 1.8 aB
P. maximum cv. Mombaça	28.6 ± 2.7 aA	18.8 ± 3.7 aB	12.0 ± 1.3 aC
P. maximum cv. Tanzânia	23.3 ± 2.3 bA	21.1 ± 2.3 aA	12.9 ± 0.8 aB
P. atratum cv. Pojuca	10.8 ± 1.6 cA	5.5 ± 0.8 bA	4.4 ± 1.3 bA

Means followed by distinct lowercase letters for the forage grass cultivars (in the column) or distinct uppercase letters for the soil water regimes (in the line) show significant differences (Scott–Knott test, $p \leq 0.05$). Values represent the mean ± mean standard error.

Petter et al. [6] reported that the growth rate of *U. brizantha* cv. Xaraés, *U. ruziziensis* cv. Comum and *P. glaucum* cv. ADR 7010 was not negatively affected by plant exposure to water stress. However, Zuffo et al. [25] showed that soil water stress inhibited the height growth rate of *U. brizantha* cv. BRS Piatã and *P. glaucum* cv. ADR 300. These results show that the adverse effects of water stress on the height growth rate of forage grasses are still inconsistent and depend on the grass development stage at which water stress occurs. Soil water stress imposed during the initial growth stage has a more significant negative impact on plant height than when imposed during the grass tillering stage [16].

The highest plant height under different soil water regimes was observed for *P. glaucum* cv. ADR 300; however, the plant height was similar to *P. maximum* cv. Aruana under an LSW regime (Table 2). These results are associated with the growth habits of forage grass cultivars. The *P. glaucum* is more extensive and has an erect growth habit, while *U. brizantha* and *P. maximum* plants are smaller, have a more tufted growth habit, and have a greater number of tillers [6,26].

Water stress significantly reduced ($p < 0.05$) the number of leaves of *U. ruziziensis* cv. Comum, *P. maximum* cv. Aruana, *P. maximum* cv. Mombaça, and *P. Atratum* cv. Pojuca, while the other forage grasses did not significantly reduce the number of leaves when grown under different soil water regimes (Table 2). Under an HSW regime, the highest number of leaves was observed for *P. atratum* cv. Pojuca, while under MSW and LSW regimes, the highest number of leaves was obtained in *U. ruziziensis* cv. Comum and *P. atratum* cv. Pojuca plants. Petter et al. [6] also reported a higher number of leaves in *U. ruziziensis* plants exposed to soil water stress. These results indicate that even the plants of *U. ruziziensis* have a significant reduction in the number of leaves when subjected to soil water stress; this species can maintain a high number of leaves under LSW regimes. The lower emergence of new leaves under water stress conditions has been considered a plant strategy to reduce the transpiration rate and increase water use efficiency [15]. However, this water stress tolerance strategy is conditioned on the specific response of the genotype [9]. The lower number of leaves has been a response of forage grasses to ensure their survival to relatively long periods of low soil water availability [8].

Leaf area was significantly ($p < 0.05$) smaller under an LSW regime for all forage grass cultivars except for *P. glaucum* cv. ADR 300 and *P. atratum* cv. Pojuca (Table 2). Under an LSW regime, the leaf area reduction of forage grasses ranged from 45% to 62% compared to plants under an HSW regime. Leaf area reduction in three forage grass cultivars (*U. brizantha* cv. BRS Piatã, *U. brizantha* cv. Marandu, *P. glaucum* cv. ADR 300) exposed to water stress conditions were also reported by Zuffo et al. [25]. The reduction in leaf area has been reported as a typical response of forage grasses when exposed to soil water stress [8–12]. One of the first processes affected in response to decreased soil water availability is cell expansion, a highly dependent process of turgidity in plants. However, with the advancement of soil water stress, other physiological processes are negatively affected, with direct effects on the photoassimilates accumulated by the forage grasses, reduction in the carbon assimilation rate, and relative growth rate [8,9]. As a result of these effects, there is a reduction in leaf area and biomass production. The reduction in leaf area occurs as a defense reaction of plants to water stress, reducing the transpiration rate and, consequently, water loss to the atmosphere [15].

Soil water regimes did not alter the root dry matter accumulation of *U. brizantha* cv. Xaraés, *P. glaucum* cv. ADR 300 and *P. atratum* cv. Pojuca, while the root dry matter accumulation of the other forage grasses was significantly lower when grown under water stress, especially under an LSW regime (Table 3). Under an HSW regime, root dry matter was higher for *U. brizantha* cv. BRS Piatã, *U. brizantha* cv. Marandu, *U. ruziziensis* cv. Comum, *P. maximum* cv. Aruana, *P. maximum* cv. Mombasa and *P. maximum* cv. Tanzania. However, when forage grass cultivars were grown under an LSW regime, there were no significant differences in root dry matter production (Table 3). The lower production of root dry matter of forage grasses under water stress conditions was also reported by Petter et al. [6] and Fariaszewska et al. [9]. Under soil water stress, plants reveal mechanisms to combat cellular tissue dehydration. The decrease in soil water availability causes an increase in the synthesis of abscisic acid (ABA) and stress proteins, which protect cell membranes and participate in osmoregulation [8]. The increase in the concentration of ABA in the cells reduces the transpiration rate by closing the stomata and results in greater water use efficiency [9]. In addition, abscisic acid inhibits shoot growth but simultaneously stimulates root growth and development, which essentially helps to overcome stress [27]. However, this stimulus in root growth caused by the higher concentration of ABA under water stress conditions has been commonly reported under field conditions [8,12]. In pot experiments, as in this study, the growth of the plant root system was limited by the soil volume in the pot.

Total dry matter accumulation and root volume were significantly lower ($p < 0.05$) under water stress conditions for all forage grass cultivars (Table 3). Under HSW regime, *P. maximum* cv. Mombaça plants have greater total dry matter accumulation and greater root volume than other forage grasses. Under soil water stress, plants of *U. ruziziensis* cv. Comum, *P. maximum* cv. Aruana, *P. maximum* cv. Mombaça, and *P. maximum* cv. Tanzânia had a greater total dry matter production, except for *P. maximum* cv. Aruana under an MSW regime (Table 3). Under water stress conditions, there was no significant difference in root volume value between forage grass cultivars, except for *P. glaucum* cv. ADR 300 and *P. atratum* cv. Pojuca under MSW, which had lower root volume than other forage grasses. The lower total dry matter production and root volume of plants exposed to soil water stress is a consequence of plant adaptation mechanisms to avoid excessive water loss [11,12], as well as the adverse effects of water stress on plant physiological metabolism, especially on the photosynthetic activity of the plants [9]. Under water stress conditions, the rate of photosynthesis decreases, which is related to a decrease in rubisco activity, a reduction in stomatal conductance, and reduced availability of CO_2 [8].

Table 3. Root dry matter, total dry matter, and root volume of nine tropical forage grass cultivars grown under different soil water regimes.

Forage Grass Cultivar	Soil Water Regimes		
	High	Middle	Low
Root dry matter (g plant^{-1})			
U. brizantha cv. BRS Piatã	23.3 ± 3.4 aA	15.0 ± 5.5 aB	5.5 ± 1.0 aC
U. brizantha cv. Marandu	19.5 ± 1.3 aA	12.5 ± 1.2 aB	7.5 ± 1.0 aB
U. brizantha cv. Xaraés	14.0 ± 1.4 bA	12.2 ± 1.1 aA	8.8 ± 1.1 aA
U. ruziziensis cv. Comum	22.8 ± 1.0 aA	14.4 ± 0.3 aB	7.1 ± 0.6 aC
P. glaucum cv. ADR 300	5.4 ± 0.6 cA	5.2 ± 0.4 bA	3.6 ± 0.2 aA
P. maximum cv. Aruana	17.5 ± 3.9 aA	8.3 ± 1.1 bB	1.2 ± 0.1 aC
P. maximum cv. Mombaça	28.4 ± 6.7 aA	13.1 ± 0.6 aB	8.3 ± 0.9 aB
P. maximum cv. Tanzânia	21.0 ± 3.6 aA	15.0 ± 3.4 aB	12.3 ± 1.2 aB
P. atratum cv. Pojuca	7.0 ± 1.5 cA	3.6 ± 0.7 bA	2.9 ± 0.3 aA
Total dry matter (g plant^{-1})			
U. brizantha cv. BRS Piatã	55.6 ± 4.3 bA	36.0 ± 1.3 bB	23.0 ± 5.9 bC
U. brizantha cv. Marandu	50.7 ± 1.6 bA	35.3 ± 2.0 bB	23.2 ± 1.7 bC
U. brizantha cv. Xaraés	42.4 ± 3.9 cA	32.5 ± 1.8 bB	22.6 ± 0.7 bC
U. ruziziensis cv. Comum	56.8 ± 1.0 bA	45.9 ± 3.5 aB	26.1 ± 1.3 aC
P. glaucum cv. ADR 300	33.3 ± 0.3 cA	25.3 ± 1.1 cB	17.7 ± 0.2 bB
P. maximum cv. Aruana	58.1 ± 4.3 bA	34.0 ± 7.4 bB	26.8 ± 0.2 aB
P. maximum cv. Mombaça	69.3 ± 7.1 aA	42.1 ± 1.2 aB	29.9 ± 0.7 aC
P. maximum cv. Tanzânia	56.3 ± 4.8 bA	43.2 ± 4.3 aB	33.7 ± 1.4 aB
P. atratum cv. Pojuca	34.4 ± 1.8 cA	18.0 ± 4.2 cB	17.3 ± 1.2 bB
Root volume (cm^3 plant^{-1})			
U. brizantha cv. BRS Piatã	88.9 ± 19.8 dA	64.4 ± 6.8 aB	38.9 ± 5.6 aC
U. brizantha cv. Marandu	106.7 ± 3.3 cA	61.8 ± 6.7 aB	37.8 ± 4.0 aB
U. brizantha cv. Xaraés	94.0 ± 6.8 cA	58.9 ± 9.5 aB	32.9 ± 0.4 aC
U. ruziziensis cv. Comum	154.4 ± 14.6 aA	85.0 ± 8.7 aB	33.2 ± 1.9 aC
P. glaucum cv. ADR 300	53.3 ± 5.7 dA	31.7 ± 1.0 bB	23.3 ± 1.9 aB
P. maximum cv. Aruana	113.0 ± 3.5 cA	85.0 ± 16.0 aB	38.9 ± 2.2 aC
P. maximum cv. Mombaça	162.8 ± 11.4 aA	71.1 ± 8.7 aB	52.2 ± 1.1 aB
P. maximum cv. Tanzânia	133.3 ± 19.0 bA	83.3 ± 3.8 aB	58.9 ± 6.8 aC
P. atratum cv. Pojuca	64.4 ± 2.2 dA	22.1 ± 9.0 bB	21.1 ± 4.0 aB

Means followed by distinct lowercase letters for the forage grass cultivars (in the column) or distinct uppercase letters for the soil water regimes (in the line) show significant differences (Scott Knott test, $p \leq 0.05$). Values represent the mean ± mean standard error.

Plants of *U. ruziziensis* vc. Comum has a higher tiller emission rate, while plants of *U. ruziziensis* cv. Comum, *P. maximum* cv. Aruana, *P. maximum* cv. Mombaça, and *P. maximum* cv. Tanzânia had a higher production of shoot dry matter (Table 4). Tillers are very important to understanding forage grass growth and regrowth. Tillers are new grass shoots made up of successive segments called phytomers. The tillering rate of grass species is controlled by the emergence rate of phytomers, genetic characteristics, and plant density in the field [5]. Thus, shoot dry matter production results from accumulated phytomers per stem and the stem density per area [11,16].

Table 4. Average value of tiller number and shoot dry matter of tropical forage grasses affected by cultivars and soil water regimes.

Causes of Variation	Number of Tillers (Units)	Shoot Dry Matter (g plant^{-1})
Forage grass cultivars		
U. brizantha cv. BRS Piatã	10.8 ± 0.8 d	20.8 ± 2.6 b
U. brizantha cv. Marandu	12.3 ± 0.9 d	23.2 ± 2.3 b
U. brizantha cv. Xaraés	11.1 ± 0.7 d	20.8 ± 2.2 b
U. ruziziensis cv. Comum	26.1 ± 3.0 a	28.2 ± 2.6 a
P. glaucum cv. ADR 300	4.2 ± 0.3 e	20.8 ± 2.0 b
P. maximum cv. Aruana	18.4 ± 1.2 b	30.6 ± 3.4 a
P. maximum cv. Mombaça	15.8 ± 1.0 c	30.5 ± 3.1 a
P. maximum cv. Tanzânia	15.6 ± 1.0 c	28.3 ± 2.2 a
P. atratum cv. Pojuca	19.8 ± 1.4 b	18.2 ± 2.6 b
Soil water regimes		
High	16.4 ± 1.2 a	33.1 ± 1.1 a
Middle	15.6 ± 1.6 a	23.6 ± 1.4 b
Low	12.6 ± 1.1 b	17.8 ± 0.8 c

Means followed by distinct lowercase letters for the forage grass cultivar and soil water regime (in the column) show significant differences (Scott Knott test, p).

The LSW regime inhibited the tiller emission of forage grasses (Table 4). Shoot dry matter production was drastically reduced when grasses were exposed to water stress conditions. The lower tillering rate and shoot dry matter accumulation of *U. brizantha* cv. BRS Piatã and P. glaucum cv. ADR 300 under water stress conditions was also reported by Zuffo et al. [25]. When water stress occurs at the initial stage of grass development, the reduction in stomatal conductance and photosynthesis rate results in lower plant tillering potential and lower shoot dry matter production [9], which results in significant forage production losses [8]. According to Fonseca and Martuscello [5], the main morphological characteristics that directly affect the forage production potential are the number of tillers, the number of leaves, and leaf size.

2.3. Interrelationship between Morphological Traits and Forage Grass Cultivars

Canonical correlation analysis was used to verify the contribution of each dependent variable measured in the tropical forage grasses as affected by soil water regimes (Figure 1). For scores to be represented in a two-dimensional graph, the percentage of retained variance must be higher than 80% [28]. In this study, variances accumulated in the two main canonical variables were 94.2%, 88.2%, and 82.9%, respectively, for each graph (Figure 1A–C), allowing an accurate interpretation.

Under the HSW regime, an angle (between vectors) less than 90° indicates a positive correlation between the dependent variable plant height (PH) with the *P. glaucum* cv. ADR 300 (T1); number of leaves (NL) with *P. atratum* cv. Pojuca (T8); and the number of tillers, leaf area, shoot dry matter, root dry matter, total dry matter, and root volume with *P. maximum* cv. Tanzania (T2), *U. ruziziensis* cv. Comum (T3), *P. maximum* cv. Aruana (T4), *P. maximum* cv. Mombaça (T5), *U. brizantha* cv. Marandu (T6), *U. brizantha* cv. BRS Piatã (T7), and *U. brizantha* cv. Xaraés (T9) (Figure 1A).

Under the MSW regime, there was a positive correlation between plant height and *P. glaucum* cv. ADR 300 (T1); the number of leaves and tillers with *U. ruziziensis* cv. Comum plants (T3) and *P. atratum* cv. Pojuca (T8); and leaf area, root volume and shoot, root and total dry matter with *P. maximum* cv. Tanzânia (T2), *P. maximum* cv. Aruana (T4), *P. maximum* cv. Mombaça (T5), *U. brizantha* cv. BRS Piatã (T7) and *U. brizantha* cv. Xaraés (T9) (Figure 1B).

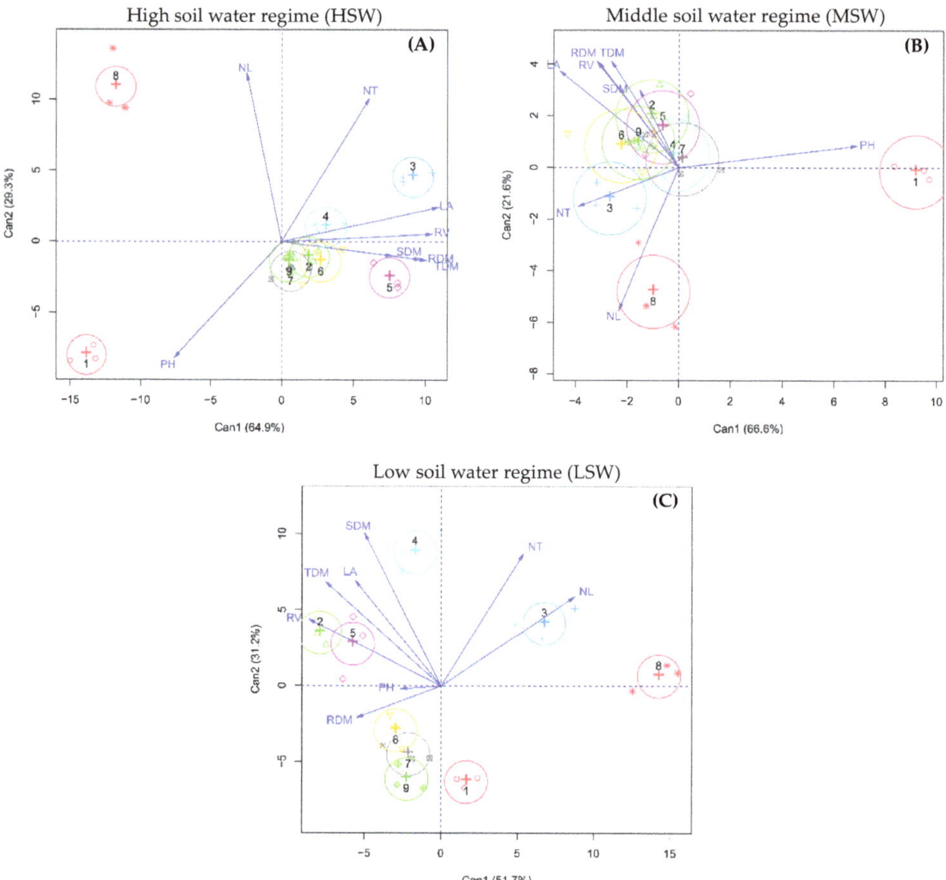

Figure 1. Canonical correlation analysis (CCA) between the morphological traits and forage grass cultivars when grown under well-irrigated control conditions (**A**) or exposed to moderate water stress (**B**) and severe water stress (**C**). The blue lines show the canonical correlation between the centroids of the first pair of canonical variates and the linear tendency line. Abbreviations: PH: plant height; NT: number of tillers; NL: number of leaves; LA: leaf area; SDM: shoot dry matter; RDM: root dry matter; TDM: total dry matter; RV: root volume. (T1) *P. glaucum* cv. ADR 300; (T2) *P. maximum* cv. Tanzânia; (T3) *U. ruziziensis* cv. Comum; (T4) *P. maximum* cv. Aruana; (T5) *P. maximum* cv. Mombaça; (T6) *U. brizantha* cv. Marandu; (T7) *U. brizantha* cv. BRS Piatã; (T8) *P. atratum* cv. Pojuca; and (T9) *U. brizantha* cv. Xaraés.

Under the LSW regime, there was a positive correlation between the number of leaves and tillers with *U. ruziziensis* cv. Comum (T3) and *P. atratum* cv. Pojuca (T8); plant height and root dry matter with *U. brizantha* cv. Marandu (T6), *U. brizantha* cv. BRS Piatã (T7) and *U. brizantha* cv. Xaraés (T9); and leaf area, root volume, shoot, and total dry matter with *P. maximum* cv. Tanzânia (T2), *P. maximum* cv. Aruana (T4) and *P. maximum* cv. Mombaça (T5) (Figure 1C).

The greater or lesser negative impact of soil water stress on the growth and development of forage grasses is determined by the genetic traits of the genotype's tolerance when exposed to water stress conditions. Each forage grass has distinct morphological characteristics that can be modified by the pasture's production environment and technical management. However, this modification of the morphological traits of grasses is limited by the phenotypic plasticity of the genotype [12]. Therefore, the cultivation environment of forage grasses results in distinct gradual, reversible changes in the morphogenic and structural characteristics of the plants [29].

The results shown in Figure 1 indicate that under non-stressful or stressful conditions, forage grasses of the species *P. maximum* have a greater capacity to produce leaf area, root volume, shoot, and total dry matter. Therefore, when the farmer aims at greater forage production, cultivars Aruana, Mombaça, and Tanzânia of *P. maximum* are excellent option for cattle feeding. Under the LSW regime, grasses of the species *U. brizantha* have a greater capacity to produce root dry matter; however, the highest root volume under LSW was observed for *P. maximum* cultivars. *U. ruziziensis* plants have a higher tillering potential under non-stressful and stressful conditions. Therefore, it can be seen that each forage grass cultivar has its intrinsic characteristics and directs the accumulation of photoassimilates to different drains, either for the growth of the stem, leaves, roots, or tillers.

The pattern of dry matter allocation among different plant organs can change throughout the plant development stages, especially when exposed to stressful environmental conditions. However, this pattern of photoassimilate allocation is essential to optimize crop growth and development under stressful conditions. This is because the pattern of photoassimilate allocation can affect plants' competitive and adaptive capacity and their responses to the stresses imposed by the cultivation environment [29]. The pattern of dry matter allocation in forage grasses is directly related to the optimization of capturing the scarcest resources from the cultivation environment. Under non-stressful conditions, grasses can allocate more photoassimilates to leaves to increase plants' light energy uptake and photosynthetic rate and increase forage production. On the other hand, grasses can allocate more photoassimilates to the roots under water stress conditions to improve water and nutrient uptake when soil water availability is low or limited [12].

2.4. Water Stress Tolerance Indices

The highest shoot biomass production under HSW regime (Y_P) was obtained for *P. maximum* cv. Aruana, *P. maximum* cv. Mombaça, and *P. maximum* cv. Tanzânia (Table 5). Under the MSW regime, the nine forage grass cultivars were grouped into the same group based on shoot biomass production (Y_S) and TOL, YSI, DI, YI, k_2STI, SSPI, and ATI indices. These results indicate that these tolerance indices did not effectively differentiate the water stress tolerance levels of forage grass cultivars exposed to moderate water stress conditions. Menezes et al. [18] also reported that the TOL and YSI indices did not differentiate water-stress-tolerant grain sorghum genotypes adequately. On the other hand, the MP, STI, GMP, and HM indices classified the forage grass cultivars into two tolerance groups, and the plants of *U. ruziziensis* cv. Comum, *P. maximum* cv. Aruana, *P. maximum* cv. Mombaça, and *P. maximum* cv. Tanzânia belonged to the group with the highest values of water stress tolerance indices (Table 5).

Under LSW regime, the TOL, YSI, and SSPI indices were not efficient in differentiating the water stress tolerance level of the nine forage grass cultivars (Table 5). On the other hand, the MP, GMP, and k_2STI indices separated the forage grass cultivars into three tolerance groups. In contrast, the DI, STI, YI, and HM indices divided the forage grass cultivars into four tolerance groups (Table 5). These results indicate that these tolerance indices were the most sensitive to differentiate forage grass cultivars regarding water stress tolerance levels. Naghavi et al. [20] showed that the STI, YI, SSPI, k_1STI, and k_2STI indices were the most suitable to identify water-stress-tolerant maize genotypes. Cabral et al. [19] reported that the MP, STI, GMP, and HM tolerance indices are the most appropriate to identify water-stress-tolerant soybean cultivars. Sánchez-Reinoso et al. [24] reported that only the SSPI index effectively identified water-stress-tolerant bean genotypes.

Table 5. Shoot biomass production and stress tolerance indices of the nine forage grass cultivars under middle or low soil water regimes.

Forage Grass Cultivar	Y_P [†]	Y_S [††]	Water Stress Tolerance Indices											
			TOL	MP	YSI	DI	STI	GMP	YI	k_1STI	k_2STI	SSPI	ATI	HM
Middle soil water regime (MSW)														
U. brizantha cv. BRS Piatã	32.3 b	21.0 a	11.3 a	26.6 b	0.65 a	0.58 a	0.62 b	26.0 b	0.89 a	0.95 b	0.89 a	23.9 a	210.4 a	25.4 b
U. brizantha cv. Marandu	31.3 b	22.8 a	8.5 a	27.0 b	0.73 a	0.71 a	0.65 b	26.7 b	0.96 a	0.89 b	0.96 a	17.9 a	161.1 a	26.3 b
U. brizantha cv. Xaraés	28.4 b	20.2 a	8.2 a	24.3 b	0.72 a	0.61 a	0.53 b	240 b	0.86 a	0.75 b	0.86 a	17.3 a	143.5 a	23.6 b
U. ruziziensis cv. Comum	34.0 b	28.1 a	5.9 a	31.1 a	0.83 a	0.99 a	0.88 a	30.9 a	1.19 a	1.06 b	1.19 a	12.4 a	131.5 a	30.8 a
P. glaucum cv. ADR 300	28.0 b	20.1 a	7.8 a	24.0 b	0.72 a	0.61 a	0.51 b	23.7 b	0.85 a	0.71 b	0.85 a	16.5 a	132.6 a	23.4 b
P. maximum cv. Aruana	40.5 a	25.7 a	14.8 a	33.1 a	0.63 a	0.82 a	0.95 a	31.3 a	1.09 a	1.50 a	1.09 a	31.4 a	270.3 a	29.8 a
P. maximum cv. Mombaça	41.0 a	29.0 a	12.0 a	35.0 a	0.72 a	0.89 a	1.08 a	34.3 a	1.23 a	1.57 a	1.23 a	25.4 a	305.4 a	33.7 a
P. maximum cv. Tanzânia	35.3 a	28.2 a	7.1 a	31.8 a	0.80 a	0.96 a	0.91 a	31.5 a	1.19 a	1.14 b	1.19 a	15.1 a	157.2 a	31.2 a
P. atratum cv. Pojuca	27.4 b	14.3 a	13.0 a	20.8 b	0.53 a	0.36 a	0.36 b	19.5 b	0.61 a	0.68 b	0.61 a	27.6 a	169.6 a	18.3 b
Mean	5.11	4.41	1.83	4.68	0.06	0.19	0.22	4.80	0.25	0.33	0.25	5.19	55.79	4.91
CV (%)	15.41	24.93	11.88	18.43	11.02	35.71	40.32	19.88	24.93	31.60	24.93	11.88	27.64	21.38
Low soil water regime (LSW)														
U. brizantha cv. BRS Piatã	32.3 b	14.8 d	17.5 a	23.6 c	0.46 a	0.39 d	0.44 d	21.9 c	0.84 d	0.95 b	0.84 c	49.4 a	203.9 b	20.3 d
U. brizantha cv. Marandu	31.3 b	15.6 d	15.6 a	23.4 c	0.50 a	0.44 d	0.45 d	22.1 c	0.88 d	0.89 b	0.88 c	44.2 a	184.6 b	20.8 d
U. brizantha cv. Xaraés	28.4 b	13.8 d	14.6 a	21.1 c	0.49 a	0.39 d	0.36 d	19.8 c	0.78 d	0.75 b	0.78 c	41.2 a	156.4 b	18.5 d
U. ruziziensis cv. Comum	34.0 b	19.0 c	15.0 a	26.5 b	0.56 a	0.61 c	0.59 c	25.4 c	1.08 c	1.06 b	1.08 c	42.3 a	204.6 b	24.4 c
P. glaucum cv. ADR 300	28.0 b	14.2 d	13.8 a	21.1 c	0.51 a	0.41 d	0.36 d	19.9 c	0.80 d	0.71 b	0.80 c	39.1 a	146.9 b	18.8 d
P. maximum cv. Aruana	40.5 a	25.6 a	14.9 a	33.1 a	0.63 a	0.91 a	0.95 a	32.2 a	1.45 a	1.50 a	1.45 a	42.3 a	257.0 a	31.4 a
P. maximum cv. Mombaça	41.0 a	21.6 b	19.4 a	31.3 a	0.53 a	0.65 c	0.82 b	29.7 a	1.22 b	1.57 a	1.22 b	54.7 a	315.7 a	28.3 b
P. maximum cv. Tanzânia	35.3 a	21.4 b	13.9 a	28.4 b	0.61 a	0.74 b	0.69 b	27.5 b	1.21 b	1.14 b	1.21 b	39.3 a	204.7 b	26.7 c
P. atratum cv. Pojuca	27.4 b	13.0 d	14.3 a	20.2 c	0.48 a	0.36 d	0.32 d	18.8 c	0.74 d	0.68 b	0.74 c	40.6 a	142.8 b	17.5 d
Mean	5.11	4.88	3.04	4.76	0.09	0.21	0.24	4.78	0.21	0.33	0.21	6.43	62.54	4.86
CV (%)	15.41	20.98	30.89	16.88	12.90	28.55	33.69	17.35	20.98	31.60	20.98	30.89	33.47	18.03

[†] Y_P represents the shoot biomass production (in grams per plant) of forage grasses grown under high soil water regime (non-stressful condition). [††] Y_S represents the shoot biomass production (in grams per plant) of forage grasses exposed to water stress conditions (middle or low soil water regime). Means followed by distinct lower case letters for the forage grass cultivar and stress tolerance indices (in the column) show significant differences (Scott Knott test, p). CV: Coefficient of variation. TOL: Tolerance, MP: Mean productivity, YSI: Yield stability index, DI: Drought resistance index, STI: GMP: Stress tolerance index, GMP: Geometric mean productivity, YI: Yield index, k_1STI: Modified stress tolerance (k1), k_2STI: Modified stress tolerance (k2), SSPI: Stress susceptibility percentage index, ATI: Abiotic tolerance index, HM: Harmonic mean.

2.5. Interrelationship between Biomass Production and Water Stress Tolerance Indices

A network diagram was constructed based on the shoot biomass production of forage grasses exposed to non-stressful and stressful conditions and on all stress tolerance indices and their respective correlations (Figure 2). The correlation network diagram shows the interactions between all the water stress tolerance indices with the shoot biomass production of the grasses. Positive and highly significant correlations were detected between water stress tolerance indices and shoot biomass production of forage plants grown under MSW (Figure 2A) or LSW (Figure 2B) regimes.

Under MSW regime, positive and significant correlations were detected between shoot biomass production under nonstressful conditions (Y_P) with all stress tolerance indices; ATI with TOL, SSPI, k_1STI, MP, GMP; k_1STI with all tolerance tolerances except YSI. Negative and significant correlations were detected between shoot biomass production under stressful conditions (Y_S) with SSPI, TOL, and ATI; between YSI and TOL, SSPI, and ATI; between TOL and MP, YI, STI, HM; between k_2STI and SSPI; and between ATI and k_2STI and YI (Figure 2A).

Under LSW regime, positive and significant correlations were detected between shoot biomass production under nonstressful (Y_P) and stressful conditions (Y_S) with all water stress tolerance indices. Negative and significant correlations were detected between the YSI with the SSPI and TOL indices (Figure 2B).

Discrimination of the water index tolerance level of the nine forage grass cultivars based on only one criterion or tolerance index can be contradictory (Table 5). Therefore, forage grasses should be differentiated and separated into different water stress tolerance levels based on all tolerance indices [20]. The ranking method has been used to classify crop genotypes into different water stress tolerance levels [21]. The ranking score of the nine forage grass cultivars in each of the 12 water stress tolerance indices under MSW and LSW regimes is shown in Table 6.

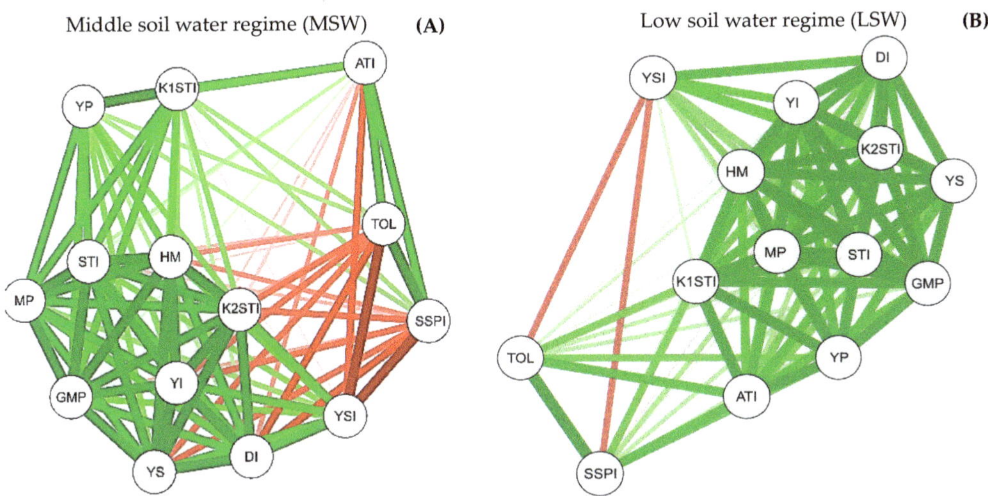

Figure 2. Correlation networks illustrate the most significant Pearson correlations between the shoot biomass production (Y_P and Y_S) and water stress tolerance indices of tropical forage grasses grown under MSW (**A**) and LSW (**B**) regime. Thicker and green lines represent the highest positive correlations (threshold set at 0.6 and p values < 0.05). Thicker and red lines represent the highest negative correlations (threshold set at 0.6 and p values < 0.05). TOL: Tolerance, MP: Mean productivity, YSI: Yield stability index, DI: Drought resistance index, STI: GMP: Stress tolerance index, GMP: Geometric mean productivity, YI: Yield index, k_1STI: Modified stress tolerance (k1), k_2STI: Modified stress tolerance (k2), SSPI: Stress susceptibility percentage index, ATI: Abiotic tolerance index, HM: Harmonic mean.

Table 6. Rank, mean rank ($\overline{R}$), and standard deviation of ranks (SD) of water stress tolerance indices of the nine forage grass cultivars under middle or low soil water regimes.

Forage Grass Cultivar	Y_P	Y_S	TOL	MP	YSI	DI	STI	GMP	YI	k_1STI	k_2STI	SSPI	ATI	HM	$\overline{R}$ (±SD)	Tolerance Level [†]
								Middle Soil Water Regime								
U. brizantha cv. BRS Piatã	5	6	6	6	7	8	6	6	6	5	6	4	3	6	5.7 (±0.8)	MS
U. brizantha cv. Marandu	6	5	5	5	3	5	5	5	5	6	5	5	5	5	5.0 (±0.3)	MT
U. brizantha cv. Xaraés	7	7	4	7	6	7	7	7	7	7	7	6	7	7	6.6 (±0.6)	MS
U. ruziziensis cv. Comum	4	3	1	4	1	1	4	4	3	4	3	9	9	3	3.8 (±1.6)	MT
P. glaucum cv. ADR 300	8	8	3	8	5	6	8	8	8	8	8	7	8	8	7.2 (±1.1)	S
P. maximum cv. Aruana	2	4	9	2	8	4	2	3	4	2	4	1	2	4	3.6 (±1.6)	MT
P. maximum cv. Mombaça	1	1	7	1	4	3	1	1	1	1	1	3	1	1	1.9 (±1.3)	T
P. maximum cv. Tanzânia	3	2	2	3	2	2	3	2	2	3	2	8	6	2	3.0 (±1.1)	T
P. atratum cv. Pojuca	9	9	8	9	9	9	9	9	9	9	9	2	4	9	8.1 (±1.5)	S
								Low Soil Water Regime								
U. brizantha cv. BRS Piatã	5	6	8	5	9	7	6	6	6	5	6	2	5	4	5.7 (±1.2)	MS
U. brizantha cv. Marandu	6	5	7	6	6	5	5	5	5	6	5	3	6	5	5.4 (±0.7)	MS
U. brizantha cv. Xaraés	7	8	4	7	7	8	8	8	8	7	8	6	7	2	6.8 (±1.2)	MS
U. ruziziensis cv. Comum	4	4	6	4	3	4	4	4	4	4	4	4	4	6	4.2 (±0.5)	MT
P. glaucum cv. ADR 300	8	7	1	8	5	6	7	7	7	8	7	9	8	3	6.5 (±1.6)	MS
P. maximum cv. Aruana	2	1	5	1	1	1	1	1	1	2	1	5	2	9	2.4 (±1.7)	T
P. maximum cv. Mombaça	1	2	9	2	4	3	2	2	2	1	2	1	1	8	2.9 (±1.8)	T
P. maximum cv. Tanzânia	3	3	2	3	2	2	3	3	3	3	3	8	3	7	3.4 (±1.2)	MT
P. atratum cv. Pojuca	9	9	3	9	8	9	9	9	9	9	9	7	9	1	7.8 (±1.8)	S

[†] T refers to a water-stress-tolerant cultivar with a mean rank ($\overline{R}$) score of 1 to 3.0; MT, moderately tolerant cultivar with an $\overline{R}$ score of 3.1 to 5.0; MS, moderately sensitive cultivar with an $\overline{R}$ score of 5.1 to 7.0; and S, water-stress-sensitive cultivar with an $\overline{R}$ score of 7.1 to 9. TOL: Tolerance, MP: Mean productivity, YSI: Yield stability index, DI: Drought resistance index, STI: GMP: Stress tolerance index, GMP: Geometric mean productivity, YI: Yield index, k_1STI: Modified stress tolerance (k1), k_2STI: Modified stress tolerance (k2), SSPI: Stress susceptibility percentage index, ATI: Abiotic tolerance index, HM: Harmonic mean.

2.6. Ranking

Considering all water stress tolerance indices, the cultivars *P. maximum* cv. Mombaça and *P. maximum* cv. Tanzânia had the highest stress tolerance indices (Table 5) and the best-ranking scores under the MSW regime (Table 6). Therefore, these two forage grass cultivars were identified as tolerant to moderate water stress. The cultivars *U. brizantha* cv. Marandu, U. ruziziensis cv. Comum, and *P. maximum* cv. Aruana were identified as moderately tolerant to moderate water stress, while *P. glaucum* cv. ADR 300 and *P. atratum* cv. Pojuca was the most sensitive cultivar to moderate water stress (Table 6).

Under LSW regime, the cultivars *P. maximum* cv. Aruana and *P. maximum* cv. Mombaça were classified as water-stress-tolerant, whereas the cultivars *U. ruziziensis* cv. Comum and *P. maximum* cv. Tanzânia were classified as moderately tolerant to severe water stress (Table 6). These results suggest that the cultivation of *P. maximum* cultivars 'Aruana', 'Mombaça', and 'Tanzânia' is an excellent option for feeding cattle during the dry season in Brazil since these forage grass cultivars were identified as tolerant to water stress and have a high capacity for forage production in low soil water availability conditions. Indeed, using tolerant cultivars in areas subject to water stress is the best solution to face the predicted climate changes in the coming years and decades.

Fonseca and Martuscello [5] reported that forage grasses belonging to the species *P. maximum* have a high potential for forage yield response to stressful environmental conditions, confirming the results of this study. Some studies with wheat [30], canola [31], and sugarcane [10] suggest that the degree of water stress tolerance is related to the ability of plants to uptake and accumulate mineral nutrients when exposed to low soil water availability conditions. In general, water stress tolerance is the characteristic of a plant species or cultivar to adapt to growing environments with low water availability. Under these stressful conditions, the extension of the root system has been a fundamental morphological characteristic to improve the ability of plants to extract water and nutrients from the soil [12,16].

Plants of *P. atratum* cv. Pojuca were identified as sensitive to water stress under MSW and LSW regimes (Table 6). Therefore, the cultivation of this forage cultivar during the dry season in tropical regions of Brazil should not be recommended due to its low capacity to produce shoot biomass in soil water stress conditions.

3. Materials and Methods

3.1. Plant Growth Conditions

The experiment was conducted at Cassilândia, Mato Grosso do Sul, Brazil ($19°05'29''$ S and $51°48'50''$ W, and altitude of 540 m) from May to August 2019, in 12 L plastic pots in a smart greenhouse with an automatic climate control system. The temperature and relative humidity inside the greenhouse were maintained at 26 °C ($\pm$2 °C) and 70% ($\pm$4%), respectively, during the trial period. The photosynthetic photon flux density (PPFD) inside the greenhouse, measured daily at midday ($\pm$12:00 h) with an Apogee MQ-500 quantum sensor, was 982 μmol m^{-2} s^{-1} ($\pm$238 μmol m^{-2} s^{-1}).

The soil used in the experiment was a typic Quartzipsamment (or Neossolo Quartzarênico Órtico latossólico) collected from the 0.0–0.30 m layer in a Cerrado native pasture area with 180 g kg^{-1} of clay, 70 g kg^{-1} of silt, and 750 g kg^{-1} of sand. The occurrence of Quartzipsamments in the eastern region of Mato Grosso do Sul state is common. This soil class has no restrictions for the use and management of agricultural soil [32].

The chemical analysis showed that the soil used in this research has a low acidity level and high fertility level, which allowed the adequate availability of nutrients for forage plants. The soil chemical analysis reported the following results: pH in CaCl$_2$ = 5.6; 20 g kg^{-1} of organic matter; 12 mg dm^{-3} of P (Mehlich^{-1}); 3.70 cmol$_c$ dm^{-3} of Ca^{2+}; 1.60 cmol$_c$ dm^{-3} of Mg^{2+}; 0.22 cmol$_c$ dm^{-3} of K$^+$; 5.80 cmol$_c$ dm^{-3} of CEC; 68% of base saturation; 1.8 mg dm^{-3} of Cu^{2+}; 1.5 mg dm^{-3} of Zn^{2+}; 73 mg dm^{-3} of Fe^{2+}; and 13 mg dm^{-3} of Mn^{2+}. All the soil chemical properties were analyzed according to Brazilian Agricultural Research Corporation standard methods described by Teixeira et al. [33].

The field capacity was measured under free-draining conditions using a water content decrease rate of 0.1 g kg^{-1} day^{-1}, as proposed by Casaroli and Lier [34], and the soil volumetric water moisture content (VWC) at pot field capacity (FC) was 218 g kg^{-1}.

The soil was placed in 12 L plastic pots and fertilized with 50 mg kg^{-1} of N (urea), 300 mg kg^{-1} of P (simple superphosphate), 150 mg kg^{-1} of K (potassium chloride), 30 mg dm^{-3} of S (gypsum), 2 mg kg^{-1} of Cu (copper sulfate), and 2 mg kg^{-1} of Zn (zinc sulfate). Fertilizers were incorporated into the entire soil volume of the pots three days before sowing the forage plants. Each plastic pot was filled with 14 kg ($\pm$10 dm^3) of air-dried soil and sieved with a 5.0 mm mesh.

3.2. Experimental Design and Treatments

The experiment was arranged in a completely randomized block design in a 3 × 9 factorial arrangement with four replicates. Treatments consisted of three soil water regimes (high soil water regime—HSW (non-stressful condition), middle soil water regime—MSW (moderate water stress), and low soil water regime—LSW (severe water stress)) and nine cultivars of tropical forage grasses (*Urochloa brizantha* cv. BRS Piatã, *U. brizantha* cv. Marandu, *U. brizantha* cv. Xaraés, *U. ruziziensis* cv. Comum, *Pennisetum glaucum* cv. ADR 300, *Panicum maximum* cv. Aruana, *P. maximum* cv. Mombaça, *P. maximum* cv. Tanzânia, *Paspalum atratum* cv. Pojuca). Some of the morphological and agronomic characteristics of the tropical forage grasses used in this study are shown in Table 7.

Table 7. Some characteristics of the nine tropical forage grass cultivars used in this study.

Forage Grass Cultivar	Common Name	Cultivar	Growth Habit	Soil Fertility Requirement	Forage Yield	Drought Tolerance
Urochloa brizantha	Palisade grass	BRS Piatã	Semierect	Medium	High	High
Urochloa brizantha	Palisade grass	Marandu	Erect	Medium	Medium	Medium
Urochloa brizantha	Palisade grass	Xaraés	Semierect	Medium	High	Medium
Panicum maximum	Guinea grass	Aruana	Erect	Medium/High	High	Medium/Low
Panicum maximum	Guinea grass	Mombaça	Erect	Medium/High	High	Medium/Low
Panicum maximum	Guinea grass	Tanzânia	Erect	Medium/High	High	Medium/Low
Pennisetum glaucum	Pearl millet	ADR 300	Erect	Medium	High	Medium/High
Urochloa ruziziensis	Ruzigrass	Comum	Semierect	Medium/High	High	Low
Paspalum atratum	Atratum	Pojuca	Erect	Low	High	Low

Source: Fonseca and Martusello [5].

Water stress treatments were applied for 25 days during the grass tillering and stalk elongation phases. Moderate and severe water stresses were achieved by simply changing the percentage volume of soil field capacity moisture in the pots, as proposed by Imakumbili [35]. This methodology often achieves severe water stress by maintaining the soil VWC in pots between 20% and 30% FC. An FC of 20% is acceptable for fine-textured soils, such as clayey soils, whereas an FC of 30% is suitable for coarse-textured soils, such as sandy soils. Moderate water stress is often achieved by maintaining soil VWC in pots at 60% FC in both coarse- and fine-textured soils, and 100% of FC will keep plants in pots under non-stressful conditions. The soil used in this study was a medium texture soil (180 g kg^{-1} of clay); therefore, the soil VWC in the pots was maintained at 60% and 25% of FC, respectively, for plants exposed to moderate water stress (middle soil water regime) or severe water stress (low soil water regime).

3.3. Plant Material and Irrigation

Seeds of nine tropical forage cultivars, three commercial cultivars of *Urochloa brizantha* (Hochst. Ex A. Rich.) R.D. Webster ('BRS Piatã', 'Marandu', and 'Xaraés'), three commercial cultivars of *Panicum maximum* Jacq. ('Aruana', 'Mombaça', and 'Tanzânia'), one commercial cultivar of *Pennisetum glaucum* (L.) R. Br. ('ADR 300'), one commercial cultivar of *Urochloa ruziziensis* (R. Germ. & C.M. Evrard) Crins ('Comum'), and a commercial cultivar of *Paspalum atratum* Swallen ('Pojuca') were sown on 8 May 2019 in 12 L pots. Ten seeds were sown at 2.0 cm depth, and five days after emergence, seedlings were thinned to two plants per pot. All plants were fertilized 30 days after emergence with 80 mg kg^{-1} of N via urea solution.

Until 40 days after sowing, the soil VWC was maintained at FC (218 g kg^{-1}) with daily irrigation. Subsequently, the experiment was divided into three groups of soil water regimes [HSW (100% of FC), MSW (60% of FC), and LSW (25% of FC)]. When initiating water stress treatments, severely and moderately stressed pots were not irrigated for 9 and 6 days until the soil VWC in the pots had dropped to a point slightly below 25% and 60% of FC, respectively. Posteriorly, the soil VWC was maintained at these water stress levels for 25 days. Soil water availability in pots was controlled daily at 9:00 a.m. and 3:00 pm using the gravimetric method [35], and the soil VWC was adjusted by adding water after weighing the pot.

3.4. Measurement of Morphological Traits

After the 25th day of exposure to water stress, the forage plants were harvested, and the plant height, number of tillers, number of leaves, leaf area, root volume, and dry matter of the plant parts were measured. The roots were put in a 1.0 mm mesh sieve and washed under running tap water to remove the adhered soil. The plants were separated into leaves, stems, and roots, oven-dried at 65 °C for three days, and then weighed. The total dry matter was obtained from all seedling parts (leaves, stems, and roots).

The plant height was determined from the soil surface until the insertion of the +1 leaf using a tape measure. The number of tillers or leaves was obtained from their count. Green leaves were detached from plants, and the leaf area was determined using Equation (1), as proposed by Benincasa [36]. Fifteen leaf discs (15.0 cm^2) detached from the basal, median, and apical leaves constituted the collected sample. Root volume was determined by water displacement method using a 1000 mL graduated cylinder.

$$LA = [(LAs \times LTDM)/DMs], \tag{1}$$

where LAs is the leaf area of the collected sample, LTDM is the leaf total dry matter, and DMs is the dry matter of the collected sample.

3.5. Calculation of Water Stress Tolerance Indices

In this study, 12 water stress tolerance indices proposed by several researchers [37–44] were used to evaluate the forage production response (i.e., shoot biomass production) of the nine forage cultivars grown under a high soil water regime (non-stressful conditions) and under middle and low soil water regimes (i.e., moderate and severe water stress). Forage biomass production data recorded for each forage cultivar in each soil water regime were used to calculate water stress tolerance indices. The 12 water stress tolerance indices used in this study are shown in Table 8.

Table 8. Stress tolerance indices used to assess water stress tolerance of nine tropical forage grass cultivars grown under different soil water regimes.

Water Stress Tolerance Index	Equation [†]	Reference
1. Tolerance	$TOL = Y_P - Y_S$	[37]
2. Mean productivity	$MP = (Y_S + Y_P)/2$	[37]
3. Yield stability index	$YSI = Y_S/Y_P$	[38]
4. Drought resistance index	$DI = [Y_S \times (Y_S/Y_P)]/\bar{Y}_S$	[39]
5. Stress tolerance index	$STI = (Y_S \times Y_P)/(\bar{Y}_P)2$	[40]
6. Geometric mean productivity	$GMP = \sqrt{(Y_S \times Y_P)}$	[40]
7. Yield index	$YI = Y_S/\bar{Y}_S$	[41]
8. Modified stress tolerance (k_1)	$k_1STI = Y_P{}^2/\bar{Y}_P{}^2$	[42]
9. Modified stress tolerance (k_2)	$k_2STI = Y_S{}^2/\bar{Y}_S{}^2$	[42]
10. Stress susceptibility percentage index	$SSPI = [(Y_P - Y_S)/2 \times \bar{Y}_P] \times 100$	[43]
11. Abiotic tolerance index	$ATI = [(Y_P-Y_S)/(\bar{Y}_P/\bar{Y}_S)] \times \sqrt{(Y_P \times YS)}$	[43]
12. Harmonic mean	$HM = [2 \times (Y_S \times Y_P)]/(Y_S + Y_P)$	[44]

[†] In the equations above, Y_P and Y_S represent the forage production of grasses grown under high soil water regime (non-stressful condition) and under middle or low soil water regimes (moderate or severe water stress) for each forage grass cultivar, respectively, whereas $\bar{Y}_P$ and $\bar{Y}_S$ represent the average forage production of all grass cultivars under high soil water regime and under middle or low soil water regimes, respectively.

3.6. Statistical Analysis

The data were previously submitted to statistical hypothesis tests to verify the homogeneity of variances (Levene test; $p > 0.05$) and normality of residues (Shapiro–Wilk test; $p > 0.05$). Then, the data were subjected to analysis of variance (ANOVA), and when significant, the means were compared by the Scott–Knott test at the 0.05 confidence level. This analysis was performed using Sisvar® version 5.6 software for Windows (Statistical Analysis Software, UFLA, Lavras, MG, Brazil).

Statistical correlations based on Pearson's correlation networks (threshold set at 0.60, $p < 0.05$) were performed between the morphological traits of tropical forage grasses. A correlation network was used to graphically illustrate Pearson's correlation analyses, in which the proximity between the nodes is proportional to the absolute correlation values between the morphological traits. The bands' relative thickness and color density indicate the strength of Pearson's correlation coefficients, and the color of each band indicates a positive or negative correlation (red for negative and green for positive).

Canonical correlation analysis (CCA) was used to study the interrelationships between sets (vectors) of independent (forage grass cultivars) and dependent (morphological traits) variables for each soil water regime (high, middle, and low). These analyses were performed using Rbio software version 140 for Windows (Rbio Software, UFV, Viçosa, MG, Brazil).

The identification of water-stress-tolerant forage grass cultivars was performed based on all stress tolerance indices using the ranking method proposed by Farshadfar et al. [21] and improved by Zuffo et al. [45]. In this method, the forage grass cultivars with the highest values for the tolerance indices Y_P, Y_S, MP, YSI, DI, STI, GMP, YI, k_1STI, k_2STI, SSPI, ATI, and HM received a ranking score of 1. Similarly, the grass cultivars with the lowest values for the TOL tolerance index were assigned a ranking score equal to 1.

The mean ranking score ($\overline{R}$) and the rank standard deviation (RSD) were calculated for all water stress tolerance indices of the nine forage grass cultivars under middle or low soil water regime. The discrimination of forage grass cultivars regarding their tolerance degree to water stress was performed based on the mean ranking score of each grass cultivar, considering the values of the quartiles that divide the nine possible ranking positions (i.e., nine forage grass cultivars) into four equal parts as idealized by Zuffo et al. [45]. A cultivar with a mean rank ($\overline{R}$) lower than the value of the first quartile (<3.0 points) is classified as tolerant (T); a cultivar with an $\overline{R}$ between the first and second quartiles (3.1 to 5.0 points) is classified as moderately tolerant (MT); a cultivar with an $\overline{R}$ between the value of the second and third quartiles (5.1 to 7.0 points) is classified as moderately sensitive (MS); and the group of water-stress-sensitive (S) cultivars is represented by grass cultivar with an $\overline{R}$ higher than the value of the third quartile ($\geq$7.1 points).

4. Conclusions

Soil water stress decreased leaf area, plant height, tillering capacity, root volume, and shoot and root dry matter production in most cultivars, with varying degrees of reduction among tropical forage grasses. *Panicum maximum* plants (cv. Mombaça and cv. Tanzânia) grown under controlled greenhouse conditions were identified as tolerant to moderate water stress, whereas the cultivars Aruana and Mombaça of *P. maximum* were identified as tolerant to severe water stress. *Panicum maximum* cv. Mombaça has greater adaptability and stability of shoot biomass production when grown under greenhouse conditions and subjected to middle and low soil water regimes; therefore, this forage grass should be tested under field conditions to confirm its forage production potential for cultivation in tropical regions with the occurrence of water stress. The MP, DI, STI, GMP, YI, k_2STI, and HM tolerance indices were the most suitable for identifying forage grass cultivars with greater water stress tolerance and a high potential for shoot biomass production under low soil water regime.

Author Contributions: Conceptualization, A.M.Z. and F.S.; methodology, A.M.Z., F.S. and R.F.R.; software, A.M.Z.; validation, A.M.Z., R.M., L.M.B. and R.F.R.; formal analysis, A.M.Z., R.M. and L.M.B.; investigation, A.M.Z., F.S., J.G.A., R.F.R. and A.S.d.S.; resources, A.M.Z.; data curation, A.M.Z., F.S. and H.H.S.G.; writing—original draft preparation, A.M.Z., R.M., L.M.B., F.S., R.F.R., H.H.S.G., P.A.C. and S.M.A.; writing—review and editing, A.M.Z., R.M., L.M.B., F.S., J.G.A., R.F.R. and A.S.d.S.; visualization, A.M.Z., J.G.A., R.F.R., A.S.d.S., H.H.S.G., P.A.C. and S.M.A.; supervision, A.M.Z.;

project administration, A.M.Z. and R.F.R.; funding acquisition, A.M.Z., H.H.S.G., P.A.C. and S.M.A. All authors have read and agreed to the published version of the manuscript.

Funding: This research received no external funding.

Institutional Review Board Statement: Not applicable.

Informed Consent Statement: Not applicable.

Data Availability Statement: All available data can be obtained by contacting the corresponding author.

Conflicts of Interest: The authors declare no conflict of interest.

Abbreviations

FC: field capacity; HSW: high soil water; ICLS: integrated crop–livestock system; LA: leaf area; LSW: low soil water; MSW: middle soil water; NT: number of tillers; NL: number of leaves; PH: plant height; RDM: root dry matter; RV: root volume; SDM: shoot dry matter; TDM: total dry matter.

References

1. USDA—United States Department of Agriculture. Foreign Agricultural Service. Available online: https://www.usda.gov/ (accessed on 5 January 2022).
2. IBGE—Instituto Brasileiro de Geografia e Estatística. Available online: https://www.ibge.gov.br/ (accessed on 5 January 2022).
3. Feltran-Barbieri, R.; Féres, J.G. Degraded pastures in Brazil: Improving livestock production and forest restoration. *Royal Soc. Open Sci.* **2021**, *8*, 201854. [CrossRef] [PubMed]
4. Steiner, F.; Zuffo, A.M.; Silva, K.C.; Lima, I.M.O.; Ardon, H.J.V. Cotton response to nitrogen fertilization in the integrated crop-livestock system. *Sci. Agrar. Paranaensis* **2020**, *19*, 211–220. [CrossRef]
5. Fonseca, D.M.; Martusello, J.A. *Plantas Forrageiras*, 2nd ed.; Editora UFV: Viçosa, Brazil, 2022; 591p.
6. Petter, F.A.; Pacheco, L.P.; Zuffo, A.M.; Piauilino, A.C.; Xavier, Z.F.; Santos, J.M.; Miranda, J.M.S. Performance of cover crops submitted to water deficit. *Semin-Cienc. Agrar.* **2013**, *34*, 3307–3319. [CrossRef]
7. Pacheco, L.P.; Petter, F.A.; Lima, M.P.D.; Alixandre, T.F.; Silva, L.M.A.; Silva, R.R.; Ribeiro, W.R.M. Development of cover crops under different water levels in the soil. *Afr. J. Agric. Res.* **2013**, *8*, 2216–2223. [CrossRef]
8. Staniak, M.; Kocoń, A. Forage grasses under drought stress in conditions of Poland. *Acta Physiol. Plant.* **2015**, *37*, 116. [CrossRef]
9. Fariaszewska, A.; Aper, J.; Huylenbroeck, J.; Swaef, T.; Baert, J.; Pecio, L. Physiological and biochemical responses of forage grass varieties to mild drought stress under field conditions. *Int. J. Plant Prod.* **2020**, *14*, 335–353. [CrossRef]
10. Silva, T.R.; Cazetta, J.O.; Carlin, S.D.; Telles, B.R. Drought-induced alterations in the uptake of nitrogen, phosphorus and potassium, and the relation with drought tolerance in sugar cane. *Cienc. Agrotec.* **2017**, *41*, 117–127. [CrossRef]
11. Macedo, L.C.P.; Dornelles, S.H.B.; Peripolli, M.; Trivisiol, V.S.; Conceição, D.Q.; Pivetta, M.; Essi, L. Phenology and dry mass production of *Urochloa plantaginea* and *Urochloa platyphylla* submitted to different water quantities in the soil. *Acta Sci. Biol. Sci.* **2019**, *41*, 46127. [CrossRef]
12. Mastalerczuk, G.; Borawska-Jarmułowicz, B. Physiological and morphometric response of forage grass species and their biomass distribution depending on the term and frequency of water deficiency. *Agronomy* **2021**, *11*, 2471. [CrossRef]
13. Abdelaal, K.; Attia, K.A.; Niedbała, G.; Wojciechowski, T.; Hafez, Y.; Alamery, S.; Alateeq, T.K.; Arafa, S.A. Mitigation of drought damages by exogenous chitosan and yeast extract with modulating the photosynthetic pigments, antioxidant defense system and improving the productivity of garlic plants. *Horticulturae* **2021**, *7*, 510. [CrossRef]
14. Basu, S.; Ramegowda, V.; Kumar, A.; Pereira, A. Plant adaptation to drought stress. *F1000Res* **2016**, *30*, 1554. [CrossRef] [PubMed]
15. Zhang, H.; Zhu, J.; Gong, Z.; Zhu, J.K. Abiotic stress responses in plants. *Nat. Rev Genet* **2022**, *23*, 104–119. [CrossRef] [PubMed]
16. Purbajanti, E.D.; Anwar, S.; Wydiati, N.; Kusmiyati, F. Drought stress effect on morphology characters, water use efficiency, growth and yield of guinea and napier grasses. *Int. Res. J. Plant Sci.* **2012**, *3*, 47–53.
17. Mohammadi, R.; Armion, M.; Kahrizi, D.; Amri, A. Efficiency of screening techniques for evaluating durum wheat genotypes under mild drought conditions. *Int. J. Plant Prod.* **2010**, *4*, 11–24. [CrossRef]
18. Menezes, C.B.; Ticona-Benavente, C.A.; Tardin, F.D.; Cardoso, M.J.; Bastos, E.A.; Nogueira, D.W.; Portugal, A.F.; Santos, C.V.; Schaffert, R.E. Selection indices to identify drought-tolerant grain sorghum cultivars. *Genet. Mol. Res.* **2014**, *13*, 9817–9827. [CrossRef]
19. Cabral, R.C.; Maekawa, S.C.E.; Zuffo, A.M.; Steiner, F. Selection indices to identify drought-tolerant soybean cultivars. *Res., Soc. Dev.* **2020**, *9*, 1–25. [CrossRef]
20. Naghavi, M.R.; Pour-Aboughadareh, A.; Khalili, M. Evaluation of drought tolerance indices for screening some of corn (*Zea mays* L.) cultivars under environmental conditions. *Not. Sci. Biol.* **2013**, *5*, 388–393. [CrossRef]
21. Farshadfar, E.; PoursiahbidI, M.M.; Abooghadareh, A.R.P. Repeatability of drought tolerance indices in bread wheat genotypes. *Intl. J. Agri. Crop Sci.* **2012**, *4*, 891–903.

22. El-Rawy, M.A.; Hassan, M.I. Effectiveness of drought tolerance indices to identify tolerant genotypes in bread wheat (*Triticum aestivum* L.). *J. Crop Sci. Biotechnol.* **2014**, *17*, 255–266. [CrossRef]
23. Gholinezhad, E.; Darvishzadeh, R.; Bernousi, I. Evaluation of drought tolerance indices for selection of confectionery sunflower (*Helianthus anuus* L.) landraces under various environmental conditions. *Not. Bot. Horti Agrobot. Cluj-Napoca* **2014**, *42*, 187–201. [CrossRef]
24. Sánchez-Reinoso, A.D.; Ligarreto-Moreno, G.A.; Restrepo-Díaz, H. Evaluation of drought indices to identify tolerant genotypes in common bean bush (*Phaseolus vulgaris* L.). *J. Integr. Agric.* **2020**, *19*, 99–107. [CrossRef]
25. Zuffo, A.M.; Ratke, R.F.; Steiner, F.; Oliveira, A.M.; Aguilera, J.G.; Lima, R.E. Silicon mitigates the effects of moderate drought stress in cover crops. *J. Agron. Crop Sci.* **2022**, *208*, 1–11. [CrossRef]
26. Rodrigues, C.S.; Nascimento-Júnior, D.; Silva, S.C.; Silveira, M.C.T.; Sousa, B.M.L.; Detmann, E. Characterization of tropical forage grass development pattern through the morphogenetic and structural characteristics. *Rev. Bras. Zootec.* **2011**, *40*, 527–534. [CrossRef]
27. Farooq, M.; Wahid, A.; Kobayashi, N.; Fujita, D.; Basra, S.M.A. Plant drought stress: Effects, mechanisms and management. *Agron. Sustain. Dev.* **2009**, *29*, 185–212. [CrossRef]
28. Mingoti, A.S. *Análise de Dados Através de Métodos de Estatística Multivariada: Uma Abordagem Aplicada*, 1st ed.; Editora UFMG: Belo Horizonte, Brazil, 2005; pp. 127–155.
29. Silva, S.C.; Sbrissia, A.F.; Pereira, L.E.T. Ecophysiology of C4 forage grasses: Understanding plant growth for optimizing their use and management. *Agriculture* **2015**, *5*, 598–625. [CrossRef]
30. Sun, M.; Gao, Z.Q.; Yang, Z.P.; He, L.H. Absorption and accumulation characteristics of nitrogen in different wheat cultivars under irrigated and dryland conditions. *Aust. J. Crop Sci.* **2012**, *6*, 613–617.
31. Ashraf, M.; Shahbaz, M.; Ali, Q. Drought-induced modulation in growth and mineral nutrients in canola (*Brassica napus* L.). *Pak. J. Bot.* **2013**, *45*, 93–98.
32. Santos, H.G.; Jacomine, P.K.T.; Anjos, L.H.C.; Oliveira, V.A.; Lumbreras, J.F.; Coelho, M.G.; Cunha, T. *Sistema Brasileiro de Classificação de Solos*, 5th ed.; Embrapa Solos: Rio de Janeiro, Brazil, 2018; 595p.
33. Teixeira, P.C.; Donagemma, G.K.; Fontana, A.; Teixeira, W.G. *Manual de Métodos de Análises de Solos*, 3rd ed.; Embrapa Solos: Rio de Janeiro, Brazil, 2017; 279p.
34. Casaroli, D.; Lier, Q.J. Critérios para determinação da capacidade de vaso. *Rev. Bras. Ci. Solo* **2008**, *32*, 59–66. [CrossRef]
35. Imakumbili, M.L.E. *Making Water Stress Treatments in Pot Experiments: An Illustrated Step-by-Step Guide*; Sokoine University of Agriculture: Morogoro, Tanzania, 2019; pp. 1–17. [CrossRef]
36. Benincasa, M.P.M. *Análise de Crescimento de Plantas: Noções Básicas*, 1st ed.; Editora FUNEP: Jaboticabal, Brazil, 2003; pp. 1–42.
37. Rosielle, A.A.; Hamblin, J. Theoretical aspects of selection for yield in stress and non-stress environments. *Crop Sci.* **1981**, *21*, 943–946. [CrossRef]
38. Bouslama, M.; Schapaugh, W.T. Stress tolerance in soybean. Part 1: Evaluation of three screening techniques for heat and drought tolerance. *Crop Sci.* **1984**, *24*, 933–937. [CrossRef]
39. Blum, A. *Plant Breeding for Stress Environments*; CRC Press: Boca Raton, FL, USA, 1988; pp. 1–12.
40. Fernández, G.C.J. Effective selection criteria for assessing plant stress tolerance. In Proceedings of the International Symposium on Adaptation of Vegetables and other Food Crops in Temperature and Water Stress, Shanhua, Taiwan, 13–18 August 1992; pp. 257–270.
41. Gavuzzi, P.; Rizza, F.; Palumbo, M.; Campaline, R.G.; Ricciardi, G.L.; Borghi, B. Evaluation of field and laboratory predictors of drought and heat tolerance in winter cereals. *Can. J. Plant Sci.* **1997**, *77*, 523–531. [CrossRef]
42. Farshadfar, E.; Sutka, J. Multivariate analysis of drought tolerance in wheat substitution lines. *Cereal Res. Commun.* **2003**, *31*, 33–39. [CrossRef]
43. Moosavi, S.S.; Samadi, B.Y.; Naghavi, M.R.; Zali, A.A.; Dashti, H.; Pourshahbazi, A. Introduction of new indices to identify relative drought tolerance and resistance in wheat genotypes. *Desert* **2008**, *12*, 165–178.
44. Jafari, A.; Paknejad, F.; Jami Al-ahmadi, M. Evaluation of selection indices for drought tolerance of corn (*Zea mays* L.) hybrids. *Int. J. Plant Prod.* **2009**, *3*, 33–38. [CrossRef]
45. Zuffo, A.M.; Steiner, F.; Sousa, T.D.O.; Aguilera, J.G.; Teodoro, P.E.; Alcântara-Neto, F.; Ratke, R.F. How does water and salt stress affect the germination and initial growth of Brazilian soya bean cultivars? *J. Agron. Crop Sci.* **2020**, *206*, 837–850. [CrossRef]

Article

Agents Affecting the Plant Functional Traits in National Soil and Water Conservation Demonstration Park (China)

Gaohui Duan [1], Zhongming Wen [1,2,*], Wei Xue [3], Yuankun Bu [4], Jinxin Lu [1], Bojin Wen [2], Boheng Wang [5] and Sihui Chen [1]

1 College of Grassland Agriculture, Northwest A&F University, Yangling 712100, China
2 Institute of Soil and Water Conservation, Chinese Academy of Sciences & Ministry of Water Resources, Yangling 712100, China
3 State Key Laboratory of Soil Erosion and Dryland Farming on the Loess Plateau, Northwest A&F University, Yangling 712100, China
4 College of Forestry, Northwest A&F University, Yangling 712100, China
5 East China Survey and Planning Institute of National Forest and Grassland Administration, Hangzhou 310019, China
* Correspondence: zmwen@ms.iswc.ac.cn

Abstract: Plant functional traits (PFTs) can reflect the response of plants to environment, objectively expressing the adaptability of plants to the external environment. In previous studies, various relationships between various abiotic factors and PFTs have been reported. However, how these factors work together to influence PFTs is not clear. This study attempted to quantify the effects of topographic conditions, soil factors and vegetation structure on PFTs. Four categories of variables were represented using 29 variables collected from 171 herb plots of 57 sites (from different topographic and various herb types) in Xindian SWDP. The partial least squares structural equation modeling showed that the topographic conditions and soil properties also have a direct effect on plant functional traits. Among the topographic conditions, slope (SLO) has the biggest weight of 0.629, indicating that SLO contributed the most to plant functional traits and vegetation structure. Among soil properties, maximum water capacity (MWC) contributes the most and is followed by soil water content (SWC), weighted at 0.588 and 0.416, respectively. In a word, the research provides new points into the quantification of the correlation between different drivers that may be important for understanding the mechanisms of resource utilization, competition and adaptation to the environment during plant recovery.

Keywords: plant functional traits; plants diversity; soil properties; random forest algorithm; PLS-SEM

Citation: Duan, G.; Wen, Z.; Xue, W.; Bu, Y.; Lu, J.; Wen, B.; Wang, B.; Chen, S. Agents Affecting the Plant Functional Traits in National Soil and Water Conservation Demonstration Park (China). *Plants* 2022, 11, 2891. https://doi.org/10.3390/plants11212891

Academic Editor: Martina Pollastrini

Received: 19 September 2022
Accepted: 26 October 2022
Published: 28 October 2022

Publisher's Note: MDPI stays neutral with regard to jurisdictional claims in published maps and institutional affiliations.

1. Introduction

The Soil and Water Conservation Demonstration Park (SWDP) is a major innovation of soil and water conservation in China. The ecosystem functions of SWDP include soil and water loss prevention, climate improvement, resource protection, etc., however, with the progress of the times, the goal of SWDP is not only limited to the improvement of park ecology through planting vegetation and construction projects but also plant restoration should become the focus of research. In order to better understand the role of vegetation in soil and water conservation projects, the changes of plant functional traits should not be ignored. The construction of Xindian SWDP can be traced back to 1952. After 68 years of continuous management, the vegetation coverage of the park has recovered from 5% to the current 75% and great changes have taken place in the ecology of the park during the 68 years of continuous management. Therefore, the study of plant functional traits reflecting the ecosystem change strategy is very important for soil and water conservation and restoration development.

As the largest terrestrial ecosystem, grassland ecosystem mainly distributes in ecologically fragile areas due to their special functional traits and strategies [1]. Functional traits are biological attributes that directly or indirectly affect species fitness and endow plants with high adaptability to environmental changes [2,3]. Functional trait variability allows plants to minimize their building costs and maximize functional efficiency. Therefore, to ascertain plant functioning and their ecological strategies in soil and water conservation measures, it is critical to explore how functional traits vary across different measures [4], especially relevant under the ongoing climatic change scenario. For instance, Scots pine (*Pinus sylvestris* L.) has the ability to adjust its leaf/sapwood area ratio, leaf-specific hydraulic conductivity and total leaf area in response to drought [5].

The functional traits of grassland communities vary greatly due to latent influencing factors, such as topographic conditions, soil properties and plant diversity [6]. On the one hand, habitat heterogeneity, due to changes in environmental factors and topography, will lead to the differences in grassland composition [7]. For example, temperature, humidity and altitude will affect grassland vegetation population and change functional traits [8]. On the other hand, there is a complex relationship between soil properties and plant. Soil which controls water and nutrients is the most critical condition for plant growth [9,10]. Soil water content is also considered to be the main factor in determining the composition of grassland vegetation [11]. Lush plants also feed back to soil nutrients and avoid soil erosion through their leaves and roots [12,13]. In addition, high vegetation coverage will create a microclimate through changes to the environment temperature and humidity and then affect soil properties [14]. In recent years, correlation or causation was explained among plant functional traits and the influencing factors. However, how these factors work together to influence plant functional traits and how they interact with each other remains unclear [15]. Hence, to resolve this major problem, the research should quantify the factors that impact on plant functional traits.

Four categories of variables were represented using 29 different observational variables collected from 171 plots of 57 sites in Xindian National Soil and Water Conservation Demonstration Park. We used random forest (RF) algorithm to identify critical indicators of plant functional traits. The RF is a machine learning classification method that has been demonstrated as an efficient algorithm to obtain key factors [16]. Moreover, structural equation modeling (SEM) was used to quantify the influence of these factors on plant functional traits [17]. In this research, we employed the partial least squares structural equation modeling (PLS-SEM). By applying these methods, the main objectives of this study were as follows: (a) Select and quantify appropriate plant functional traits in the study area; (b) Discuss the interaction between topographic conditions, soil properties and plant diversity affecting plant functional traits; (c) Evaluate the effects of topographic conditions and diversity indexes on plant functional traits. The study on the above issues is helpful to understand the status quo and change rules of plant functional traits in the special ecosystem of Xindian National Soil and Water Conservation Demonstration Park; it is of great significance to understand the mechanism of plant community construction under the interaction of different factors.

2. Materials and Methods

2.1. Study Area

Xindian Soil and WATER Conservation Demonstration Park, located on the left bank of the middle reaches of Wuding River, was built in 1952. It covers an area of 1.44 km^2 and is all composed of hills. The terrain is very broken, with 31 gullies that are 200 m long and cultivated land above 25 degrees accounting for 49% of the total cultivated land area (Figure S1). According to the observed data in 1952, the annual average soil loss amount was 19,900 t. Since 1952, 24 silt DAMS have been built in the demonstration park. At present, the control degree (through engineering measures and vegetation engineering) of the demonstration park has reached 80%, the forest and grass coverage rate has promoted from 5% to 75% and the sand blocking rate has reached 98%.

The soil texture is sandy loam with dense gullies, which has typical loess hilly and gully landform (Figure 1). It is a temperate continental semi-arid climate, with a monthly mean temperature, ranging from −7.5° in January to 24° in July, mean annual temperature is 8.3° and the mean annual precipitation is 486 mm from 2010 to 2020, most of which occurs in the form of rainstorms from July to September [18].

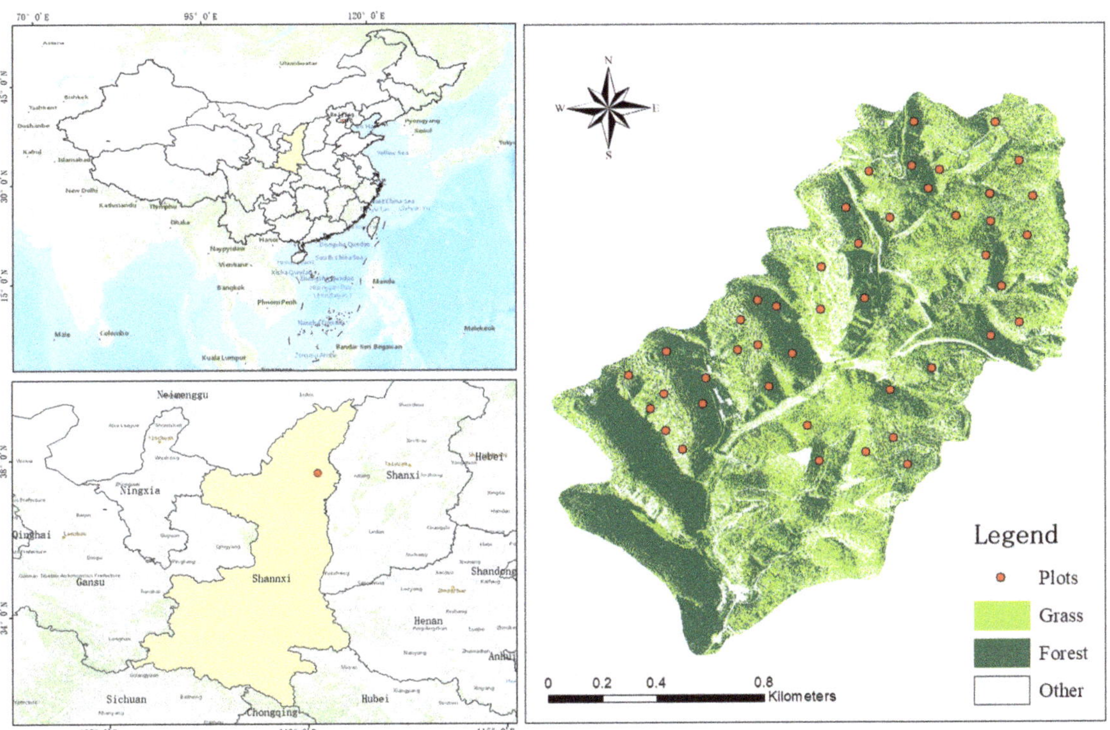

Figure 1. Research area.

The study area is dominated by grassland and accounts for more than 80% of the total vegetation area. Our research plots are located at an altitude of 850 m to 1287 m, with a slope range from 3 °C to 40 °C.

2.2. Plot Survey, Sample Collection and Analysis

There were one hundred and seventy-one herb plots (1 m × 1 m) from fifty-seven sampling sites in the study area (Figure 2). On each plot, we recorded the names, number, coverage, proportion of all herb species and plant height of each herb. The vegetation coverage was captured by Canon fisheye lens camera and processed using ArcGIS 10.6 to get the total coverage and dominant species coverage. After the investigation, ten well-lit and developed leaves were collected from dominant species for measuring the leaf thickness, area, and dry leaf weight of the species (Table S1). The remaining leaves were taken back to the laboratory for chemical element determination after drying.

Soil samples from the fifty-seven sites were also obtained at 11:00 a.m. to 15:00 p.m. from 29 July to 31 July 2020. Seven points were selected along an S-shape line in each plot (Figure 2). The sampling was collected from 0–30 cm and mixed. Then, the soil samples were required to pass a 2 mm sieve to remove impurities, the samples were taken back to the laboratory for subsequent analysis. Soil bulk density and soil water content were obtained from three samples along the diagonal in each grassland plot, using a cylindrical metal sampler (100 mm^2) [19].

The measurement of leaf traits was mainly carried out according to the literature [20,21]. A scanner was used to obtain the leaf area (10 replicates) and a vernier card was used to measure and record the leaf thickness. Then, the measured leaves were put into envelopes and the dry matter content of the leaves was obtained by drying method. The soil bulk density was determined by soil–core method. The SWC was calculated as the ratio of soil water mass to oven-dry weight [22,23]. The organic matter was assayed by dichromic oxidation method. The total nitrogen content was measured by Kjeldahl method using a FOSS Kjeltec 8400 Analyzer Unit (FOSS, Hillerod, Denmark) [24]. The total phosphorus was digested by H_2SO_4-$HClO_4$ and measured by spectrophotometer [25]. Total carbon was measured from 1 mm screened samples using Liaui TOC II analyzer (ELMENTAR, Langenselbold, Germany).

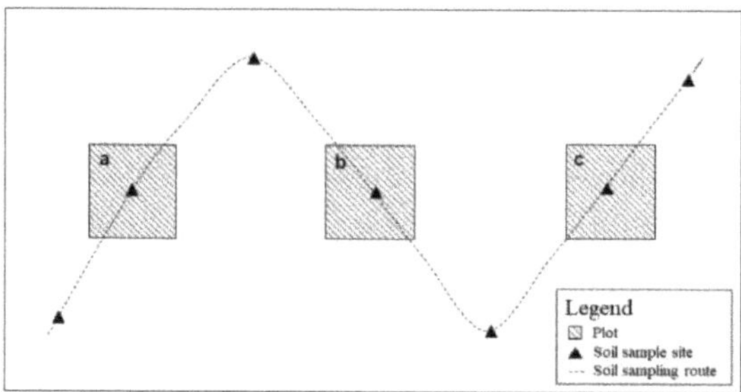

Figure 2. Grassland plots position (a, b, and c) and soil collected in sampling plots.

2.3. Variables

The SEM analysis method is a comprehensive technique that uses covariance matrix to analyze the relationships in multivariate data and identify the causality between observed variables and latent variables. In this study, we explained the relationships between the various influencing factors that pertain to plant functional traits by using latent variables and observational variables. The PLS-SEM was constructed by selecting three explanatory latent variables (topographic conditions, soil properties and vegetation structure) and one latent dependent variable (plant functional traits). Six explanatory observation variables were used to represent the latent variable characteristics of soil, namely soil organic matter content (SOM), soil water content (SWC), soil bulk density (BD), maximum soil water content (MWC), soil total phosphorus content (TP) and soil total nitrogen content (TN). Four topographic variables, namely altitude, slope, slope position and aspect, were chosen due to their effect on the hydrothermal conditions of sampling sites. In order to ensure the integrity of the vegetation diversity information, research needs of the integration of multiple levels and multiple diversity indexes [26]. We selected five different vegetation diversity indexes and two features as latent variables to illustrate vegetation structure, including the total number of plants (N), plants species index (S), Shannon–Wiener index (SHA), Simpson index (SIM), Margalef index (MAR) and Gleason index (GLE). Finally, 12 indexes, including leaf thickness, leaf dry weight, organic matter of leaves, total nitrogen content, total phosphorus content, nitrogen-to-phosphorus ratio, leaf tissue density, leaf area, specific leaf area, vegetation coverage, average plant height and ratio of dominant species were used to represent the observed variables of plant functional traits (Tables S2 and S3).

2.4. Calculation of Variables

The calculation and treatment of some specific variables were as follows:

(1) Margalef species richness index (MAR)

$$MAR = \frac{S-1}{\ln N}$$

where S is the total number of species in the community and N is the total number of individuals of all observed species.

(2)   Shannon diversity index (SHA)

$$SHA = -\sum p_i \ln p_i \text{ and } p_i = \frac{N_i}{N}$$

where p_i is the proportion of the number of species i to the total number.

(3)   Simpson diversity index (SIM)

$$SIM = 1 - \sum p_i^2 \text{ and } p_i = \frac{N_i}{N}$$

where p_i is the proportion of the number of species i to the total number.

(4)   Pielou's evenness index (PIE)

$$PIE = \frac{N(N-1)}{\sum N_i(N_i - 1)}$$

where N_i is the number of individuals of species i and N is the sum of individuals of all species in the community.

(5)   Gleason richness index (GLE)

$$GLE = \frac{S}{\ln A}$$

where A is the total area investigated and S is the total number of species in the community.

(6)   Community weighted means (CWM)

The functional characteristics of plant communities were characterized by community weighting method [27]. CWM was calculated based on the plant functional traits of each species in the plots. In addition, the relative above-ground biomass (AGB) of each species in the plot was weighted. Additionally, the CWM mainly reflects the attribute and strategy of dominant species in the community [28,29]. The *CWM* units are the same as functional traits units involved in the calculation. The formula is as follows:

$$CWM = \sum_{i=1}^{n} trait_i \times P_i$$

where $trait_i$ is the plant functional trait value of species i, P_i is the relative AGB of species i in the plot, and n is the number of species in the plot.

2.5. Selection of Variables

In the research, RF algorithm was used to filter important dependent variables to reduce the latent redundancy of dependent variables [30]. RF is an ensemble algorithm that classifies by voting on multiple unbiased classifier decision trees [31,32]. The algorithm was based on the Boruta package in R 4.1.2. The Boruta feature selection, providing important values to indicate whether features are important or not, which are obtained by mixing original attribute values between objects [33].

Plant functional traits can be reflected by multiple indices with different correlations with explanatory observed variables. Therefore, we used a random forest algorithm to screen for the indices of plant functional traits that were more correlated with explanatory observed variables. The important values of 12 plant functional traits indexes were calculated using 17 explanatory observational variables. All the plant functional traits indexes were used to classify the explanatory observed variables and an important classification value was obtained. The results for each important value were summarized and the functional trait indicators that showed statistical significance in a classification were voted on. Such voting was conducted to select more relevant and key functional traits. Random forest and voting results were used as input parameters to support latent dependent variables in the PLS-SEM model. In our research, confidence levels and maximum runs of RF algorithm were set as 0.01 and 100, respectively [19].

2.6. Establish of Preliminary Model and Paths Determination

Structural equation model (SEM) is a method to establish, estimate and test causality models, which contain both observable variables and latent variables that cannot be directly observed [34]. It is mainly used in PLS-SEM and structural equation modeling based on covariance. Compared with CB-PLS, which focuses on parameter evaluation, PLS-SEM has a more accurate prediction accuracy [35]. In exploratory research, we should focus on PLS-SEM when the relationship between variables is complex and unclear [36]. In addition, according to the research, when the sample number is limited and does not follow the normal distribution, PLS has a wider tolerance than CB [37,38]. Therefore, PLS-SEM was chosen in the research.

There were three causal hypothesis to establish the preliminary model: (a) The variables of topographic conditions, soil properties and vegetation structure directly influenced dependent variables of plant functional traits; (b) The latent variables of topographic conditions indirectly impacted the latent variables of plant functional traits by influencing hydrothermal conditions, species distribution and ecological processes; (c) Vegetation and soil influence each other. The latent variables of vegetation structure indirectly impacted the latent variables of plant functional traits by influencing the latent variables of soil properties, and the latent variable of soil properties can also affect the functional character of vegetation by improving the vegetation structure (Figure 3).

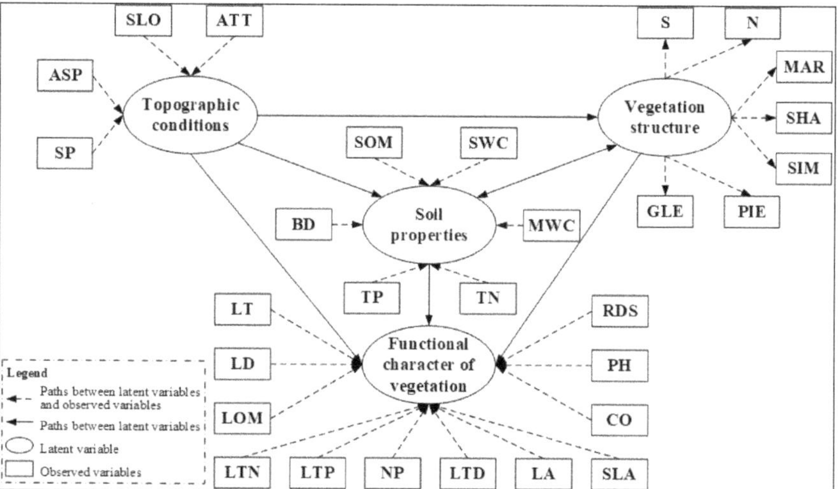

Figure 3. A PLS-SEM model was established to show the relationship between latent variables. The arrow represents the influence of the weight of the latent variable or observed variable on the latent variable. The ovals represent the latent variables; the rectangles represent the observed variables. SLO, slope; ALT, altitude; ASP, aspect; SP, slope position; MAR, Margalef species richness index; SHA, Shannon diversity index; SIM, Simpson diversity index; PIE, Pielou's evenness index; GLE, Gleason richness index; N, Total number of shrubs; S, Total shrub species; SOM, Soil organic matter; BD, Soil bulk density; MWC, Maximum water capacity; TP, Total phosphorus content; TN, Total nitrogen content; SWC, Soil water content; LT, Blade thickness; LD, Leaf dry weight; LOM, Organic matter of leaves; LTN, Total nitrogen content; LTP, Total nitrogen content; NP, Nitrogen-to-phosphorus ratio; LTD, Leaf tissue density; LA, Leaf area; SLA, Specific leaf area; CO, Vegetation coverage; PH, Plant height; RDS, Ratio of dominant species.

To improve the reliability of model, correlation tests were carried out on different latent variables. We used CCA analysis to correlate each set of variables. Seventeen explanatory observed variables were divided into three groups of latent variables. Then, six pairs of CCA were calculated for each variable in pairs. The secondary PLS-SEM is determined by analyzing the significance of six sets of data and eliminating non-significant paths in the model. The CCA was carried out using the R package, Vegan.

2.7. Evaluation of PLS-SEM

We carried out factor analysis and path analysis in PLS-SEM after obtaining the results of RF and CCA analysis. In the final structural equation model, compound reliability values (CR) and average variance extraction values (AVE) are used to evaluate the structural reliability and internal model validity of SEM. The CR value > 0.7 indicates good internal consistency and reliability of the model [39]. The AVE value > 0.5 indicates that the model fits well and converges effectively. The identification validity should be evaluated with a matrix, according to the Fornell–Larcker criterion. The square root of the AVE should be greater than the relevant value of other variables [40]. We performed a variance inflation factor (VIF) analysis to avoid multicollinearity, among observing variables [41]. To avoid model misjudgment, standardized root means square residual (SRMR) was used as a fitting measure. Bootstrap programs were used to obtain T-statistics for significance tests of structural paths in PLS-SEM [42]. The path coefficient is significant when the p-value < 0.05 [36]. In this work, the maximum number of iterations of the PLS algorithm was set to 5000. The threshold value, which determines the maximum of the difference of the external weight, is set to 10^{-7}. In addition, the number of guide sub-samples was set as 10,000 and the significance level of guide was set as 0.05 to ensure reliability. Smart PLS 3 was used for all PLS-SEM statistical analyses.

3. Results

Fifty-one species and 10,899 individuals of herbs were observed in 171 plots from 57 sites. *Lespedeza davurica* was the most common herb observed at 57 sites, followed by *Setaria viridis* (L.) *Beauv* at 25 sites. The twelve dependent observed variables used to explain plant functional traits were as follows: LT:0.085–0.412; LH:5.74–60.261; LOM:404.453–516.743; LTN:18.981–71.337; LTP:1.296–6.457; NP:8.078–33.559; LTD:2.423–21.617; LV:0.0655–11.479; SLA:2.994–21.071; CO:0.477–0.977; PH:10.15–55; PDS:0.4–0.975.

3.1. Observed Variables of Plant Functional Traits

Seventeen explanatory observation variables among soil properties, vegetation structure and topographic conditions were used to classify 12 causal observation variables by RF, considering the contribution of dependent variables to explanatory variables as important values. Based on the regression analyses, we obtained 17 important values (Figure 4). The importance values of N and SOM groups showed the best results, and only four values were rejected in the group. CO was rejected by both N and SOM groups, and their importance values were −3.20 and −1.44. The importance values in MAR all indicate low importance, and all variables in this group are rejected. Eleven, fourteen and sixteen importance values are derived from topographic conditions, soil properties and diversity, respectively.

We ranked the 12 dependent variables by the 42 importance values obtained. In descending order, the results are as follows: PH (eight votes) < LTD and LTN (seven votes each) < PDS (six votes) < LOSM and SLA (five votes each) < LTP (three votes) < LT (one vote) < LD, NP, LA and CO (zero votes). PH received the highest number of votes, which were based on soil properties. LTN received four votes from soil properties, three votes from diversity, and zero votes from topographic conditions; LOM received four votes from diversity and one vote from topographic conditions; and SLA received zero votes from diversity. PH, RDS and LTD obtained votes from three latent variables, respectively, while LD, NP, LA and CO did not obtain any votes. According to the total number of votes for each dependent variable, six variables—PH, PDS, LTN, LTD, LOM and SLA—were chosen as observed variables of plant functional traits in PLS-SEM.

	LT	LD	LOM	LTN	LTP	NP	LTD	LA	SLA	CO	PH	PDS	
SOM	− 0.25	− 0.75	− 0.40	3.68*	0.87	0.16	8.53*	0.91	7.33*	− 1.44	2.92*	2.27*	Soil properties
TP	− 0.60	− 0.20	0.64	0.37	− 0.91	− 2.08	0.13	− 1.38	0.52	− 0.69	5.69*	− 0.17	
TN	− 0.11	− 0.38	− 0.14	2.29*	− 0.31	− 0.68	3.01*	− 0.29	3.74*	− 0.79	2.93*	0.29	
SWC	0.44	− 1.16	− 2.21	2.07*	− 0.71	− 0.30	1.05	0.81	0.54	2.39	− 0.84	0.44	
BD	− 0.16	− 0.44	− 1.46	0.23	0.09	− 1.10	0.46	0.50	2.23*	− 0.97	1.98*	0.57	
MWC	0.23	− 1.41	− 0.45	1.75*	1.84	0.11	2.06*	0.72	1.71	− 0.79	1.00	− 0.16	
ALT	0.87	− 0.52	− 0.72	− 0.13	2.80*	1.12	6.32*	− 0.55	4.98*	− 0.65	− 0.65	− 0.21	Topographic conditions
SLO	− 1.13	− 1.77	− 0.40	0.24	3.58*	0.42	2.06*	0.13	4.61*	− 0.92	6.19*	2.32*	
ASP	-0.91	− 1.26	9.18*	1.11	0.73	0.39	0.62	− 0.67	− 0.42	− 0.05	0.37		
SP	− 0.55	− 0.94	− 0.70	− 0.79	0.08	0.01	− 0.50	− 0.86	0.61	− 0.09	2.26*	0.86	
MAR	− 0.39	− 1.24	− 0.56	− 1.75	− 1.11	− 1.02	1.08	0.11	0.54	− 1.24	0.19	0.66	Vegetation structure
GLE	0.16	− 0.42	3.96*	1.53*	0.54	− 0.24	0.68	− 0.41	0.43	− 2.99	− 0.43	− 0.56	
SIM	− 0.06	− 0.80	0.12	0.53	0.35	0.70	0.39	− 0.36	− 0.54	− 1.08	2.25*	4.40*	
SHA	− 0.46	− 1.22	1.19	− 0.85	− 1.57	0.10	− 0.11	− 0.97	− 0.04	− 1.98	− 0.02	6.13*	
PIE	− 0.16	− 1.66	4.11*	0.57	0.29	0.09	− 0.55	− 0.86	0.42	0.38	− 1.72	3.78*	
S	0.61	− 0.94	4.15*	1.88*	0.34	− 0.88	1.91*	0.52	0.42	− 2.16	− 0.11	0.44	
N	4.97*	− 0.62	9.82*	4.54*	4.19*	0.56	3.20*	1.14	− 1.08	− 3.20	4.35*	0.63	

Figure 4. The RF is adopted for factor significance, where the red part represents acceptance and the blue part represents rejection, and * represents the significance of explanatory observation variables selected due to observation variables. SLO, Slope; ALT, Altitude; ASP, Aspect; SP, Slope position; MAR, Margalef species richness index; SHA, Shannon diversity index; SIM, Simpson diversity index; PIE, Pielou's evenness index; GLE, Gleason richness index; N, Total number of shrubs; S, Total shrub species; SOM, Soil organic matter; BD, Soil bulk density; MWC, Maximum water capacity; TP, Total phosphorus content; TN, Total nitrogen content; SWC, Soil water content; LT, Blade thickness; LD, Leaf dry weight; LOM, organic matter of leaves; LTN, Total nitrogen content; LTP, Total nitrogen content; NP, Nitrogen-to-phosphorus ratio; LTD, Leaf tissue density; LA, Leaf area; SLA, Specific leaf area; CO, Vegetation coverage; PH, Plant height; RDS, Ratio of dominant species.

3.2. Possible Path and Model

CCA analysis showed that the relationship between topographic conditions and soil properties was not statistically significant. This may be because topographic conditions have little direct effect on the underlying variables of soil properties. The relations between vegetation structure and plant functional traits were highly statistically significant, which was consistent with the interaction between vegetation structure and functional traits. In addition, CCA results showed significant correlation between vegetation structure vs. soil properties, topographic conditions vs. vegetation structure, topographic conditions vs. plant functional traits and soil properties vs. plant functional traits. These results show the interaction among the vegetation structure, soil properties, topographic conditions and plant functional traits (Table 1).

Table 1. The significance of paths between latent variables.

Latent Variables Group	Canonical Correlations 1	Eigenvalue	*p*-Value	Wilk's	DF
Vegetation structure vs. Soil properties	0.660	0.771	0.028 *	0.285	42
Topographic conditions vs. Vegetation structure	0.651	0.737	0.042 *	0.430	28
Topographic conditions vs. Soil properties	0.551	0.435	0.101	0.518	24
Vegetation structure vs. Functional character of vegetation	0.612	0.600	0.002 **	0.226	42
Topographic conditions vs. Functional character of vegetation	0.592	0.540	0.024 *	0.457	24
Soil properties vs. Functional character of vegetation	0.668	0.808	0.012 *	0.309	36

* indicates $p < 0.05$; ** indicates $p < 0.01$.

If CCA results between the two sets of variables are not significant, we conclude there is no latent path between the two pairs in model. Therefore, we preliminarily determined the path of the PLS-SEM based on CCA results of different variable groups by eliminating a path from topographic conditions to soil properties. The following figure describes the optimized secondary structure equation model (Figure 5).

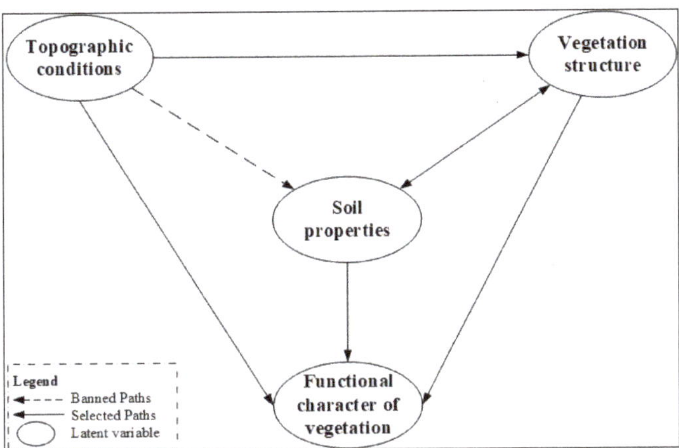

Figure 5. PLS-SEM based on CCA.

3.3. Model Fit

The optimal model diagram is shown below (Figure 6). In this model, the CR value is 0.795, indicating that the model has good precision and interpretability. The AVE (value = 0.660) shows that the SEM finally obtained has good fitting convergence. Identification validity was assessed through the Fornell–Larcker standard matrix. AVE square root of plant functional traits was 0.812 in the matrix, which is greater than its correlation with other variables. VIF values of all variables were less than three. The SRMR value of 0.093 is acceptable in PLS-SEM.

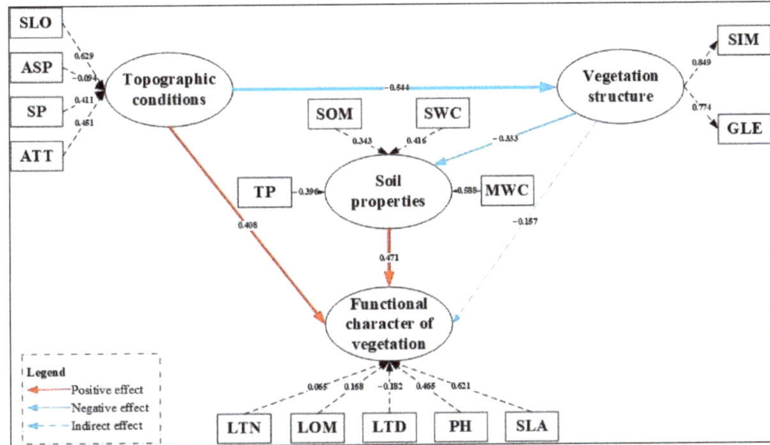

Figure 6. The final SEM describing the relationship between variables. The red arrow indicates positive impact and the blue indicates negative impact; the thickness of the arrow indicates the weight.

3.4. Evaluation of Influencing Factors of Plant Functional Traits

In the final PLS-SEM, the results indicate the weight of each observed variable among potential variables. The weights of PH and SLA were 0.465 and 0.621, indicating that they contributed 94.9% to latent variables of plant functional traits. For latent variables of plant species diversity, the weights of SIM and GLE were 0.849 and 0.774, respectively. SLO had the highest contribution value of latent quality variable, which was 0.629. ASP has the lowest contribution value of -0.094. The contribution rates of SOM, TP, SWC and MWC to soil properties were 0.343, 0.396, 0.416 and 0.588, respectively; they have similar contribution rates.

The bootstrapping results demonstrated an acceptable t-test and p-value, which implies that the four total effects of the final model are significant. The effects of topographic conditions on plant functional traits ($t = 3.245$, $p < 0.01$), topographic conditions on plant diversity ($t = 5.324$, $p < 0.001$), plant diversity on soil properties ($t = 2.214$, $p < 0.05$) and soil properties on plant functional traits ($t = 3.593$, $p < 0.001$) were 0. 408, -0.544, -0.333 and 0.471, respectively. In addition, the indirect effect of latent variables of plant diversity on latent variables of plant functional traits ($t = 1.967$, $p < 0.05$) was -0.157, which was acceptable according to the bootstrapping results. Finally, the remaining paths (from soil properties to vegetation structure) in the prediction model (Figure 5) were deleted because their $p > 0.05$ and would affect the overall effect of the model.

4. Discussion

4.1. Results of Screening Plant Functional Traits by RF

According to the results of RF, there are finally five indicators that get high votes, they are LTN, LOM, LTD, PH and SLA. PH received the most votes, which was mostly from soil properties. Soil is the most important factor affecting plant growth, and TP is the highest score among the four votes obtained through soil properties. The reason is that phosphorus is an indispensable factor for plant growth, which can promote the rapid growth of roots and aboveground parts and improve the ability of plants to adapt to external environmental conditions [43]. Secondly, LTN and LTD both obtained the majority of seven votes from soil properties and vegetation structure. Soil can provide nutrients for plants, and reasonable vegetation structure can provide good hydrothermal conditions for plants, LOM and SLA get higher votes for the same reason. Finally, LT received only one vote from vegetation structure, LT did not get the topographical status vote because the study area was located in SWDP, and many engineering measures changed the small topographical area of the watershed, resulting in a low impact of topography on LT. The reason for not obtaining soil voting is that the change of soil properties has limited influence on LT. As an important leaf shape component, LT is mainly affected by environmental factors, especially light intensity, which is the reason why LT can obtain vegetation structure voting.

4.2. Direct Effects of Topographic Conditions on Vegetation Structure and Plant Functional Traits

In the research, the direct effects of topographic conditions on plant diversity and plant functional traits [36] were evaluated. The results showed that the effects of topographic conditions on plant diversity and functional traits were mainly reflected by the differences in altitude (ALT), slope aspect (ASP) and slope position (SP), as well as the different water, heat, and light conditions, thus, leading to the formation of life strategies with different combinations of functional traits in plants. Some scholars also confirmed the same problem, the study showed that topographic factors strongly influenced the pattern of community, ecosystem, and landscape, thus, affecting plant diversity (some studies pointed out that topographic factors were the main factor, limiting the change of vegetation distribution in the loess hilly region) [44].

According to the results of PLS-SEM, topographic conditions have a negative impact on vegetation structure. SLO index had the greatest negative impact on vegetation structure. Perhaps the main reason is that with the increase in slope, due to the special natural geographical conditions of the Loess Plateau gully region, the soil required to form surface runoff when the rainfall intensity is beyond a certain range cannot effectively replenish groundwater, and thus cannot form a soil dry layer of a certain depth, eventually affecting the plant growth and reducing the plant community diversity. In the model, SP and ALT have similar contributions to vegetation structure, The main reason is that they share a similar principle: water and heat change with height, which ultimately affects the structure of the vegetation [45,46]. Finally, slope aspect had the least impact on plant diversity. Slope aspect mainly affected the light received by plants, but in the Loess Plateau region, regardless of the slope aspect, the plant received sufficient light, which resulted in a small impact of slope aspect factors on plant diversity [47].

According to the results of our research in the SWDP, the variation of SLO degree significantly affects plant functional characteristics [48]. The possible reason is that the plant growth is mainly restricted by water condition, and the change of the slope degree affects the slope surface runoff and soil erosion intensity and indirectly leads to differences in soil nutrient and moisture [49]. A study from Chinese scholars also showed that the changes of certain functional traits in plants were mainly determined by slope changes and plant functional traits gradually changed with the increase in slope value [50]. Altitude is another major variable in determining the functional traits of vegetation, which affects the temperature, water and carbon dioxide required for plant growth [51]. In addition, our results suggested that slope position is also critical to the latent variable, affecting plant functional traits. These results are consistent with the report, which showed that slope position has a significant impact on the vegetation at microtopography [52]. The main reason for this result is that the poor water and fertilizer retention ability of the upper slope position causes a large amount of water to accumulate at the bottom of the slope, resulting in significant differences in water and fertilizer conditions at different slope positions [53]. At the same time, the temperature at the bottom of the slope was higher than that at the top due to the geographical advantage, and the water and heat were locally redistributed due to the microtopography, which ultimately led to the differences in plant diversity and functional traits [54]. In this study, slope aspect contributed very little to plant functional traits (especially LTN), which was inconsistent with the results of Bennie et al. [55]. This may be because the slope aspect has little influence on soil moisture and heat in the Loess Plateau due to the low slopes and extreme drought [56].

4.3. Direct Effects of Soil Properties on Plant Functional Traits

Based on the results of model, in addition to topographic conditions, soil properties also directly affect the plant functional traits [57]. Some scholars also reported similar results, pointing out the availability of underground nutrients plays a key role in plant growth [58].

For target variables, our results indicate MWC and SWC are the main factors affecting plant functional traits. The biggest contribution of MWC and SWC is mainly because area is located in the Loess Plateau, where soil water retention is poor, precipitation is low and soil erosion is serious, so water is the main factor limiting plant functional traits [59]. There are other studies [60] reported that changes in plant functional traits are mainly influenced by MWC and SWC. In addition, our study showed that TN is also critical to latent variables of plant functional traits, which is consistent with Chinese scholars findings that phosphorus is an important factor, affecting plant traits in arid regions [61]. In this study, SOM contribution is 0.343. Organic matter is one of the main sources of plant nutrition, which can promote plant growth and have a great impact on plant functional traits [62]. According to the study along nutrient gradients on a provincial scale, SOM is proportional to SLA LTN, LTP, etc. [63].

4.4. Indirect Effect of Vegetation Structure on Plant Functional Traits

The model shows that vegetation structure (SIM and GLE) directly affect soil properties and indirectly affect plant functional traits (through soil properties), which showed there is no evidence for a direct causal correlation of species diversity and plant functional traits.

Vegetation structure has a direct negative impact on soil properties, which may be due to sub-standard soil in the Loess Plateau due to low soil moisture and nutrient content; the lack of soil nutrient accumulation and recruitment was due to excessive uptake of soil nutrients in species-rich plots. Additionally, the study area will have heavy rain that produces surface runoff and leaching from July to September, which further affects soil properties. SIM index has a great influence on soil properties. The important reason is that SIM index is an index indicating species dominance. Different species require different nutrients. Plots with high dominance indicate that the proportion of a certain species in the area is too high, which may lead to excessive consumption of certain elements in the soil and eventually have a negative impact on soil properties. For example, studies have shown that cropland (a single type of land) reduces soil organic matter content by 40% and has the worst soil structure, compared to forest and grassland [64]. Second, the GLE index of species richness based on plot area also had a negative impact on soil properties. When the plot area is fixed, there is fierce competition among vegetation in the plot with the increase in SIM index. Soil moisture is the biggest limiting factor for vegetation growth in the loess hilly–gully region. The competition between vegetation will accelerate the consumption of soil water and eventually affect soil properties. For example, the influence of vegetation structure on the spatial variation of soil's physical and chemical properties is sometimes even more significant than that of climate factors [65]. Finally, there are two main reasons why other vegetation diversity is not used in the model. On the one hand, RF results showed that PIE, MAR and SHA had low correlations with selected plant functional traits. On the other hand, N and S indexes have collinearity with SIM and GLE in PLS-SEM. Therefore, the SIM and GLE indices were finally selected.

According to our results, plant functional traits were significantly affected by the composition and distribution of herbaceous layers [66]. Vegetation structure have a direct impact on soil properties, and the vegetation structure index can also indirectly reflect the strength of inter-vegetation competitiveness [67]. For the target variables, the model results showed that the SLA index of plant functional traits was indirect impacted by vegetation structure. SLA is a functional trait that represents the interaction between plants and the environment [68]. In the same community, the specific leaf area of plants was larger when light was decreased. When the species diversity is high, the competition among plants will lead to the lack of light for some plants, thus, affecting SLA index [69]. The change of species diversity also had a significant impact on PH, mainly because of the high light intensity and hot climate in the Loess Plateau region, but the shortage of water resources and soil nutrients. The competition among species is mainly through the root system to compete for soil water and nutrients, while the aboveground part mainly uses conservative strategies to reduce light exposure. For example, dwarf plants are more conducive to growth in the Loess Plateau. The research of scholars has also proved this point. Vegetation on the Loess Plateau generally has developed roots to obtain soil water and nutrients, while the aboveground parts are generally short [70]. In addition, changes in species diversity had a certain impact on LTN and LOM indices, and the lack of soil nutrient accumulation and recruitment was due to the excessive uptake of soil nutrients in species-rich plots. Additionally, the study area will have heavy rains that produce surface runoff and leaching from July to September, leading to a decrease in soil nutrient content, and ultimately a comprehensive impact on LTN and LOM indicators [71]. Finally, the results of PLS-SEM showed that the change of species diversity would affect the LTD; in order to adapt to the harsh living environment, plants will put more dry matter synthesized by plants into the structure of leaves (LTD increases) and improve the resistance to drought environment by increasing the distance of internal water diffusion to leaves [72].

5. Conclusions

Quantitative description of the relationship between the factors that may affect plant functional traits is very important for understanding the heterogeneity of vegetation habitats in Xindian National Soil and Water Conservation Demonstration Park, which can provide new insights for vegetation management and restoration in soil and water conservation area. For example, we should pay attention to vegetation structure in the future measures; reasonable vegetation structure can play a greater role in the same terrain and soil conditions. In the research, PLS-SEM was built to quantify the influence between influencing drivers on grassland vegetation functional traits, and the results showed that: (1) Topographic conditions and plant diversity were the main factors affecting plant functional traits. (2) The effect of topographic conditions on latent variables of soil properties was limited, although some topographic conditions variables were strongly correlated with it. (3) Species diversity indirectly affects latent variables of plant functional traits by influencing soil properties. In general, our results show that topographic conditions and soil properties are still the most important influencing factors, although other factors will also have an impact on the functional characteristics of vegetation; in the future, park management should continue to adhere to the policy of engineering measures in the main, with vegetation measures as the auxiliary. Finally, the results of the model are relatively reasonable for the short-term mechanism, but the time scale of the existence of the ecosystem is longer. In the future, we can continue to study SWDP to obtain the long-term mechanism and causal relationship between the components of the ecosystem in SWDP.

Supplementary Materials: The following supporting information can be downloaded at: https://www.mdpi.com/article/10.3390/plants11212891/s1, Figure S1: Slope aspect map, slope map and elevation map of the study area; Table S1: Dominant species information; Table S2: Details of variables used in the research; Table S3: The data details.

Author Contributions: Conceptualization: Z.W.; Methodology: G.D., Y.B. and B.W. (Bojin Wen); data collection and curation: G.D., B.W. (Boheng Wang) and J.L.; field investigation: G.D., W.X. and S.C.; writing—review and editing: G.D., Z.W. and W.X. All authors have read and agreed to the published version of the manuscript.

Funding: This study was funded by the National Natural Science Foundation of China (No. 41977077).

Institutional Review Board Statement: Not applicable.

Informed Consent Statement: Not applicable.

Data Availability Statement: All available data can be obtained by contacting the corresponding author.

Conflicts of Interest: The authors declare that they have no know financial interests or personal relationships that could have influenced the information reported in this paper.

References

1. Yan, Y.; Zhao, C.; Quan, Y.; Lu, H.; Rong, Y.; Wu, G. Interrelations of Ecosystem Services and Rural Population Wellbeing in an Ecologically-Fragile Area in North China. *Sustainability* **2017**, *9*, 709. [CrossRef]
2. Violle, C.; Navas, M.L.; Vile, D.; Kazakou, E.; Fortunel, C.; Hummel, I.; Garnier, E. Let the concept of trait be functional! *Oikos* **2007**, *116*, 882–892. [CrossRef]
3. Lavorel, S.; McIntyre, S.; Landsberg, J.; Forbes, T. Plant functional classifications: From general groups to specific groups based on response to disturbance. *Trends Ecol. Evol.* **1997**, *12*, 474–478. [CrossRef]
4. Crowther, T.W.; Maynard, D.S.; Peccia, T.R.; Smith, J.R.; Bradford, M.A. Untangling the fungal niche: The trait-based approach. *Front. Microbiol.* **2014**, *2014*, 5. [CrossRef]
5. Seidel, H.; Matiu, M.; Menzel, A. Compensatory Growth of Scots Pine Seedlings Mitigates Impacts of Multiple Droughts Within and Across Years. *Front. Plant Sci.* **2019**, *10*, 519. [CrossRef]
6. Zhou, P.; Geng, Y.; Ma, W.-H.; He, J.-S. Linkages of functional traits among plant organs in the dominant species of the inner mongolia grassland, China. *Chin. J. Plant Ecol.* **2010**, *34*, 7.
7. Propastin, P. Multiscale analysis of the relationship between topography and aboveground biomass in the tropical rainforests of Sulawesi, Indonesia. *Int. J. Geogr. Inf. Sci.* **2011**, *25*, 455–472. [CrossRef]
8. Niu, Y.; Zhou, J.; Yang, S.; Chu, B.; Ma, S.; Zhu, H.; Hua, L. The effects of topographical factors on the distribution of plant communities in a mountain meadow on the Tibetan Plateau as a foundation for target-oriented management. *Ecol. Indic.* **2019**, *106*, 105532.1–105532.12. [CrossRef]
9. Reinhart, K.; Vermeire, L. Power and limitation of soil properties as predictors of variation in peak plant biomass in a northern mixed-grass prairie. *Ecol. Indic.* **2017**, *80*, 268–274. [CrossRef]

10. Zak, D.R.; Holmes, W.E.; White, D.C.; Peacock, A.D.; Tilman, D. Plant diversity, soil microbial communities, and ecosystem function: Are there any links? *Ecology* **2003**, *84*, 2042–2050. [CrossRef]

11. Bui, N.E. Soil salinity: A neglected factor in plant ecology and biogeography. *J. Arid. Environ.* **2013**, *92*, 14–25. [CrossRef]

12. Butt, M.J.; Waqas, A.; Mahmood, R. The Combined Effect of Vegetation and Soil Erosion in the Water Resource Management. *Water Resour. Manag.* **2010**, *24*, 3701–3714. [CrossRef]

13. Dong, L.; Li, J.; Zhang, Y.; Bing, M.; Liu, Y.; Wu, J.; Hai, X.; Li, A.; Wang, K.; Wu, P.; et al. Effects of vegetation restoration types on soil nutrients and soil erodibility regulated by slope positions on the Loess Plateau. *J. Environ. Manag.* **2021**, *302*, 113985. [CrossRef] [PubMed]

14. Wang, C.; Liang, W.; Yan, J.; Jin, Z.; Zhang, W.; Li, X. Effects of vegetation restoration on local microclimate on the Loess Plateau. *J. Geogr. Sci.* **2022**, *32*, 291–316. [CrossRef]

15. Díaz, S.; Cabido, M. Vive la différence: Plant functional diversity matters to ecosystem processes. *Trends Ecol. Evol.* **2001**, *16*, 646–655. [CrossRef]

16. Ho, T.K. Random Decision Forests. In Proceedings of the 3rd International Conference on Document Analysis and Recognition, Montreal, QC, Canada, 14–16 August 1995; Volume 1, pp. 278–282.

17. Kaplan, D. *Structural Equation Modeling: Foundations and Extensions*; Sage Publications: Thousand Oaks, CA, USA, 2008; Volume 10.

18. Wang, Z.-J.; Jiao, J.-Y.; Rayburg, S.; Wang, Q.-L.; Su, Y. Soil erosion resistance of "grain for green" vegetation types under extreme rainfall conditions on the loess plateau, China. *Catena* **2016**, *141*, 109–116. [CrossRef]

19. Wang, B.; Bu, Y.; Li, Y.; Li, W.; Zhao, P.; Yang, Y.; Qi, N.; Gou, R. Quantifying the Relationship among Impact Factors of Shrub Layer Diversity in Chinese Pine Plantation Forest Ecosystems. *Forests* **2019**, *10*, 781. [CrossRef]

20. De Bello, F.; Carmona, C.P.; Dias, A.T.; Götzenberger, L.; Moretti, M.; Berg, M.P. *Handbook of Trait-Based Ecology: From Theory to R Tools*; Cambridge University Press: Cambridge, UK, 2021.

21. Perez-Harguindeguy, N.; Diaz, S.; Garnier, E.; Lavorel, S.; Poorter, H.; Jaureguiberry, P.; Bret-Harte, M.; Cornwell, W.; Craine, J.; Gurvich, D. New handbook for standardised measurement of plant functional traits worldwide. *Aust. Bot.* **2013**, *61*, 167–234. [CrossRef]

22. Dane, J.H.; Topp, C.G. *Methods of Soil Analysis, Part 4: Physical Methods*; John Wiley & Sons: Hoboken, NJ, USA, 2020.

23. Sparks, D.L.; Page, A.L.; Helmke, P.A.; Loeppert, R.H. *Methods of Soil Analysis, Part 3: Chemical Methods*; John Wiley & Sons: Hoboken, NJ, USA, 2020.

24. Bremner, J.M. Nitrogen-Total. In *Methods of Soil Analysis: Part 3 Chemical Methods*; American Society of Agronomy, Inc.: Madison, WI, USA, 1996; Volume 5, pp. 1085–1121.

25. Parkinson, J.A.; Allen, S.E. A wet oxidation procedure suitable for the determination of nitrogen and mineral nutrients in biological material. *Commun. Soil Sci. Plant Anal.* **1975**, *6*, 1–11. [CrossRef]

26. Lyashevska, O.; Farnsworth, K.D. How many dimensions of biodiversity do we need? *Ecol. Indic.* **2012**, *18*, 485–492. [CrossRef]

27. Garnier, E.; Cortez, J.; Billès, G.; Navas, M.-L.; Roumet, C.; Debussche, M.; Laurent, G.; Blanchard, A.; Aubry, D.; Bellmann, A. Plant functional markers capture ecosystem properties during secondary succession. *Ecology* **2004**, *85*, 2630–2637. [CrossRef]

28. Díaz, S.; Lavorel, S.; de Bello, F.; Quétier, F.; Grigulis, K.; Robson, T.M. Incorporating plant functional diversity effects in ecosystem service assessments. *Proc. Natl. Acad. Sci. USA* **2007**, *104*, 20684–20689. [CrossRef] [PubMed]

29. He, K.; Huang, Y.; Qi, Y.; Sheng, Z.; Chen, H. Effects of nitrogen addition on vegetation and soil and its linkages to plant diversity and productivity in a semi-arid steppe. *Sci. Total Environ.* **2021**, *778*, 146299. [CrossRef] [PubMed]

30. Tuv, E.; Borisov, A.; Runger, G.; Torkkola, K. Feature selection with ensembles, artificial variables, and redundancy elimination. *J. Mach. Learn. Res.* **2009**, *10*, 1341–1366.

31. Hair, J.F.; Anderson, R.E.; Tatham, R.L.; Black, W.C. *Multivariate Data Analysis*, 5th ed.; Prentice Hall: New York, NY, USA, 1998.

32. Breiman, L. Random forests. *Mach. Learn.* **2001**, *45*, 5–32. [CrossRef]

33. Kursa, M.B.; Rudnicki, W.R. Feature selection with the boruta package. *J. Stat. Softw.* **2010**, *36*, 1–13. [CrossRef]

34. Grace, J.B. *Structural Equation Modeling and Natural Systems*; Cambridge University Press: Cambridge, UK, 2006.

35. Samani, S.A. Steps in research process (partial least square of structural equation modeling (pls-sem)). *Int. J. Soc. Sci. Bus.* **2016**, *1*, 55–66.

36. Wong, K.K.-K. Partial least squares structural equation modeling (pls-sem) techniques using smartpls. *Mark. Bull.* **2013**, *24*, 1–32.

37. Bacon, L.D. Using Lisrel and Pls to Measure Customer Satisfaction. In Proceedings of the Sawtooth Software Conference Proceedings, La Jolla, CA, USA, 2–5 February 1999; pp. 305–306.

38. Wong, K. Handling small survey sample size and skewed dataset with partial least square path modelling. *Vue Mag. Mark. Res. Intell. Assoc.* **2010**, *20*, 20–23.

39. Hair, J.F., Jr. *Successful Strategies for Teaching Multivariate Statistics*; Kennesaw State University: Kennesaw, GA, USA, 2006; pp. 1–5.

40. Fornell, C.; Larcker, D.F. Evaluating structural equation models with unobservable variables and measurement error. *J. Mark. Res.* **1981**, *18*, 39–50. [CrossRef]

41. Hair, J.F.; Ringle, C.M.; Sarstedt, M. Pls-sem: Indeed a silver bullet. *J. Mark. Theory Pract.* **2011**, *19*, 139–152. [CrossRef]

42. Henseler, J.; Dijkstra, T.K.; Sarstedt, M.; Ringle, C.M.; Diamantopoulos, A.; Straub, D.W.; Ketchen, D.J., Jr.; Hair, J.F.; Hult, G.T.M.; Calantone, R.J. Common beliefs and reality about pls: Comments on rönkkö and evermann (2013). *Organ. Res. Methods* **2014**, *17*, 182–209. [CrossRef]

43. Schöb, C.; Armas, C.; Guler, M.; Prieto, I.; Pugnaire, F.I. Variability in functional traits mediates plant interactions along stress gradients. *J. Ecol.* **2013**, *101*, 753–762. [CrossRef]
44. Sánchez-González, M.; Stiti, B.; Chaar, H.; Cañellas, I. Dynamic dominant height growth model for spanish and tunisian cork oak (*Quercus suber* L.) forest. *For. Syst.* **2010**, *19*, 285–298.
45. Bruns, T.D.; Kennedy, P.G. Individuals, populations, communities and function: The growing field of ectomycorrhizal ecology. *New Phytol.* **2009**, *182*, 12–14. [CrossRef] [PubMed]
46. Whittaker, R.H.; Niering, W.A. Vegetation of the Santa Catalina Mountains, Arizona. V. Biomass, production, and diversity along the elevation gradient. *Ecology* **1975**, *56*, 771–790. [CrossRef]
47. Kutiel, P.; Lavee, H. Effect of slope aspect on soil and vegetation properties along an aridity transect. *Isr. J. Plant Sci.* **1999**, *47*, 169–178. [CrossRef]
48. Zhang, Q.-P.; Wang, J.; Gu, H.-L.; Zhang, Z.-G.; Wang, Q. Effects of continuous slope gradient on the dominance characteristics of plant functional groups and plant diversity in alpine meadows. *Sustainability* **2018**, *10*, 4805. [CrossRef]
49. Zheng, Z.; He, S.; Wu, F. Effects of soil surface roughness on sheet erosion and change under different rainfall conditions. *Trans. Chin. Soc. Agric. Eng.* **2010**, *26*, 139–145.
50. Zhang, L.; Wen, Z.M.; Miao, L.P. Source of variation of plant functional traits in the yanhe river watershed: The influence of environment and phylogenetic background. *Acta Ecol. Sin.* **2013**, *33*, 6543–6552. [CrossRef]
51. Calvet, J.-C.; Wigneron, J.-P.; Mougin, E.; Kerr, Y.; Brito, J. Plant water content and temperature of the Amazon forest from satellite microwave radiometry. *IEEE Trans. Geosci. Remote Sens.* **1994**, *32*, 397–408. [CrossRef]
52. Palmer, M.I. *The Effects of Microtopography on Environmental Conditions, Plant Performance, and Plant Community Structure in Fens of the New Jersey Pinelands*; Rutgers The State University of New Jersey: New Brunswick, NJ, USA, 2005.
53. Hao, Y.; Lal, R.; Owens, L.; Izaurralde, R.; Post, W.; Hothem, D. Effect of cropland management and slope position on soil organic carbon pool at the North Appalachian Experimental Watersheds. *Soil Tillage Res.* **2002**, *68*, 133–142. [CrossRef]
54. Qin, J.; Bai, H.; Li, S.; Wang, J.; Gan, Z.; Huang, A. Differences in growth response of larix chinensis to climate change at the upper timberline of southern and northern slopes of mt. Taibai in central qinling mountains, china. *Acta Ecol. Sin.* **2016**, *36*, 5333–5342.
55. Bennie, J.; Huntley, B.; Wiltshire, A.; Hill, M.O.; Baxter, R. Slope, aspect and climate: Spatially explicit and implicit models of topographic microclimate in chalk grassland. *Ecol. Model.* **2008**, *216*, 47–59. [CrossRef]
56. Huang, Y.-M.; Liu, D.; An, S.-S. Effects of slope aspect on soil nitrogen and microbial properties in the chinese loess region. *Catena* **2015**, *125*, 135–145. [CrossRef]
57. Helm, J.; Dutoit, T.; Saatkamp, A.; Bucher, S.F.; Leiterer, M.; Rmermann, C. *Recovery of Mediterranean Steppe Vegetation after Cultivation: Legacy Effects on Plant Composition, Soil Properties and Functional Traits*; John Wiley & Sons, Ltd: Hoboken, NJ, USA, 2019.
58. Sadiq, S.A.; Baloch, D.M.; Ahmed, N. Role of coal-derived humic acid in the availability of nutrients and growth of sunflower under calcareous soil. *J. Anim. Plant Sci.* **2014**, *24*, 1737–1742.
59. Vilagrosa, A.; Chirino, E.; Peguero-Pina, J.-J.; Barigah, T.S.; Cochard, H.; Gil-Pelegrin, E. Xylem Cavitation and Embolism in Plants Living in Water-Limited Ecosystems. In *Plant Responses to Drought Stress*; Springer: Berlin/Heidelberg, Germany, 2012; pp. 63–109.
60. Rosado, B.H.P.; Joly, C.A.; Burgess, S.S.O.; Oliveira, R.; Aidar, M.P.M. Changes in plant functional traits and water use in Atlantic rainforest: Evidence of conservative water use in spatio-temporal scales. *Trees* **2015**, *30*, 47–61. [CrossRef]
61. Atkin, O.K.; Bloomfield, K.J.; Reich, P.B.; Tjoelker, M.G.; Asner, G.P.; Bonal, D.; Bönisch, G.; Bradford, M.G.; Cernusak, L.A.; Cosio, E.G. Global variability in leaf respiration in relation to climate, plant functional types and leaf traits. *New Phytol.* **2015**, *206*, 614–636. [CrossRef]
62. Elgala, A.M.; Amberger, A. Effect OP PH, orgrnic matter and plant growth on the movement of iron in soils. *J. Plant Nutr.* **1982**, *5*, 841–855. [CrossRef]
63. Hernández-Vargas, G.; Sánchez-Velásquez, L.R.; López-Acosta, J.C.; Noa-Carrazana, J.C.; Perroni, Y. Relationship between soil properties and leaf functional traits in early secondary succession of tropical montane cloud forest. *Ecol. Res.* **2019**, *34*, 213–224. [CrossRef]
64. Celik, I. Land-use effects on organic matter and physical properties of soil in a southern Mediterranean highland of Turkey. *Soil Tillage Res.* **2005**, *83*, 270–277. [CrossRef]
65. Goderya, F.S. Field Scale Variations in Soil Properties for Spatially Variable Control: A Review. *Soil Sediment Contam. Int. J.* **1998**, *7*, 243–264. [CrossRef]
66. de Castro, H.I.F. *Effects of Land Use Change on Plant Composition and Ecosystem Functioning in an Extensive Agro-Pastoral System: Plant Functional Traits and Ecosystems Processes*; Universidade de Coimbra: Coimbra, Portugal, 2008.
67. Thakur, S.; Negi, V.S.; Dhyani, R.; Bhatt, I.; Yadava, A. Influence of environmental factors on tree species diversity and composition in the Indian western Himalaya. *For. Ecol. Manag.* **2022**, *503*, 119746. [CrossRef]
68. Kittelson, P.; Maron, J.; Marler, M. An invader differentially affects leaf physiology of two natives across a gradient in diversity. *Ecology* **2008**, *89*, 1344–1351. [CrossRef] [PubMed]
69. Wright, A.J.; Kroon, H.; Visser, E.J.W.; Buchmann, T.; Ebeling, A.; Eisenhauer, N.; Fischer, C.; Hildebrandt, A.; Ravenek, J.; Roscher, C.; et al. Plants are less negatively affected by flooding when growing in species-rich plant communities. *New Phytol.* **2016**, *213*, 645–656. [CrossRef] [PubMed]

70. Wang, G.; Liu, G.; Xu, M. Above- and belowground dynamics of plant community succession following abandonment of farmland on the Loess Plateau, China. *Plant Soil* **2008**, *316*, 227–239. [CrossRef]
71. Jia, Z.; Meng, M.; Li, C.; Zhang, B.; Zhai, L.; Liu, X.; Ma, S.; Cheng, X.; Zhang, J. Rock-Solubilizing Microbial Inoculums Have Enormous Potential as Ecological Remediation Agents to Promote Plant Growth. *Forests* **2021**, *12*, 357. [CrossRef]
72. Van De Weg, M.J.; Meir, P.; Grace, J.; Atkin, O.K. Altitudinal variation in leaf mass per unit area, leaf tissue density and foliar nitrogen and phosphorus content along an Amazon-Andes gradient in Peru. *Plant Ecol. Divers.* **2009**, *2*, 243–254. [CrossRef]

plants

Article

Biomass and Leaf Nutrition Contents of Selected Grass and Legume Species in High Altitude Rangelands of Kashmir Himalaya Valley (Jammu & Kashmir), India

Javed A. Mugloo [1], Mehraj ud din Khanday [2], Mehraj ud din Dar [1], Ishrat Saleem [1], Hesham F. Alharby [3,4], Atif A. Bamagoos [3], Sameera A. Alghamdi [3], Awatif M. Abdulmajeed [5], Pankaj Kumar [6] and Sami Abou Fayssal [7,8,*]

1 Division of Silviculture and Agro Forestry, Faculty of Forestry, Sher-e-Kashmir University of Agricultural Sciences and Technology of Kashmir, Kashmir 190025, India; drjaveed72@gmail.com (J.A.M.); mihraj.dar@gmail.com (M.u.d.D.); ishratsaleem1992@gmail.com (I.S.)
2 Division of Soil Science, Faculty of Horticulture, Sher-e-Kashmir University of Agricultural Sciences and Technology of Kashmir, Kashmir 190025, India; mehraj197@gmail.com
3 Department of Biological Sciences, Faculty of Science, King Abdulaziz University, Jeddah 21589, Saudi Arabia; halharby@kau.edu.sa (H.F.A.); abamagoos@kau.edu.sa (A.A.B.); saalghamdy1@kau.edu.sa (S.A.A.)
4 Plant Biology Research Group, Department of Biological Sciences, Faculty of Science, King Abdulaziz University, Jeddah 21589, Saudi Arabia
5 Biology Department, Faculty of Science, University of Tabuk, Umluj 46429, Saudi Arabia; awabdulmajeed@ut.edu.sa
6 Agro-Ecology and Pollution Research Laboratory, Department of Zoology and Environmental Science, Gurukula Kangri (Deemed to Be University), Haridwar 249404, India; rs.pankajkumar@gkv.ac.in
7 Department of Agronomy, Faculty of Agronomy, University of Forestry, 10 Kliment Ohridski Blvd, 1797 Sofia, Bulgaria
8 Department of Plant Production, Faculty of Agriculture, Lebanese University, Beirut 1302, Lebanon
* Correspondence: sami.aboufaycal@st.ul.edu.lb

Citation: Mugloo, J.A.; Khanday, M.u.d.; Dar, M.u.d.; Saleem, I.; Alharby, H.F.; Bamagoos, A.A.; Alghamdi, S.A.; Abdulmajeed, A.M.; Kumar, P.; Abou Fayssal, S. Biomass and Leaf Nutrition Contents of Selected Grass and Legume Species in High Altitude Rangelands of Kashmir Himalaya Valley (Jammu & Kashmir), India. *Plants* **2023**, *12*, 1448. https://doi.org/10.3390/ plants12071448

Academic Editors: Bingcheng Xu and Zhongming Wen

Received: 12 November 2022
Revised: 10 February 2023
Accepted: 20 March 2023
Published: 25 March 2023

Abstract: The yield and nutritional profile of grass and legume species in Kashmir Valley's rangelands are scantly reported. The study area in this paper included three types of sites (grazed, protected, and seed-sown) divided into three circles: northern, central, and southern Kashmir. From each circle, three districts and three villages per district were selected. Most sites showed higher aboveground biomass (AGB) compared to belowground biomass (BGB), which showed low to moderate effects on biomass. The comparison between northern, central, and southern Kashmir regions revealed that AGB (86.74, 78.62, and 75.22 t. ha^{-1}), BGB (52.04, 51.16, and 50.99 t. ha^{-1}), and total biomass yield (138.78, 129.78, and 126.21 t. ha^{-1}) were the highest in central Kashmir region, followed by southern and northern Kashmir regions, respectively. More precisely, AGB and total biomass yield recorded the highest values in the protected sites of the central Kashmir region, whereas BGB scored the highest value in the protected sites of southern Kashmir region. The maximum yield (12.5 t. ha^{-1}) recorded among prominent grasses was attributed to orchard grass, while the highest crude fiber and crude protein contents (34.2% and 10.4%, respectively), were observed for Agrostis grass. The maximum yield and crude fiber content (25.4 t. ha^{-1} and 22.7%, respectively), among prominent legumes were recorded for red clover. The highest crude protein content (33.2%) was attributed to white clover. Those findings concluded the successful management of Kashmir rangelands in protected sites, resulting in high biomass yields along with the considerable nutritional value of grasses and legumes.

Keywords: biomass production; chemical composition; ecological management; ecosystem diversity; grazing

1. Introduction

The Jammu and Kashmir (J and K) region (India) is well known for its alpine and subalpine pastures [1], locally named "Margs" or "Bahaks". These pastures constitute a crucial ecological resource and play a significant role in the socioeconomic state of the Himalayas Valley. The total area of Kashmir pasturelands is around 9595 km^2 [2]. These pasturelands are rural areas that involve around 97% of the population in the agricultural sector. The rearing of sheep, goats, and cattle constitutes the locals' subsidiary occupation. In addition, a huge population of nomadic Bakerwals, Gujjars, Chopans, Changpas, and Gaddies depends directly on meadow products and pasturelands for herds of maintained livestock.

The term "grassland" (also designated by "pastureland") can be defined as land (and the vegetation growing on it) devoted to the production of introduced or indigenous forage for harvest via grazing, cutting, or both. The grassland's vegetation includes grasses, legumes, and scantly woody species [3]. Thus, grassland is a highly dynamic ecosystem that supports fauna, flora, and human populations worldwide. It also encloses fodder crops that covered approximately 3.5 billion hectares in 2000 and contains around 20% of the world's soil carbon stocks [4]. These stocks can be well enriched through the good management of grasslands via Voisin's rational grazing (VRG), resulting in an increased milk production by ruminants [5]. This system allows the maximization of pasture growth associated with ruminant intake while maintaining a sustainable circular cycle [6]. The adoption of such a system is a must, especially in the Himalayan pasturelands, which was a scene of overgrazing over decades and centuries [7]. The literature reported a decrease in the available grazing area in the alpine and subalpine pasturelands of Kashmir from 0.15 ha. animal^{-1} to 0.10 ha. animal^{-1} between 1977 and 1982 [8]. Unfortunately, it is still at a continuous decrease rate according to a recent report [9].

As the globe is still facing a rise in climatic change, there is an increased demand to monitor, forecast and predict its effect on the productivity and quality of pasturelands [10]. However, the prediction of pasturelands' biomass is sometimes not very reliable, while the accurate and precise estimation consists of traditional methods that are mainly costly, time-consuming, and non-environmentally friendly. In addition to this, the traditional influencing factors (i.e., high wind velocity, low temperature, snowstorms, and so on) and the increase in solar and ultraviolet (UV) radiations—as a direct response to climate change—have resulted in the decreased biomass production of Himalaya pasturelands [11]. Similar observations were acknowledged in Nepal [12] and in the African pasturelands of Niger and Zambia [10,13]. Thus, since the crucial role of pastureland in ecosystem functioning and soil stability is endangered, the safeguarding of the Himalayan alpine and subalpine vegetation exerts its weight [14].

Although earlier studies focused on the aboveground herbaceous species production in a very limited area in Kashmir Valley [15], none have explored the whole region's above and belowground biomass (AGB and BGB, respectively), nor documented reliably the prominent grasses and legumes species growing there. These grasses and legumes are the main diet for sheep, goats, and cattle reared in the region. Thus, as the local population mainly relies on milk production, the compositional quality evaluation of biomass used for sheep, goats, and cattle nutrition is crucial. Therefore, the current study aimed to (a) investigate and estimate the biomass yield and leaf nutritional profile of grasses and legumes in high-altitude pasturelands of the northern, central and southern Kashmir Himalaya Valley (Jammu and Kashmir (J&K)), India; and (b) detect any possible effect (danger i.e., overgrazing) on grassland biomass in the studied zones.

2. Results

2.1. Analysis of Above (AGB), Below Ground Biomass (BGB) and Total Biomass Yield in Northern Kashmir

In northern Kashmir region, the average AGB/BGB ratio was the highest in the grazed and seed-sown sites of Kupwara district (1.61 and 1.85, respectively), and in the protected sites of Baramulla district (1.38) (Table 1). The average ratio of protected sites biomass over grazed sites biomass (R1) was comparable between districts, whereas the average ratio of seed-sown sites' biomass over grazed sites' biomass (R2) was the highest in Kupwara district. The percentages of villages within districts, where AGB > BGB (P1), and seed-sown sites biomass > grazed sites biomass (P3), were the highest in Kupwara district (100%). All districts showed a comparable percentage (100%) of villages where protected sites biomass > grazed sites biomass (P2).

Table 1. Comparison between northern Kashmir districts in terms of biomass production.

Parameter	Kupwara			Baramulla			Bandipora		
	Grazed Sites	Protected Sites	Seed-Sown Sites	Grazed Sites	Protected Sites	Seed-Sown Sites	Grazed Sites	Protected Sites	Seed-Sown Sites
AGB/BGB	1.61a	1.32b	1.85a	1.31b	1.38a	1.65c	1.29b	1.15c	1.81b
R1		2.37a			2.33a			2.36a	
R2		1.48a			1.12b			1.43b	
P1 (%)		100.00a			77.78b			66.67c	
P2 (%)		100.00a			100.00a			100.00a	
P3 (%)		100.00a			33.33c			66.67b	

AGB: above-ground biomass; BGB: below-ground biomass; R1: average ratio of protected sites biomass over grazed sites biomass (protected sites biomass/grazed sites biomass); R2: average ratio of seed-sown sites biomass over grazed sites biomass (seed-sown sites biomass/grazed sites biomass); P1: percentage of villages within district where ABG > BGB; P2: percentage of villages within district where protected sites biomass > grazed sites biomass; P3: percentage of villages within district where seed-sown sites biomass > grazed sites biomass. Values are means; means within the same row followed by different letters are significantly different at $p < 0.05$ according to Duncan's multiple range test.

All northern Kashmir districts showed higher AGB by 32.0%–43.2% in grazed sites compared to BGB, except in Firozpur (Baramulla district) and Ketsan (Bandipora district) where BGB was higher ($p < 0.05$) by 40.0% than AGB (Table 2). In protected sites, AGB was also higher ($p < 0.05$) by a range of 1.4%–40.3% than BGB, except in Ketsan (Bandipora district) where the latter was higher ($p < 0.05$) by 12.3% than the former. In all northern Kashmir districts, seed-grown sites showed higher ($p < 0.05$) AGB by a range of 20.1%–56.4% than BGB, whereas BGB was more abundant than AGB ($p < 0.05$) by 15.3% and 12.1% in Firozpur (Baramulla district) and Ketsan (Bandipora district), respectively. It was also depicted that the average biomass was higher ($p < 0.05$) for AGB than BGB in all districts (24.4%–43.2%), except for Firozpur (Baramulla district) and Ketsan (Bandipora district) (14.8% and 18.6%, respectively). Rajwar (Kupwara district), Gulmarg (Baramulla district), and Cithernaar (Bandipora district) showed the highest AGB and BGB among all districts and studied sites of northern Kashmir region. On the other hand, Gulmarg (Baramulla district) had the highest ($p < 0.05$) AGB among all studied sites, while Rajwar (Kupwara district) and Cithernaar (Bandipora district) had the highest ($p < 0.05$) BGB.

Table 2. Above (AGB) and below Ground Biomass (BGB) (t/ha) in the pasture of northern Kashmir.

District	Village	Biomass Type	Grazed Sites	Protected Sites	Seed-Sown Sites	Average Biomass
Kupwara	Bangas Valley	AGB	0.75Da	1.46Da	1.34Ea	1.19Ea
		BGB	0.51Eb	1.06Gb	1.07Eb	0.80Eb
	Wogubal	AGB	1.11Ca	3.58Ca	1.81Ca	2.16Da
		BGB	0.63Db	2.53Cb	0.81Fb	1.32Cb
	Rajwar	AGB	4.26Ba	5.98Ba	4.83Aa	5.02Ba
		BGB	2.66Bb	5.17Ab	2.32Ab	3.38Ab

Table 2. *Cont.*

District	Village	Biomass Type	Grazed Sites	Protected Sites	Seed-Sown Sites	Average Biomass
Baramulla	Firozpur	AGB	0.51Eb	1.38Ea	1.05Fb	0.98Fb
		BGB	0.85Ca	1.36Fb	1.24Da	1.15Da
	Dragbah	AGB	1.11Ca	3.58Ca	1.65Da	2.11Da
		BGB	0.63Db	2.43Db	0.85Fb	1.30Cb
	Gulmarg	AGB	4.46Aa	6.98Aa	4.81Aa	5.41Aa
		BGB	2.82Ab	4.17Bb	2.22Bb	3.07Bb
Bandipora	Ketsan	AGB	0.51Eb	1.28Fb	1.09Fb	0.96Fb
		BGB	0.85Ca	1.46Ea	1.24Da	1.18Da
	Awathwooth	AGB	1.11Ca	3.58Ca	1.95Ba	2.21Ca
		BGB	0.63Db	2.53Cb	0.85Fb	1.33Cb
	Cithernaar	AGB	4.26Ba	5.98Ba	4.81Aa	5.01Ba
		BGB	2.82Ab	5.17Ab	2.12Cb	3.37Ab

Values are means (3 replicates of each biomass type); means within the same column (comparison between villages in terms of AGB/BGB) followed by different capital letters are significantly different at $p < 0.05$ according to Duncan test; means within the same column (comparison between AGB and BGB within each village) followed by different lowercase letters are significantly different at $p < 0.05$ according to Student's t test.

2.2. Analysis of Above (AGB), Below Ground Biomass (BGB) and Total Biomass Yield in Central Kashmir

In central Kashmir region, the average AGB/BGB ratio was the highest in all studied sites of Budgam district (2.08, 1.72, and 2.11 at grazed, protected and seed-sown sites, respectively), (Table 3). R1 was comparable between Ganderbal and Srinagar districts (2.70), whereas R2 was the highest in Srinagar district (1.53). P1, P2, and P3 were comparable between all central Kashmir districts (100%).

Table 3. Comparison between central Kashmir districts in terms of biomass production.

Parameter	Ganderbal			Budgam			Srinagar		
	Grazed Sites	Protected Sites	Seed-Sown Sites	Grazed Sites	Protected Sites	Seed-Sown Sites	Grazed Sites	Protected Sites	Seed-Sown Sites
AGB/BGB	1.70b	1.46c	1.93b	2.08a	1.72a	2.11a	1.45c	1.57b	1.83c
R1		2.70a			2.62b			2.70a	
R2		1.47b			1.32c			1.53a	
P1 (%)		100.00a			100.00a			100.00a	
P2 (%)		100.00a			100.00a			100.00a	
P3 (%)		100.00a			100.00a			100.00a	

AGB: above-ground biomass; BGB: below-ground biomass; R1: average ratio of protected sites biomass over grazed sites biomass (protected sites biomass/grazed sites biomass); R2: average ratio of seed-sown sites biomass over grazed sites biomass (seed-sown sites biomass/grazed sites biomass); P1: percentage of villages within district where ABG > BGB; P2: percentage of villages within district where protected sites biomass > grazed sites biomass; P3: percentage of villages within district where seed-sown sites biomass > grazed sites biomass. Values are means; means within the same row followed by different letters are significantly different at $p < 0.05$ according to Duncan's multiple range test.

All central Kashmir districts showed a higher AGB ($p < 0.05$) by 19.7%–70.2% in grazed sites than BGB (Table 4). In protected sites, AGB was also higher ($p < 0.05$) by a range of 12.8%–47.9% than BGB. In all central Kashmir districts, seed-grown sites showed higher ($p < 0.05$) AGB by a range of 18.7%–57.1% than BGB. Moreover, the average biomass was higher ($p < 0.05$) for AGB than BGB in all districts (20.8%–55.6%). Narang (Ganderbal district), Kanidajan (Budgam district), and Chirenbal (Srinagar district) showed the highest AGB and BGB among all districts and studied sites of central Kashmir region. On the other hand, Kanidajan (Budgam district) showed significantly higher ($p < 0.05$) AGB and BGB in all sites except protected ones.

Table 4. Above (AGB) and belowground biomass (BGB) (t/ha) in the pasture of central Kashmir.

District	Village	Biomass Type	Grazed Sites	Protected Sites	Seed-Sown Sites	Average Biomass
Ganderbal	Rayil	AGB	0.85Ea	1.56Ga	1.34Fa	1.25Ga
		BGB	0.54Gb	1.36Fb	1.09Db	0.99Gb
	Wangeth	AGB	1.11Ca	4.58Da	1.95Da	2.54Ea
		BGB	0.63Fb	2.93Db	0.85Fb	1.47Fb
	Naranag	AGB	4.26Ba	6.98Ba	4.81Ba	5.35Ca
		BGB	2.42Cb	4.17Cb	2.13Cb	2.91Db
Budgam	Yousmarg	AGB	0.94Da	1.88Ea	1.19Ga	1.33Fa
		BGB	0.28Hb	0.98Hb	0.51Gb	0.59Hb
	Dodhpathri	AGB	1.18Ca	4.68Ca	1.98Da	2.61Da
		BGB	0.72Eb	2.97Db	0.87Fb	1.52Eb
	Kanidajan	AGB	4.76Aa	7.99Aa	5.88Aa	6.21Aa
		BGB	3.82Ab	4.80Bb	3.42Ab	4.01Ab
Srinagar	Astanmarg	AGB	0.73Fa	1.78Fa	1.41Ea	1.31Fa
		BGB	0.55Gb	1.16Gb	1.09Db	0.94Gb
	Zahgemarg	AGB	1.11Ca	4.61Ca	2.09Ca	2.61Da
		BGB	0.83Db	2.83Eb	0.95Eb	1.54Eb
	Chirenbal	AGB	4.26Ba	7.98Aa	4.85Ba	5.71Ba
		BGB	2.52Bb	5.17Ab	2.42Bb	3.37Bb

Values are means (3 replicates of each biomass type); means within the same column (comparison between villages in terms of AGB/BGB) followed by different capital letters are significantly different at $p < 0.05$ according to Duncan test; means within the same column (comparison between AGB and BGB within each village) followed by different lowercase letters are significantly different at $p < 0.05$ according to the Student's t-test.

2.3. Analysis of Above (AGB), Below Ground Biomass (BGB) and Total Biomass Yield in Southern Kashmir

In southern Kashmir region, the average AGB/BGB ratio was the highest in the grazed and seed-sown sites of Anantnag district (1.71 and 1.82, respectively), and the protected sites of Shopian district (1.38) (Table 5). R1 and R2, and P1 and P3 scored the highest values in Anantnag and Kulgam districts, respectively, (2.38 and 1.47, and 100%, respectively). P3 was comparable between all southern Kashmir districts (100%).

Table 5. Comparison between southern Kashmir districts in terms of biomass production.

Parameter	Anantnag			Shopian			Kulgam		
	Grazed Sites	Protected Sites	Seed-Sown Sites	Grazed Sites	Protected Sites	Seed-Sown Sites	Grazed Sites	Protected Sites	Seed-Sown Sites
AGB/BGB	1.71a	1.18b	1.82a	1.31c	1.38a	1.69c	1.43b	1.13b	1.71b
R1		2.38a			2.19c			2.23b	
R2		1.43b			1.31c			1.47a	
P1 (%)		100.00a			66.67b			66.67b	
P2 (%)		100.00a			100.00a			100.00a	
P3 (%)		66.67b			66.67b			100.00a	

AGB: above-ground biomass; BGB: below-ground biomass; R1: average ratio of protected sites biomass over grazed sites biomass (protected sites biomass/grazed sites biomass); R2: average ratio of seed-sown sites biomass over grazed sites biomass (seed-sown sites biomass/grazed sites biomass); P1: percentage of villages within district where ABG > BGB; P2: percentage of villages within district where protected sites biomass > grazed sites biomass; P3: percentage of villages within district where seed-sown sites biomass > grazed sites biomass. Values are means; means within the same row followed by different letters are significantly different at $p < 0.05$ according to Duncan's multiple range test.

Particularly, all southern Kashmir districts showed a higher AGB ($p < 005$) by 31.6%–51.9% in grazed sites than BGB, except in Dabjan (Shopian district) and Astanmarg (Kulgam district) where BGB was higher ($p < 0.05$) by 30.6% and 42.0% than AGB, respectively

(Table 6). The same trend was observed regarding protected sites where AGB were higher ($p < 0.05$) by a range of 2.0%–40.3% than BGB, except in Dabjan (Shopian district) and Astanmarg (Kulgam district) where BGB was higher ($p < 0.05$) by 5.4% and 18.0% than AGB, respectively. In all southern Kashmir districts, seed-grown sites showed higher ($p < 0.05$) AGB by a range of 17.4%–56.6% than BGB, whereas BGB was more abundant ($p < 0.05$) by 15.3% and 12.8% than AGB in Dabjan (Shopian district) and Astanmarg (Kulgam district), respectively. Also, the average biomass was higher ($p < 0.05$) for AGB over BGB in all districts (14.9%–44.3%), except for Dabjan (Shopian district) and Astanmarg (Kulgam district) (15.1% and 22.0%, respectively). Aru Valley (Anantnag district), Kaller (Shopian district), and Chirenbal (Kulgam district) showed the highest AGB and BGB among all districts and studied sites. On the other hand, AGB was the highest ($p < 0.05$) in Kaller (Shopian district) in grazed and protected sites, and in Chirenbal (Kulgam district) among seed-sown sites. BGB was the highest ($p < 0.05$) in Chirenbal (Kulgam district) in grazed and seed-grown sites, and in Aru Valley (Anantnag district) in protected sites.

Table 6. Above (AGB) and belowground biomass (BGB) (t/ha) in the pasture of southern Kashmir.

District	Village	Biomass Type	Grazed Sites	Protected Sites	Seed-Sown Sites	Average Biomass
Anantnag	Daksum	AGB	0.79Ga	1.51Fa	1.32Fa	1.21Ea
		BGB	0.54Fb	1.48Gb	1.09Eb	1.03Eb
	Alhan	AGB	1.25Ea	3.74Ca	1.98Da	2.32Ca
		BGB	0.65Eb	2.58Db	0.91Fb	1.38Cb
	Aru Valley	AGB	4.69Aa	5.98Ba	4.81Ba	5.16Ba
		BGB	2.68Bb	5.57Ab	2.32Bb	3.52Ab
Shopian	Dabjan	AGB	0.59Hb	1.41Gb	1.05Gb	1.01Fb
		BGB	0.85Ca	1.49Ga	1.24Da	1.19Da
	Hirpora	AGB	1.21Fa	3.68Da	1.85Ea	2.24Da
		BGB	0.77Db	2.43Eb	0.92Fb	1.37Cb
	Kaller	AGB	4.70Aa	6.98Aa	4.91Ba	5.51Aa
		BGB	2.82Ab	4.17Cb	2.22Cb	3.07Bb
Kulgam	Astanmarg	AGB	0.51Hb	1.28Hb	1.09Gb	0.96Gb
		BGB	0.88Ca	1.56Fa	1.25Da	1.23Da
	Zahgemarg	AGB	1.31Da	3.58Ea	2.05Ca	2.31Ca
		BGB	0.63Eb	2.53Db	1.05Eb	1.40Cb
	Chirenbal	AGB	4.56Ba	5.98Ba	5.81Aa	5.45Aa
		BGB	2.82Ab	5.19Bb	2.52Ab	3.51Ab

Values are means (3 replicates of each biomass type); means within the same column (comparison between villages in terms of AGB/BGB) followed by different capital letters are significantly different at $p < 0.05$ according to Duncan test; means within the same column (comparison between AGB and BGB within each village) followed by different lowercase letters are significantly different at $p < 0.05$ according to the Student's t test.

2.4. Comparison of Above (AGB), Belowground Biomass (BGB) and Total Biomass Yield between Kashmir Regions

The comparison between northern, central, and southern Kashmir regions revealed that AGB (86.74, 78.62, and 75.22 t. ha^{-1}), BGB (52.04, 51.16, and 50.99 t. ha^{-1}), and total biomass yield (138.78, 129.78, and 126.21 t. ha^{-1}) were the highest in central Kashmir region, followed by southern and northern Kashmir ones, respectively, (Figure 1). More precisely, AGB and total biomass yield recorded the highest values in the protected sites of central Kashmir region (42.04 and 68.41 t. ha^{-1}, respectively), whereas BGB scored the highest value in the protected sites of southern Kashmir region (27.00 t. ha^{-1}).

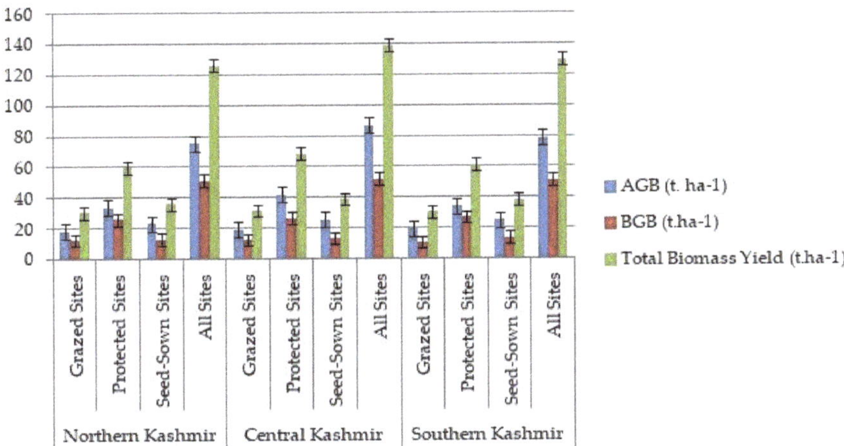

Figure 1. AGB, BGB and total biomass yield (t. ha^{-1}) in studied circles.

2.5. Yield and Nutrient Profile of Prominent Grasses and Legumes

The results of prominent grasses' yield and nutrient profile evaluation are shown in Table 7. Orchard grass (*Dactylis glomerata*) showed the highest yield ($p < 0.05$) (12.50 $\pm$ 0.6 t. ha^{-1}) compared to other grasses, whereas, the highest nitrogen content ($p < 0.05$) was detected in Timothy grass (1.81 $\pm$ 0.05%). Phosphorus and crude protein contents were the most abundant ($p < 0.05$) in Agrostis grass (*Agrostis alba*) (0.34 $\pm$ 0.04% and 10.40 $\pm$ 0.5%, respectively), whereas perennial ryegrass (*Lolium perenne*) enclosed the highest ($p < 0.05$) potassium and crude fiber contents (0.44 $\pm$ 0.05% and 35.12 $\pm$ 0.8%, respectively). It should be noted that no significant difference ($p > 0.05$) and very low standard deviations (SDs) (0.2 < SD < 0.6, and 0.01 < SD < 0.8) were observed in terms of grass yield, and leaf nutrient content, respectively, between all studied sites in the three Kashmir regions.

Table 7. Yield and leaf nutrient profile of prominent grasses.

Grasses	Yield (t. ha^{-1})	N (%)	p (%)	K (%)	Crude Fiber (%)	Crude Protein (%)
Dactylis glomerata (Orchard grass)	12.50 $\pm$ 0.6a	1.51 $\pm$ 0.04b	0.07 $\pm$ 0.01d	0.31 $\pm$ 0.04c	15.30 $\pm$ 0.6b	10.30 $\pm$ 0.5a
Festuca arundinacea (Tall fescue grass)	11.80 $\pm$ 0.6ab	1.22 $\pm$ 0.03c	0.18 $\pm$ 0.02c	0.41 $\pm$ 0.05ab	11.20 $\pm$ 0.6b	7.50 $\pm$ 0.3b
Lolium perenne (Perennial rye grass)	10.23 $\pm$ 0.5b	1.29 $\pm$ 0.03c	0.29 $\pm$ 0.03b	0.44 $\pm$ 0.05a	35.12 $\pm$ 0.8a	5.10 $\pm$ 0.2c
Phleum pratense (Timothy grass)	8.57 $\pm$ 0.4c	1.81 $\pm$ 0.05a	0.21 $\pm$ 0.03c	0.39 $\pm$ 0.04b	31.20 $\pm$ 0.8a	9.50 $\pm$ 0.4a
Agrostis alba (Agrostis grass)	3.51 $\pm$ 0.2d	1.23 $\pm$ 0.03c	0.34 $\pm$ 0.04a	0.23 $\pm$ 0.03d	34.20 $\pm$ 0.8a	10.40 $\pm$ 0.5a

Values are means (3 replicates of each grass); means within the same column followed by different letters are significantly different at $p< 0.05$ according to Duncan's multiple range test.

Table 8 showed that red clover (*Trifolium pretense*) had the highest ($p < 0.05$) yield, nitrogen, potassium, and crude protein contents (25.40 $\pm$ 0.6 t. ha^{-1}, 1.69 $\pm$ 0.05%, 0.45 $\pm$ 0.05%, and 22.70 $\pm$ 0.6%, respectively), among prominent legumes found in Kashmir Valley, whereas the phosphorus content was the most abundantly found ($p < 0.05$) in white clover (*Trifolium repens*) and sainfoin (*Onobrychis viciifolia*) (0.34 $\pm$ 0.04%). In addition, the crude fiber content scored its highest value in white clover (33.12 $\pm$ 0.8%) compared to other prominent legumes studied. It should be noted that no significant difference ($p > 0.05$) and very low SDs (0.3 < SD < 0.6, and 0.02 < SD < 0.8) were observed in terms of legume yield and leaf nutrient content, respectively, between all studied sites in the three Kashmir regions.

Table 8. Yield and leaf nutrient profile of prominent legumes.

Legumes	Yield (t. ha^{-1})	N (%)	*p* (%)	K (%)	Crude Fiber (%)	Crude Protein (%)
Trifolium pratense (Red clover)	25.40 ± 0.6a	1.69 ± 0.05a	0.33 ± 0.04a	0.45 ± 0.05a	29.89 ± 0.7a	22.70 ± 0.6a
Trifolium repens (White clover)	24.20 ± 0.6ab	1.23 ± 0.03c	0.34 ± 0.04a	0.25 ± 0.03c	33.12 ± 0.8a	21.10 ± 0.6a
Medicago sativa L. (Lucerne)	23.40 ± 0.6b	1.67 ± 0.05a	0.31 ± 0.04a	0.43 ± 0.05a	11.50 ± 0.6b	18.80 ± 0.5b
Onobrychis viciifolia (Sainfoin)	7.50 ± 0.3d	1.19 ± 0.02c	0.34 ± 0.04a	0.23 ± 0.03c	10.20 ± 0.5b	15.40 ± 0.5c
Securigera varia (Crown vetch)	9.30 ± 0.4c	1.47 ± 0.04b	0.11 ± 0.02b	0.35 ± 0.04b	9.23 ± 0.4b	11.56 ± 0.5d

Values are means (3 replicates of each grass); means within the same column followed by different letters are significantly different at $p < 0.05$ according to Duncan's multiple range test.

3. Discussion

3.1. Analysis of Above (AGB), Belowground Biomass (BGB), and Total Biomass Yield

The healthy functioning of an ecosystem can be estimated by the overall plant biomass it yields. Over-grazing is one of the most critical issues facing plant biomass in Kashmiri grasslands. The current study detected higher AGB compared to BGB and a positive AGB/BGB ratio in all districts of northern, central, and southern Kashmir regions. This simulates that the studied locations were not over-grazed. However, the high BGB in Aru Valley, Kaller and Chirenbal (northern Kashmir), Rajwar, Gulmarg, and Cithernaar (southern Kashmir), and Naranag, Kanidajan, and Chirenbal (central Kashmir) might reveal possible moderate grazing within these regions. Such regions might be abandoned after being heavily grazed in elder decades. In this regard, Dai et al. [16] reported that the moderate grazing promoted the root biomass of *Kobresia* meadow (BGB) in the northern Qinghai-Tibet pastures, which corroborates with our findings. Thus, the adaptive response of plants might occur in which they tend to increase their root development to survive [17]. Also, during the April-May period, snow melting occurs in the pasturelands of Kashmir Valley which naturally favors and promotes the growth and development of BGB. Furthermore, the protected areas (sites) by local authorities may have helped in the preservation of plant biomass away from rearing and over-grazing, thus resulting in increased biomass yields. This is consistent with the report of Lone and Pandit [18] on the Langate Forest division of Kashmir. Tittonell et al. [19] proposed an agroecological research agenda, suggesting species breeding to preserve diversity and a co-innovation of large-scale farming with farmers, policymakers, and value chains. Although site protection resulted in substantial improvements in rangeland grazing management in Namibia, it did not enhance cattle productivity nor rangeland health [20]. The interesting observation in the present study is that all studied species were found in grazed, protected, and seed-sown sites, which outlines again that grasslands were not over-grazed in the studied locations. Two decades back, it was reported that dry matter biomass yield ranged between 1.41 and 6.23 t. ha^{-1} in temperate pastures of the northwestern Himalayas [21], which is far below that observed in the current study. This could be explained by the fact that Kashmiri citizens are now more conscious and aware of the risks on grassland biomass and the disequilibrium of the environmental balance associated with overgrazing. On the other hand, climatic contrasts, species heterogeneity, and anthropogenic disturbances were reported to affect the biomass yield in the lesser Himalayan foothills, and northwestern regions of Kashmir Valley [22,23].

This simulates a possible inclusion of non-native species to the studied sites, resulting in the variation of soil organic carbon (SOC), and thus a variation in biomass yields [24]. Our findings outlined that AGB, BGB, and total biomass yield were the highest in central Kashmir, followed by southern and northern Kashmir. Such a variation in biomass yields was reported to be correlated with the variation in carbon sequestration potential (CSP) [25]. Panwar et al. [25] reported a high CSP in northern India (J & K as a whole state), associated with high biomass yields (AGB: 6.7–159.4 t. ha^{-1}; BGB: 1.6–71.5 t. ha^{-1}; total biomass yield: 15.9–202.6 t. ha^{-1}). However, the study was very general and did not take into consideration the difference between Kashmiri regions in terms of vegetative populations nor site types (grazed, protected and seed-sown sites). Despite that, the AGB, BGB, and total biomass yields observed in the present study fall within the ranges stated in the aforementioned ones.

3.2. Yield and Nutrient Profile of Prominent Grasses and Legumes

The nutritional composition of grasses used for animal forage is a factor determining their growth, reproduction, and livestock production. Additionally, the climate, soil type, and degree of maturity of grasses influence their nutritional composition [26,27]. Sampling was performed during July which means that the studied grasses and legumes were at their harvest stage [28]. This simulates that their nutrient richness may have started to decline [29]. On the other hand, Hao and He [30] outlined an increased biomass yield once a nutrient loss occurs. This statement partly agrees with our findings as biomass yield was satisfying while grasses were highly nutritious. Leaf N, P, and K contents in orchard grass was several folds higher than outlined in other grasslands (N: 0.22–0.26%, p: 0.002–0.004%, K: 0.007–0.02%) [31]. A previous study on Chinese seed-sown pastures outlined leaf P and K contents in tall fescue grass higher by 1.7-fold and 4.9-fold than our findings [32]. Generally, leaf nitrogen content in grasses should not exceed 3.5% [33]; thus, our findings showed safe values. Leaf phosphorus content in grasses does not usually exceed 0.7–0.8% [34], which agrees with our findings, whereas leaf potassium content can range between 1.2 and 2.0% [35]. This simulates that the selected grasses may show some K deficiency. Moreover, Chang et al. [32] depicted a crude protein content in the range of 11–14%, being 1.5–1.9-fold higher than observed in the present study. On the other hand, the crude protein and crude fiber contents in selected legumes was promising. It was recommended that 12–19% of crude protein would be suitable for cattle feed [36]; which means that some of the selected grasses can be mixed with red or white clover to improve the protein requirements for rearing cattle. Thus, the high management of grasslands is proposed to increase the crude protein and crude fiber contents in selected grasses. In this context, Berauer et al. [37] reported that crude protein in grasses increased by 22–30% after high land management. Furthermore, the moderate nitrogen percentages in the studied grasses revealed the absence of an over-grazing activity in the studied sites. Dong et al. [38] explained that increased N rates, N mineralization, and nitrification processes occur when associated with over-grazing. It is worth noting that animals have different behaviors based on their preference. For instance, cattle and sheep diets are mainly based on grasses and legumes while goats' diet is more related to herbal biomass [39]. Therefore, the present grasslands studied enclose nutritious grasses and legumes highly abundant for cattle and sheep foraging. This cannot be achieved without the inhibition of biomass species' eradication unless correct and serious management of lands is performed associated with new technology that is timesaving and has a lower negative impact on the environment.

4. Materials and Methods

4.1. Study Area and Sites Description

The study area enclosed the whole Jammu and Kashmir (J and K) Valley, which was divided into three zone circles: northern, central, and southern. Within each circle, three districts were selected, and subsequently, three sites within each district were chosen (Table 9). All studied sites varied in elevation between 1450 and 4800 m above mean sea level. Studied areas enclosed: (a) Grazed Sites, (b) Protected Sites, and (c) Seed-Sown Sites. Grazed sites are grassland areas covered by grasses and legumes that are suitable for livestock grazing. Protected sites are grassland areas where carbon emissions from land use change are limited, and nutrient sapping is avoided. This included the cultivation of trees as windbreaks to reduce soil erosion and crop rotation to keep good nutrient availability in soil. Seed-sown sites are grassland areas which were seeded with native engendered species due to overgrazing or extensive agricultural exploitation. Seeded species included: orchard, tall fescue, perennial rye, and Agrostis grasses, and legumes such as: red clover, white clover, lucerne, sainfoin, and crown vetch. These species were sown in a randomized complete block design (RCBD) between mid-August and mid-September at a rate of 36 g seeds/m^2. A harrowing process was also practiced for initial soil preparation in order to optimize seedling establishment.

Table 9. Spatial distribution of studied sites.

Zone Name	District Name	Village Name	District Coordinates	District Altitude
Northern Kashmir	Kupwara	Bangus Valley, Wogubal, Rajwar	34°18′–34°47′ N, 73°45′–74°30′ E	2000–3500 m
	Baramulla	Firozpur, Dragbah, Gulmarg	34°11′–34°19′ N, 74°21′–74°36′ E	1630–2085 m
	Bandipora	Ketsan, Awathwooth, Cithernaar	34°25′–34°41′ N, 74°39′–74°65′ E	2700–4800 m
Central Kashmir	Srinagar	Baedhmargh Reshipora, Astanmarg Dara, Syedpora Bla	34°05′–34°08′ N, 74°50′–74°83′ E	1450–3942 m
	Budgam	Yousmarg, Dodhpathri, Kanidajan	33°93′–34°02′ N, 74°69′–74°79′ E	2000–2730 m
	Ganderbal	Rayil, Wangeth, Naranag	33°44′–33°73′ N, 75°09′–75°15′ E	1716–3397 m
Southern Kashmir	Anantnag	Daksum, Ahlan, Aru Valley	33°36′–34°25′ N, 75°02′–75°59′ E	1600–1723 m
	Shopian	Dabjan, Hirpora, Kaller	33°43′–33°72′ N, 74°50′–74°83′ E	1650–4720 m
	Kulgam	Astanmarg, Zahgemarg, Chirenbal	33°55′–33°78′ N, 74°90′–75°17′ E	1740–4800 m

4.2. Climate

The union territory of J and K, India (33°17′–37°20′ N latitude, 73°25′–80°30′ E longitude) comprises two main physical regions: Outer Himalayas with sub-tropical and intermediate climate (Jammu), and Inner Himalayas with a temperate climate (Kashmir). The climate varies considerably with altitude; it is mild and salubrious in lower altitudes but very cold in higher-ups. Spring is cool and rather wet. Regarding the Outer Himalayas (Jammu), average minimum and maximum temperatures vary between −11 °C and 33 °C during winter and summer, respectively. Autumn is bright and pleasant, while winter is extremely cold and experiences heavy snowfalls. Frost is experienced from the middle of November onwards. The main form of precipitation is snow in winter and some stray rains, and showering in spring. The Jammu region receives an average annual precipitation of about 1103 mm in the form of rain and snow for about 70 days. Unlike the Outer Himalayas, there is no distinct rainy season in the Inner Himalayas. The minimum temperature of the Kashmir region falls within −7 °C in winter and the maximum goes up to 35 °C in summer. It is characterized by a mean minimum temperature below 8 °C for more than six consecutive months per year. Mean annual minimum and maximum temperatures range between 6.68 and 19.31 °C, respectively, and mean annual soil temperature ranges between 8 and 15 °C, thus the area belongs to the mesic temperature regime. The mean annual rainfall in the Inner Himalayas is 710 mm and the soil in the studied area does not remain generally dry for more than 90 cumulative days. Hence, it belongs to the udic moisture regime.

4.3. Vegetation Diversity

Several types of grasses and legumes growing in the studied Kashmir regions are outlined in Table 10.

Table 10. Diversity of grasses and legumes in the studied Kashmir regions.

Zone	Grasses	Legumes
Northern Kashmir	*Dactylis glomerata*	*Trifolium pratense*
	Festuca arundinacea	*Trifolium repens*
	Lolium perenne	*Onobrychis viciifolia*
	Phleum pratense	*Medicago sativa*
	Bromus unioloides	*Securigera varia*
	Phalaris spp.	
	Poa pratensis	
	Lolium multiflorum	
	Agrostis alba	
	Avena sativa	
Central Kashmir	*Dactylis glomerata*	*Trifolium alexandrinum*
	Festuca arundinacea	*Stylosanthus hamata*
	Lolium perenne	*Macroptilium atropupreum*
	Dicanthium annulatum	*Trifolium pratense*
	Chloris gayana	*Trifolium repens*
	Chrysopogon fulvus	*Onobrychis viciifolia*
	Heteropogon contortus	*Medicago sativa*
	Agrostis alba	*Securigera varia*
	Setaria spp.	
	Avena sativa	
	Phleum pratense	
Southern Kashmir	*Dactylis glomerata*	*Trifolium pratense*
	Festuca arundinacea	*Trifolium repens*
	Lolium perenne	*Onobrychis viciifolia*
	Agrostis alba	*Medicago sativa*
	Phleum pratense	*Securigera varia*
	Avena sativa	
	Setaria spp.	

4.4. Sampling

Samples were collected following a direct field plot harvest method [40]. Briefly, three transects, divided into two blocks (100 m distant from each other) were performed. Moreover, three quadrants of 1 m^2 each were performed within each block. All grasses and legumes, within these quadrants, were collected from their roots by digging 5 cm^2 pits up to a depth of 30 cm. Then, they were packed in ice-cooled-bags, transported directly to the laboratory for identification, and stored at a cool temperature for further analyses.

4.5. Compositional Analyses

Before analyses, the vegetative components including roots were washed thoroughly under a jet of running tap water to remove the attached soil. Then, they were dipped in diluted HCl (1 mL concentrated HCL.L^{-1} water) [41]. Further washing was performed with de-ionized water. Sampled species were first identified as grasses or legumes, then the above- and belowground biomass yields were estimated in t. ha^{-1} [41]. Afterward, the roots were removed, and the clean leaf samples were dried in a hot air-circulating oven at 105 °C for 24 h until a constant weight is obtained. Then, they were girded for nutrient analysis [41].

The nitrogen content was estimated using the micro Kjeldahl method [42]. The phosphorus content was determined following the vanado-molybdo-phosphoric Acid yellow colorimetric method [43]. Briefly, the dissolved reactive phosphorus reacted with ammonium molybdate under acid conditions. Hence, the molybdo-phosphoric acid was formed, and a yellow vanado-molybdo-phosphoric acid was obtained in the presence of vanadium. This corresponds to the phosphorus concentration. Such concentration was detected at a wavelength of 470 nm using Thermo Spectronic Helios Gamma UV (Hellma model:

178.712-QS, flow cell: 10 mm light path, inner optical volume: 30 L), connected to a Kipp & Zonen BD112 recorder [43].

The potassium content was determined using the FLAPHO flame photometer method [44]. Briefly, the instrument was warmed up for 10 min and distilled water was fed to the instrument. Then, the indicators were adjusted to 0 (reading). The concentrated standard solution was aspirated, and the readout was adjusted to 90 (on the uppermost scale). Afterward, the distilled water was aspirated, and the instrument read 0. All standards and standard solutions were aspirated, and results were recorded. Then, the calibration curves were drawn; the potassium concentration corresponded to the abscissa, whereas the instrument readouts corresponded to the ordinate. Finally, the potassium concentration was noted.

Using the acid–alkali digestion method, the residue after acid and alkaline digestion (determined gravimetrically) corresponded to the crude fiber content [45]. The crude protein content was calculated after the determination of the leaf nitrogen via the micro Kjeldahl method. The leaf nitrogen content value was multiplied by a coefficient factor of 6.25 [46]. All compositional elements were expressed as percentage (%) dry matter.

4.6. Statistical Analysis

One-way ANOVA, Duncan, and Student's t tests were applied for data analysis using the SPSS 25® program. A confidence level of 95% ($p = 0.05$) was adopted for all statistical tests. A comparison between regions districts was performed (different lower-case letters refer to a significant difference) in terms of AGB/BGB, R1, R2, P1, P2, and P3 (Tables 1, 3, and 5). A comparison between villages of different districts was performed (different capital letters refer to a significant difference) as well as between AGB and BGB (different lowercase letters refer to a significant difference (Tables 2, 4, and 6). In a similar vein, a comparison between grasses and legume types in terms of yield, and compositional elements (N, P, K, crude fiber and crude protein), was performed (different letters refer to a significant difference) (Tables 7 and 8).

5. Conclusions

Kashmir Valley rangelands (northern, central and southern Kashmir regions) were investigated for their AGB, BGB, and total biomass (grasses/legumes) yields in three site types (grazed, protected, and seed-sown) along with their leaf nutritional profiles (N, P, K, crude fiber, and crude protein contents). Results showed an overall moderate grazing with a low to moderate effect on biomass. AGB, BGB, and total biomass yields were the highest in central Kashmir, followed by southern, and northern Kashmir. AGB and total biomass yields recorded the highest values in the protected sites of central Kashmir region, whereas, BGB yield scored the highest value in the protected sites of southern Kashmir region. On the other hand, Agrostis grass showed the highest crude fiber and crude protein contents among grasses found in the studied regions, whereas the highest crude fiber and crude protein contents among prominent legumes were recorded for red clover and white clover, respectively. Those findings concluded the successful management of Kashmir rangelands in protected sites, resulting in high biomass yields along with the considerable nutritional value of grasses and legumes.

Author Contributions: Conceptualization, J.A.M. and M.u.d.K.; methodology, J.A.M. and M.u.d.K.; software, P.K. and S.A.F.; validation, M.u.d.D., I.S., H.F.A., A.A.B., S.A.A., A.M.A., P.K. and S.A.F.; formal analysis, S.A.F.; investigation, S.A.F.; resources, H.F.A., A.A.B., S.A.A. and A.M.A.; data curation, S.A.F.; writing—original draft preparation, M.u.d.K. and S.A.F.; writing—review and editing, I.S., P.K. and S.A.F.; visualization, M.u.d.D., I.S., H.F.A., P.K. and S.A.F.; supervision, M.u.d.K. and S.A.F.; project administration, S.A.F.; funding acquisition, H.F.A. and S.A.F. All authors have read and agreed to the published version of the manuscript.

Funding: This work was funded by Institutional Fund Projects under grant No. (IFPIP: 394-130-1443), Ministry of Education in Saudi Arabia.

Institutional Review Board Statement: Not applicable.

Informed Consent Statement: Not applicable.

Data Availability Statement: All data used in the present study are included in the manuscript.

Acknowledgments: This research work was funded by Institutional Fund Projects under grant No. (IFPIP: 394-130-1443). The authors gratefully acknowledge technical and financial support provided by the Ministry of Education and King Abdulaziz University, DSR, Jeddah, Saudi Arabia.

Conflicts of Interest: The authors declare no conflict of interest.

References

1. Haq, S.M.; Hassan, M.; Jan, H.A.; Al-Ghamdi, A.A.; Ahmad, K.; Abbasi, A.M. Traditions for Future Cross-National Food Security–Food and Foraging Practices among Different Native Communities in the Western Himalayas. *Biology* **2022**, *11*, 455. [CrossRef] [PubMed]
2. Singh, J.P.; Dev, I.; Deb, D.; Chaurasia, R.S.; Radotra, S. Identification and characterization of pastureland and other grazing resources of Jammu & Kashmir using GIS and satellite remote sensing technique. In *Proceedings of the XXIII International Grassland Congress, Sustainable Use of Grassland Resources for Forage Production, Biodiversity and Environmental Protection*; Roy, M.M., Malaviya, D.R., Yadav, V.K., Singh, T., Sah, R.P., Vijay, D., Radhakrishna, A., Eds.; Range Management Society of India: New Delhi, India, 2015.
3. Louhaichi, M.; Gamoun, M.; Hassan, S.; Abdallah, M.A.B. Characterizing Biomass Yield and Nutritional Value of Selected Indigenous Range Species from Arid Tunisia. *Plants* **2021**, *10*, 2031. [CrossRef]
4. FAO. Food and Agriculture Organization of the United Nations—Statistic Division (FAOSTAT). 2015. Available online: http://faostat3.fao.org/home/E (accessed on 20 August 2022).
5. Seó, H.L.S.; Machado Filho, L.C.P.; Brugnara, D. Rationally Managed Pastures Stock More Carbon than No-Tillage Fields. *Front. Environ. Sci.* **2017**, *5*, 87. [CrossRef]
6. Pinheiro Machado Filho, L.C.; Seó, H.L.S.; Daros, R.R.; Enriquez-Hidalgo, D.; Wendling, A.V.; Pinheiro Machado, L.C. Voisin Rational Grazing as a Sustainable Alternative for Livestock Production. *Animals* **2021**, *11*, 3494. [CrossRef] [PubMed]
7. Apollo, M.; Andreychouk, V.; Bhattarai, S. Short-Term Impacts of Livestock Grazing on Vegetation and Track Formation in a High Mountain Environment: A Case Study from the Himalayan Miyar Valley (India). *Sustainability* **2018**, *10*, 951. [CrossRef]
8. Misri, B. *Improvement of Sub-Alpine and Alpine Himalayan Pastures. Research Centre*; Indian Grassland and Fodder Research Institute, HPKV Campus: Palampur, India, 2003.
9. Qureshi, R.A.; Ghufran, M.A.; Gilani, S.A.; Sultana, K.; Ashraf, M. Ethnobotanical studies of selected medicinal plants of Sudhan Gali and Ganga Chotti hills, district Bagh, Azad Kashmir. *Pak. J. Bot.* **2007**, *39*, 2275–2283.
10. Clementini, C.; Pomente, A.; Latini, D.; Kanamaru, H.; Vuolo, M.R.; Heureux, A.; Fujisawa, M.; Schiavon, G.; Del Frate, F. Long-Term Grass Biomass Estimation of Pastures from Satellite Data. *Remote Sens.* **2020**, *12*, 2160. [CrossRef]
11. Sulzberger, B.; Austin, A.T.; Cory, R.M.; Zepp, R.G.; Paul, N.D. Solar UV radiation in a changing world: Roles of cryosphere-land-water-atmosphere interfaces in global biogeochemical cycles. *Photochem. Photobiol. Sci.* **2019**, *18*, 747–774. [CrossRef]
12. Paudel, K.P.; Andersen, P. Assessing rangeland degradation using multi temporal satellite images and grazing pressure surface model in Upper Mustang, Trans Himalaya, Nepal. *Remote Sens. Environ.* **2010**, *114*, 1845–1855. [CrossRef]
13. Schucknecht, A.; Meroni, M.; Kayitakire, F.; Rembold, F.; Boureima, A. Biomass estimation to support pasture management in Niger. *Int. Arch. Photogramm. Remote Sens. Spat. Inf. Sci.* **2015**, *XL-7/W3*, 109–114. [CrossRef]
14. Kuniyal, J.C.; Maiti, P.; Kumar, S.; Kumar, A.; Bisht, N.; Sekar, K.C.; Arya, S.C.; Rai, S.; Nand, M. Dayara bugyal restoration model in the alpine and subalpine region of the Central Himalaya: A step toward minimizing the impacts. *Sci. Rep.* **2021**, *11*, 16547. [CrossRef] [PubMed]
15. Saleem, I.; Mugloo, J.A.; Anjum Baba, A.; Buch, K. Biomass estimation of herbaceous species of Benhama area, Kashmir. *J. Pharmacogn. Phytochem.* **2019**, *8*, 2926–2929.
16. Dai, L.; Guo, X.; Ke, X.; Zhang, F.; Li, Y.; Peng, C.; Shu, K.; Li, Q.; Lin, L.; Cao, G.; et al. Moderate grazing promotes the root biomass in *Kobresia* meadow on the northern Qinghai-Tibet Plateau. *Ecol. Evol.* **2019**, *9*, 9395–9406. [CrossRef] [PubMed]
17. Qu, K.; Cheng, Y.; Gao, K.; Ren, W.; Fry, E.L.; Yin, J.; Liu, Y. Growth-Defense Trade-Offs Induced by Long-term Overgrazing Could Act as a Stress Memory. *Front. Plant Sci.* **2022**, *13*, 917354. [CrossRef]
18. Lone, H.A.; Pandit, A.K. Impact of grazing on community features and biomass of herbaceous species in Langate Forest Division of Kashmir. *Indian For.* **2007**, *133*, 93–100.
19. Tittonell, P.; Piñeiro, G.; Garibaldi, L.A.; Dogliotti, S.; Olff, H.; Jobbagy, E.G. Agroecology in Large Scale Farming—A Research Agenda. *Front. Sustain. Food Syst.* **2020**, *4*, 584605. [CrossRef]
20. Coppock, D.L.; Crowley, L.; Durham, S.L.; Groves, D.; Jamison, J.C.; Karlan, D.; Norton, B.E.; Ramsey, R.D. Community-based rangeland management in Namibia improves resource governance but not environmental and economic outcomes. *Commun. Earth Environ.* **2022**, *3*, 32. [CrossRef]
21. Sharma, B.R.; Koranne, K.D. Present status and management strategies for increasing biomass production in North-Western Himalayan rangelands. In *Rangelands—Resources and Management*; Singh, P., Pathak, P.S., Eds.; Range Management Society of India, Indian Grassland and Fodder Research Institute: Jhansi, India, 1988; pp. 138–147.

22. Dad, J.M. Organic carbon stocks in mountain grassland soils of northwestern Kashmir Himalaya: Spatial distribution and effects of altitude, plant diversity and land use. *Carbon Manag.* **2019**, *10*, 149–162. [CrossRef]

23. Khan, R.W.A.; Shaheen, H.; Awan, S.N. Biomass and soil carbon stocks in relation to the structure and composition of Chir Pine dominated forests in the lesser Himalayan foothills of Kashmir. *Carbon Manag.* **2021**, *12*, 429–437. [CrossRef]

24. Hoffmann, U.; Hoffmann, T.; Jurasinski, G.; Glatzel, S.; Kuhn, N.J. Assessing the spatial variability of soil organic carbon stocks in an alpine setting (Grindelwald, Swiss Alps). *Geoderma* **2014**, *232*, 270–283. [CrossRef]

25. Panwar, P.; Mahalingappa, D.G.; Kaushal, R.; Bhardwaj, D.R.; Chakravarty, S.; Shukla, G.; Thakur, N.S.; Chavan, S.B.; Pal, S.; Nayak, B.G.; et al. Biomass Production and Carbon Sequestration Potential of Different Agroforestry Systems in India: A Critical Review. *Forests* **2022**, *13*, 1274. [CrossRef]

26. Ravhuhali, K.E.; Mlambo, V.; Beyene, T.S.; Palamuleni, L.G. Effect of soil type on spatial distribution and nutritive value of grass species growing in selected rangelands of South Africa. *S. Afr. J. Plant Soil* **2021**, *38*, 361–371. [CrossRef]

27. Mokgakane, T.J.; Mlambo, V.; Ravhuhali, K.E.; Magoro, N. Contribution of Soil Type to Quantify and Nutritiional Value of Grass Species on the South African Highveld. *Resources* **2021**, *10*, 106. [CrossRef]

28. Boob, M.; Elsaesser, M.; Thumm, U.; Hartung, J.; Lewandowski, I. Harvest Time Determines Quality and Usability of Biomass from Lowland Hay Meadows. *Agriculture* **2019**, *9*, 198. [CrossRef]

29. Mwangi, F.W.; Charmley, E.; Adegboye, O.A.; Gardiner, C.P.; Malau-Aduli, B.S.; Kinobe, R.T.; Malau-Aduli, A.E.O. Chemical Composition and In Situ Degradability of *Desmanthus* spp. Forage Harvested at Different Maturity Stages. *Fermentation* **2022**, *8*, 377. [CrossRef]

30. Hao, Y.; He, Z. Effects of grazing patterns on grassland biomass and soil environments in China: A meta-analysis. *PLoS ONE* **2019**, *14*, e0215223. [CrossRef]

31. Pavlů, L.; Gaisler, J.; Hejcman, M.; Pavlů, V.V. What is the effect of long-term mulching and traditional cutting regimes on soil and biomass chemical properties, species richness and herbage production in Dactylis glomerata grassland? *Agric. Ecosyst. Environ.* **2016**, *217*, 13–21. [CrossRef]

32. Chang, S.; Xie, K.; Du, W.; Jia, Q.; Yan, T.; Yang, H.; Hou, F. Effects of Mowing Times on Nutrient Composition and In Vitro Digestibility of Forage in Three Sown Pastures of China Loess Plateau. *Animals* **2022**, *12*, 2807. [CrossRef] [PubMed]

33. Gislum, R.; Griffith, S.M. Tiller Production and Development in Perennial Ryegrass in Relation to Nitrogen Use. *J. Plant Nutr.* **2005**, *27*, 2135–2148. [CrossRef]

34. Zhao, X.; Lyu, Y.; Jin, K.; Lambers, H.; Shen, J. Leaf Phosphorus Concentration Regulates the Development of Cluster Roots and Exudation of Carboxylates in *Macadamia integrifolia*. *Front. Plant Sci.* **2021**, *11*, 610591. [CrossRef] [PubMed]

35. Jin, Y.; Lai, S.; Chen, Z.; Jian, C.; Zhou, J.; Niu, F.; Xu, B. Leaf Photosynthetic and Functional Traits of Grassland Dominant Species in Response to Nutrient Addition on the Chinese Loess Plateau. *Plants* **2022**, *11*, 2921. [CrossRef]

36. Katongole, C.B.; Yan, T. Effect of Varying Dietary Crude Protein Level on Feed Intake, Nutrient Digestibility, Milk Production, and Nitrogen Use Efficiency by Lactating Holstein-Friesian Cows. *Animals* **2020**, *10*, 2439. [CrossRef] [PubMed]

37. Berauer, B.J.; Wilfahrt, P.A.; Reu, B.; Schuchardt, M.A.; Garcia-Franco, N.; Zistl-Schlingmann, M.; Dannenmann, M.; Kiese, R.; Kühnel, A.; Jentsch, A. Predicting forage quality of species-rich pasture grasslands using vis-NIRS to reveal effects of management intensity and climate change. *Agric. Ecosyst. Environ.* **2020**, *296*, 106929. [CrossRef]

38. Dong, J.; Tian, L.; Zhang, J.; Liu, Y.; Li, H.; Dong, Q. Grazing Intensity Has More Effect on the Potential Nitrification Activity Than the Potential Denitrification Activity in An Alpine Meadow. *Agriculture* **2022**, *12*, 1521. [CrossRef]

39. Ayele, J.; Tolemariam, T.; Beyene, A.; Tadese, D.A.; Tamiru, M. Biomass composition and dry matter yields of feed resource available at Lalo kile district of Kellem Wollega Zone, Western Ethiopia. *Heliyon* **2022**, *8*, e08972. [CrossRef] [PubMed]

40. Ray, D.K.; Sloat, L.L.; Garcia, A.S.; Davis, K.F.; Ali, T.; Xie, W. Crop harvests for direct food use insufficient to meet the UN's food security goal. *Nat. Food* **2022**, *3*, 367–374. [CrossRef]

41. Ruan, L.; Cheng, H.; Ludewig, U.; Li, J.; Chang, S.X. Root Foraging Strategy Improves the Adaptability of Tea Plants (*Camellia sinensis* L.) to Soil Potassium Heterogeneity. *Int. J. Mol. Sci.* **2022**, *23*, 8585. [CrossRef]

42. Hicks, T.D.; Kuns, C.M.; Raman, C.; Bates, Z.T.; Nagarajan, S. Simplified Method for the Determination of Total Kjeldahl Nitrogen in Wastewater. *Environments* **2022**, *9*, 55. [CrossRef]

43. Desai, S.; Amaresan, N. Qualitative and Quantitative Estimation of Phosphate Solubilizing Actinobacteria. In *Methods in Actinobacteriology*; Dharumadurai, D., Ed.; Springer Protocols Handbooks Humana: New York, NY, USA, 2022. [CrossRef]

44. Junsomboon, J.; Jakmunee, J. Determination of potassium, sodium, and total alkalies in portland cement, fly ash, admixtures, and water of concrete by a simple flow injection flame photometric system. *J. Anal. Methods Chem.* **2011**, *2011*, 742656. [CrossRef]

45. Dhingra, D.; Michael, M.; Rajput, H.; Patil, R.T. Dietary fibre in foods: A review. *J. Food Sci. Technol.* **2012**, *49*, 255–266. [CrossRef]

46. Liu, T.; Ren, T.; White, P.J.; Cong, R.; Lu, J. Storage nitrogen co-ordinates leaf expansion and photosynthetic capacity in winter oilseed rape. *J. Exp. Bot.* **2018**, *69*, 2995–3007. [CrossRef] [PubMed]

Article

Intraspecific Variation in Functional Traits of *Medicago sativa* Determine the Effect of Plant Diversity and Nitrogen Addition on Flowering Phenology in a One-Year Common Garden Experiment

Yue Ma [1,2], Xiang Zhao [1,*], Xiaona Li [2], Yanxia Hu [2] and Chao Wang [2,*]

1 College of Grassland Science, Shanxi Agricultural University, Jinzhong 030801, China
2 Institute of Grassland, Flowers and Ecology, Beijing Academy of Agriculture and Forestry Sciences, Beijing 100097, China; sunny3629@126.com (Y.H.)
* Correspondence: sxndzhaox@126.com (X.Z.); wangchao@grass-env.com (C.W.)

Abstract: Nitrogen deposition and biodiversity alter plant flowering phenology through abiotic factors and functional traits. However, few studies have considered their combined effects on flowering phenology. A common garden experiment with two nitrogen addition levels (0 and 6 g N m^{-2} year^{-1}) and five species richness levels (1, 2, 4, 6, and 8) was established. We assessed the effects of nitrogen addition and plant species richness on three flowering phenological events of *Medicago sativa* L. via changes in functional traits, soil nutrients, and soil moisture and temperature. The first flowering day was delayed, the last flowering day advanced, and the flowering duration shortened after nitrogen addition. Meanwhile, the last flowering day advanced, and flowering duration shortened along plant species richness gradients, with an average of 0.64 and 0.95 days change per plant species increase, respectively. Importantly, it was observed that plant species richness affected flowering phenology mainly through changes in plant nutrient acquisition traits (i.e., leaf nitrogen and carbon/nitrogen ratio). Our findings illustrate the non-negligible effects of intraspecific variation in functional traits on flowering phenology and highlight the importance of including functional traits in phenological models to improve predictions of plant phenology in response to nitrogen deposition and biodiversity loss.

Keywords: functional traits; abiotic factors; flowering phenology; nitrogen addition; plant diversity; common garden

Citation: Ma, Y.; Zhao, X.; Li, X.; Hu, Y.; Wang, C. Intraspecific Variation in Functional Traits of *Medicago sativa* Determine the Effect of Plant Diversity and Nitrogen Addition on Flowering Phenology in a One-Year Common Garden Experiment. *Plants* **2023**, *12*, 1994. https://doi.org/10.3390/plants12101994

Academic Editor: Luca Vitale

Received: 13 February 2023
Revised: 27 April 2023
Accepted: 8 May 2023
Published: 16 May 2023

1. Introduction

Human activities such as agricultural production and fossil fuel combustion add ~200 Tg/year of nitrogen (N) to global ecosystems, approximately equal to that provided by natural N fixation [1]. Land use changes, climate changes, and N deposition increase are the major causes of biodiversity loss [2,3]. Numerous studies showed that plant functional traits and ecosystem functions changed significantly under N deposition increase and biodiversity loss [4–6]. Phenology, the periodic events in the life cycles of plants, which is sensitive to global changes [7], and thus understanding the influence of N addition, plant diversity, and its interactions on plant phenology is essential.

Flowering phenology, including flowering date and duration, is an important developmental stage in plant phenology and is known to be determined by numerous abiotic factors, such as temperature, photoperiod, and resource availability [8,9]. Recent research revealed that plant functional traits, such as plant height and leaf traits, may serve as direct determinants for flowering phenology [10–12]. For example, plant height is critical for a species' competitive ability to capture light; taller species with higher maximized light interception and faster relative growth rate often flower later than shorter ones [13,14]; specific leaf area (SLA) is linked to growth rates, and higher SLA was found to be associated with stronger phenological shifts [15]; leaf carbon and N content are measures for the plants' investment in structural components and photosynthetic accumulation [16,17], whereas shifts in flowering phenology could be due to a trade-off between the investment between vegetative and reproductive growth [18]. Although the links between plant functional traits and flowering phenology have been reported, little is known about which traits are better for predicting the changes in flowering phenology.

Functional traits are measurable biotic properties related to the adaptation of plants to the environment [19,20]. Nitrogen addition stimulates plant growth through increasing soil-available nitrogen, resulting in changes in functional traits, such as plant height, leaf area, and nitrogen content, but varied with plant species [21,22]. Multiple changes in the biotic and abiotic environments along plant diversity gradients require a coordinated response in numerous traits to achieve a balance among different functions [4,23]. For instance, plants with slow growth rate may adjust morphological traits to tolerate low light availability through the formation of longer and thinner leaves [24]; variation in leaf nitrogen content (LNC) may serve as a physiological adjustment to improve photosynthetic carbon gain [25,26]. Hence, plant flowering phenology was significantly affected by N addition [27,28] and plant diversity [5,29], while our knowledge regarding how plant diversity and N deposition affect flowering phenology through plant functional traits was essential to step forward.

Here, we report on a common garden experiment investigating the influence of N addition and plant species richness on three flowering phenologies of *Medicago sativa* L. in an assemblage grassland (Figure S1). Specifically, we explored the effects of N addition and plant species richness on the first flowering day (FFD), last flowering day (LFD), flowering duration (FD), flower numbers (FN), environmental factors, and a series of functional traits. We hypothesized that: (a) increased N inputs and plant species richness would delay flowering phenology; (b) increased N inputs promote light acquisition traits (e.g., SLA and plant height), while decreased plant species richness promotes nutrient acquisition traits (e.g., leaf N concentration); and (c) intraspecific variation in functional traits and changes in abiotic factors would co-drive the response of flowering phenology to N addition and plant species richness.

2. Results

2.1. The Effects of N Addition and Plant Diversity on Soil Environmental Conditions

In our common garden experiment, soil temperature and soil moisture at a depth of 10 cm were significantly influenced by nitrogen addition and plant species richness (Table 1). Soil temperature significantly increased, but soil moisture did not change after nitrogen addition (Table 1 and Figure 1). Soil temperature and soil moisture changed markedly following plant species richness increase (Figure 1c,d). Soil temperature significantly decreased along plant species richness gradients under nitrogen addition (Figure 1c). Soil moisture showed no significant relationship with plant species richness under control and nitrogen addition treatments (Figure 1d).

Table 1. Linear mixed-effects model results showing the effects of plant species richness and nitrogen addition treatments on biotic and abiotic factors.

Parameters	Nitrogen Addition			Species Richness			Nitrogen Addition × Species Richness		
	numDF	*F*	*P*	*denDF*	*F*	*P*	*denDF*	*F*	*P*
Soil moisture	1	1.99	0.16	4	7.61	**<0.01**	4	1.42	0.23
Soil temperature	1	44.62	**<0.01**	4	4.52	**<0.01**	4	6.33	**<0.01**
Soil available nitrogen	1	0.15	0.71	4	5.94	**0.01**	4	0.36	0.55
Leaf mass	1	4.07	0.06	4	0.17	0.68	4	0.71	0.41
Leaf area	1	7.15	**<0.01**	4	1.43	0.24	4	0.41	0.53
Specific leaf area	1	1.61	0.21	4	8.51	**<0.01**	4	0.02	0.97
Leaf carbon content	1	0.01	0.97	4	0.26	0.61	4	0.05	0.82
Leaf nitrogen content	1	0.41	0.53	4	4.15	**<0.01**	4	0.41	0.53
Leaf carbon/nitrogen ratio	1	0.25	0.62	4	3.91	**<0.01**	4	0.24	0.63
First flowering day	1	2.52	0.09	4	2.01	0.15	4	0.01	0.91
Last flowering day	1	2.23	0.09	4	20.05	**<0.01**	4	0.46	0.49
Flowering duration	1	8.62	**<0.01**	4	13.09	**<0.01**	4	0.23	0.64
Flowering numbers	1	6.38	**0.01**	4	14.31	**<0.01**	4	0.02	0.89

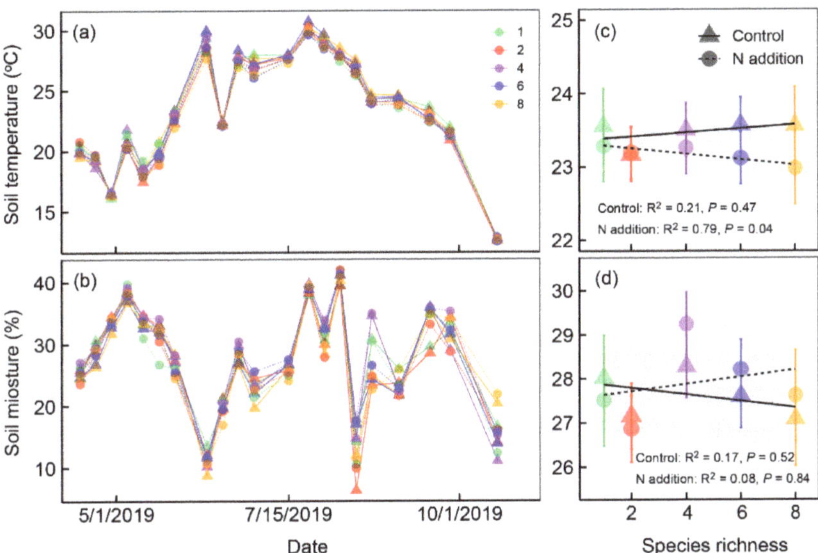

Figure 1. Soil temperature (°C, at a depth of 10 cm) (**a**) and soil moisture (%, at a depth of 10 cm) (**b**) under different plant species richness and nitrogen addition levels from April to October 2019. Mean (±*se*, *n* = 4 or 8) soil temperature (**c**) and soil moisture (**d**) for the whole growing season along the plant species richness gradients under different nitrogen (N) addition levels in 2019. We applied the ordinary linear regression analysis to test the dynamic of soil temperature and soil moisture along plant species richness gradients. Points and lines with different shapes represent different nitrogen addition levels, and points with different colors represent different plant species richness levels.

2.2. The Effects of N Addition and Plant Diversity on Flowering Phenology

The flowering phenology of *M. sativa* was significantly affected by nitrogen addition and plant species richness (Table 1), but the effects varied in different phenological phases (Figure 2). Nitrogen addition delayed the first flowering day of *M. sativa* by 1.9 ± 0.39 days, but plant species richness had no effects on it (Figure 2a). The last flowering day advanced by 1.2 ± 0.21 days after nitrogen addition (Figure 2b) and advanced by an average of 0.51 and 0.77 days per plant species increase under control and nitrogen addition treatments, respectively (Figure 2b). The delay of the first flowering day and the advancement of the last flowering day inevitably led to the shortening of the flowering duration. Nitrogen addition shortened the flowering duration of *M. sativa* by 3.2 ± 0.31 days, and it was shortened by 1.08 and 0.82 days per species increase under control and nitrogen addition treatments, respectively (Figure 2c). Moreover, our study also showed that flower numbers significantly decreased after nitrogen addition but increased along the plant species richness gradients (Figure 2d).

2.3. The Effects of N Addition and Plant Diversity on Functional Traits

Nine different leaf traits of *M. sativa* were measured in our study. For light acquisition traits, nitrogen addition significantly increased leaf area but did not change leaf mass and specific leaf area (Table 1 and Figure 3). Specific leaf area significantly decreased with increasing plant species richness (Figure 3), but not for other traits (Table 1 and Figure 3). For nitrogen acquisition traits, nitrogen addition had minor effects on leaf carbon content, leaf nitrogen content, and leaf carbon/nitrogen ratio (Table 1). Leaf nitrogen content significantly decreased, but the leaf carbon/nitrogen ratio significantly increased with increasing plant species richness (Figure 3d,e). There is no interaction effect between nitrogen addition and plant species richness on functional traits (Table 1).

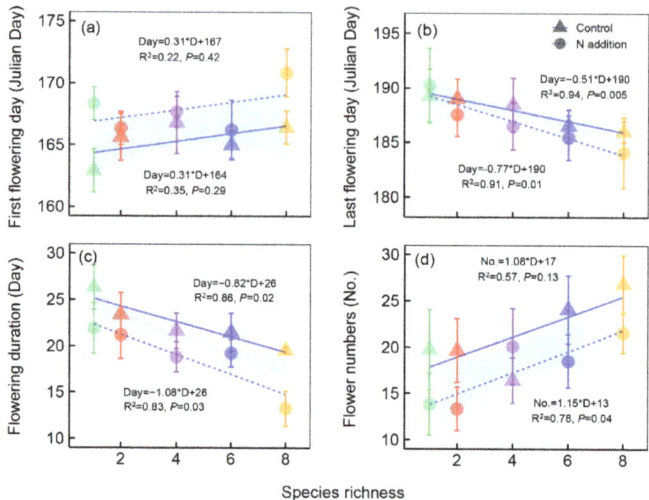

Figure 2. The first flowering day (**a**), last flowering day (**b**), flowering duration (**c**), and flower numbers (**d**) of *Medicago sativa* along the plant species richness gradients under different nitrogen (N) addition levels. All the analyses were performed using the ordinary linear regression analysis to test the dynamic of flower phenological events along plant species richness gradients. Points and lines with different shapes represent different N addition levels, and points with different colors represent different species richness levels. Shaded areas show 95% confidence intervals of the fit.

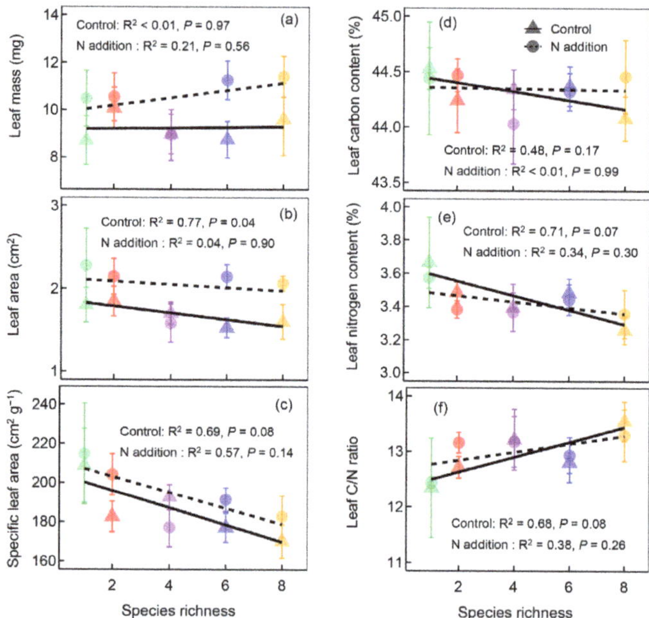

Figure 3. The leaf mass (**a**), leaf area (**b**), specific leaf area (**c**), leaf carbon content (**d**), leaf nitrogen content (**e**), and leaf carbon/nitrogen ratio (**f**) of *Medicago sativa* along the plant species richness gradients under different nitrogen (N) addition levels. Points and lines with different shapes represent different nitrogen addition levels, and points with different colors represent different plant species richness levels.

2.4. Ecological Factors Influencing Flowering Phenology

The last flowering day was positively correlated with the first flowering day significantly (Figure 4a); flowering duration was negatively correlated with the first flowering day and positively correlated with the last flowering day (Figure 4b,d); flower numbers were positively correlated with last flowering day and flowering duration (Figure 4e,f).

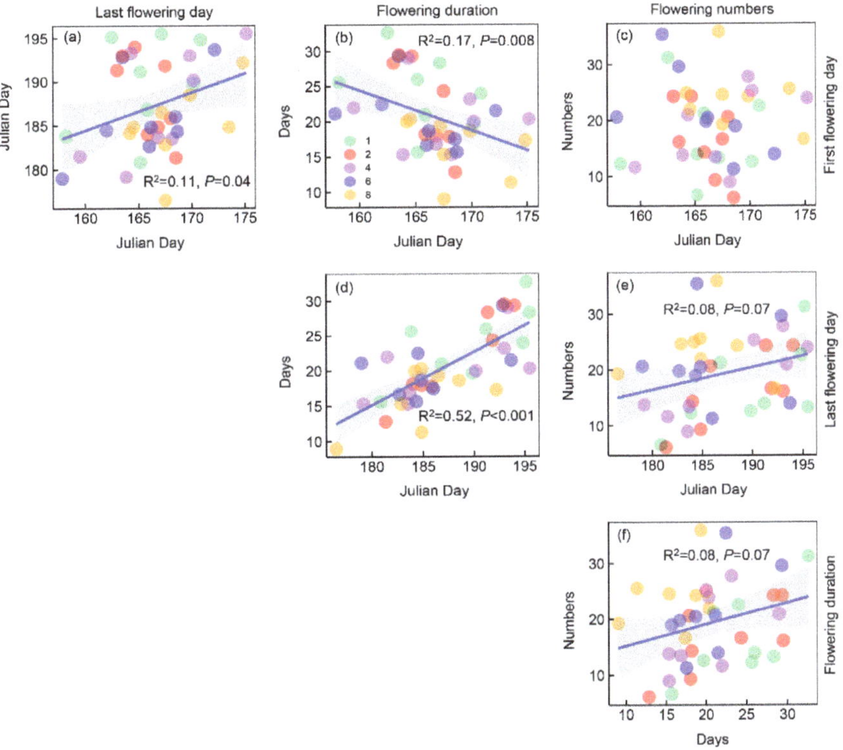

Figure 4. Plots showing the relationships between the four metrics of phenology. Points with different colors represent different plant species richness levels. Shaded areas show 95% confidence intervals of the fit.

Variation partitioning analysis indicated that nutrient acquisition traits and abiotic factors explained a much greater portion of the variance in the first flowering day (31% and 15%, Figure S5a), last flowering day (23% and 44%, Figure S5b), flowering duration (33% and 46%, Figure S5c), and flower numbers (32% and 44%, Figure S5d), respectively. The structural equation models showed that the changes in the first flowering day were positively correlated with soil-available nitrogen but negatively correlated with leaf nitrogen content and leaf carbon/nitrogen ratio (Figure 5a). The negative effects of plant species richness on the last flowering day and flowering duration were mainly through its negative effects on leaf nitrogen content (Figure 5b,c). Plant species richness impacted flower numbers directly or through its negative effects on leaf carbon/nitrogen ratio indirectly (Figure 5d), and nitrogen addition impacted flower numbers directly or through its positive effects on leaf mass indirectly (Figure 5d).

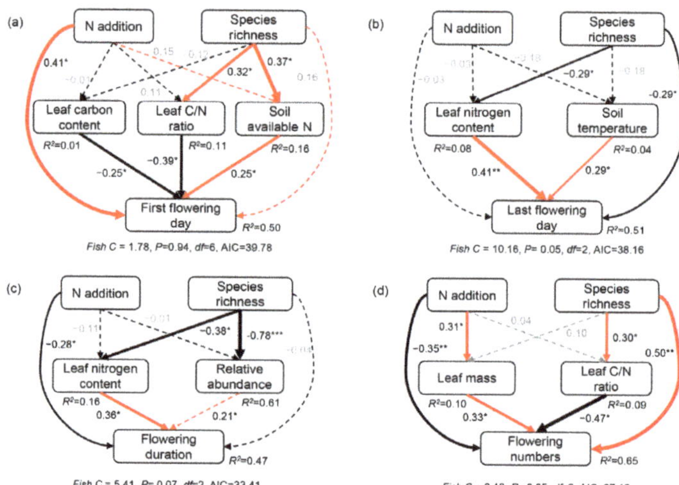

Figure 5. Structural equation modellings of plant species richness and nitrogen (N) addition impact on the first flowering day (**a**), last flowering day (**b**), flowering duration (**c**), and flower number (**d**). Red and black arrows represent significant positive and negative pathways, respectively. Solid and dashed arrows indicate significant and non-significant pathways, respectively. Numbers near the arrow indicate the standardized path coefficients indicating the effect size of the relationship with * indicating $p < 0.05$, ** indicating $p < 0.01$, and *** indicating $p < 0.001$. The arrow width is proportional to the strength of the relationship. R^2 represents the proportion of variance explained for each dependent variable. The goodness-of-fit statistics for the structural equation modeling are shown below each model. C, carbon.

3. Discussion

3.1. The Effects of N Addition and Plant Diversity on Flowering Phenology

The flowering phenology is one of the more important factors determining the reproductive success of plants because the timing of flowering represents when plants expose their reproductive organs to the changing biotic and abiotic environments [30,31]. Flowering phenology changed after N addition in our study was consistent with the results from natural grasslands [27,32]. Soil-available N increased following N addition, which delayed the plant switches from vegetative to reproductive growth and shortened the reproductive duration [28], resulting in a delay in FFD and advancement in LFD. FD was positively correlated with LFD and negatively correlated with FFD, and then the FD was shortened after N addition. A previous study found that plants delayed the switches from vegetative to reproductive growth due to the stimulation of growth through enhancement in soil N availability [28], which led to lower investment in plant reproduction, resulting in a reduction in FD in the N addition treatment.

Similar to the effects of N addition, plant species richness did not change FFD but advanced LFD, which resulted in a shortening in FD. The effects of plant species richness on FFD was inconsistent with the results (ranging from advancement of 1.8 d per species lost to a delay of 0.7 d per species lost; the average is 0.6 d advance) reported from the serpentine grasslands in North America [5], suggesting that the effects of plant species richness on FFD varied among plant species. Moreover, we noticed that the amplitude of changes (0.64 d) in LFD was larger than that in FFD after per species change, which revealed that LFD might be more sensitive to plant species richness. Soil-available N increased along the plant species richness gradients (Figure S4), which delayed the plant switches from vegetative to reproductive growth and shortened reproductive duration [28], resulting in advancement in LFD. FD was positively correlated with LFD, resulting in a shortening in FD.

3.2. The Effects of N Addition and Plant Diversity on Functional Traits

We found evidence to support our second hypothesis that increased N inputs promote light acquisition traits. Specifically, the leaf area of *M. sativa* was promoted after N addition may vary due to the increasing demand for light [33]. These results were inconsistent with that reported on the grasses and forbs from the alpine grasslands in China [6,34], which may be partly caused by the differences in plant species. *M. sativa*, legumes, can fix N using symbionts to relieve their N limitation [35], and thus N addition had no effects on the nutrient acquisition traits of *M. sativa*, which may lead to no interactions with plant species richness.

Our findings for plant species richness also support our second hypothesis that both light (SLA) and nutrient (LNC and relative abundance) acquisition traits were promoted with decreasing plant species richness. These results were inconsistent with the results reported from the grasslands in Germany [4], which may be caused by the differences in plant species. Closed canopies are characterized by pronounced gradients in spectral light quality and quantity [36]. In the present study, plant species richness impacts on morphological traits associated with light acquisition were mainly attributable to the changes in the relative height of *M. sativa*. Relative height increased along the plant species richness gradients, indicating less effort for light acquisition should be allocated by plants [37,38]. Hence, *M. sativa* exhibited low SLA in adjustment to decreasing competition for light. Moreover, nitrogen is not the limited resource of *M. sativa*. Lower light acquisition traits are accompanied by lower nutrient acquisition traits [33], resulting in a decrease in LNC along with plant species richness gradients.

3.3. Intraspecific Variation in Functional Traits Determines Flowering Phenology

Our findings did not support our third hypothesis that intraspecific variation in functional traits and alteration in abiotic factors co-drove the response of flowering phenology to N addition and plant species richness; the contributions of intraspecific variation in functional traits for changes in flowering phenology was far greater than abiotic factors (Figures 5 and S5). Specifically, the delay in FFD along the plant species richness gradients was mainly induced by an increase in soil-available N, which was consistent with the result from Wolf et al. in American serpentine grasslands [5]. Higher soil available N delayed the plant switches from vegetative to reproductive growth [28] and then led to the delay in FFD in the monocultures. However, the changes in LFD and FD were mainly driven by the intraspecific variation in LNC. Leaf nutrients could influence the strength of shifts in flowering phenology as plant performance is positively associated with phenological shifts [15,39] and the alteration in the natural selection of plants [12,40,41]. In the monocultures, lower intraspecific competition for resources and light [38] led to higher LNC and SLA and minimized light interception [42], resulting in advancing FFD, delaying LFD, and extending FD (Figure 2). As plant species richness increases, the intraspecific competition decreases, but the interspecific competition among species increases [43,44]. *M. sativa* is more conservative in resource utilization, with low SLA, LNC, and maximized light interception, and thus the "slow" growth strategy [45–47] led to advancement in LFD and shortened in FD. Therefore, the intraspecific variation in leaf traits caused the alteration in the growth strategy of *M. sativa* due to interspecific competition increased with increasing plant species richness, which resulted in changes in flowering phenology. Furthermore, the survival pressure of *M. sativa* increased due to higher interspecific competition in the mixed communities, which may cause *M. sativa* to invest more resources in reproductive growth [10], resulting in a positive relationship between FN and plant species richness.

4. Materials and Methods

4.1. Study Site and Experimental Design

Our study site (40°10′ 45″ N, 116°26′ 13″ E, and 50 m above sea level) is located at the Station of the Institute of Grassland, Flowers and Ecology on Xiaotang Mountain in Beijing, China. The mean annual precipitation is 526 mm, the mean annual temperature is

11.8 °C (2000–2018) in our study site. The common garden experiment was established in 2019, using a split-plot experiment design with N addition (0 and 6 g N m^{-2} year^{-1}) as the main treatment factors and plant species richness (species richness is 1, 2, 4, 6, and 8) as the subfactors (Figure S2 and Table S1). The experiment comprised the following: 2 N addition levels × 4 replicate blocks × 17 combinations of species per N treatment/replicate (four combinations per level of species richness with 1, 2, 4, and 6 species and one with 8 species) = 136 pipes. Each pipe (30 cm in diameter and 50 cm in height polyvinylchloride sealed pipes) was filled with uniformly mixed soil (natural soil with stones and roots sieved out) with sand (in a 3:1 soil–sand ratio) and then buried into the ground without any shelters. Eight plant species were selected in our study to build assembled grassland communities [48]. Different species of seeds were mixed up well and distributed randomly in the pipes, maintained at around 60 individuals in each pipe at the beginning of the experiment, and did not re-sow later. Nitrogen was added as urea (12.86 g m^{-2} year^{-1}), dissolved in N-free water, and then applied by spraying on 1 May of each year; the control treatment received equal N-free water. Additional information regarding the experimental design was provided by Wang et al. [48].

4.2. Phenology Monitoring

To track the flowering phenology of *M. sativa*, phenology was monitored every 3–4 days during the growing season from May to September 2019. Three individuals in each pipe were randomly selected, marked, and monitored across the growing season. The first and last date of a flower observed for each marked individual was recorded as the first and last flowering day, and the periods between the first and last flowering day were recorded as flowering duration. Flowering numbers were counted for each marked individual. Flowering phenological events and flowering numbers were averaged for three individuals in each pipe.

4.3. Functional Traits and Abiotic Factors Measurements

Light acquisition traits (plant height and relative height, leaf mass and area, leaf length and width, and specific leaf area) and nutrient acquisition traits (leaf carbon content, leaf nitrogen content, leaf C/N ratio, abundance and relative abundance of plant species) are closely related to plant phenology [49,50]. Consequently, we determined these traits to explore the mechanism underlying regulating the response of flowering phenology to experimental N addition and plant species richness in the peak growing season in 2019. Before the measurements, we investigated the abundance and height of each plant species in the pipes. *M. sativa* is the predominant species (relative abundance >40% in each pipe) (Figure S3), three healthy individuals, which we measure flowering phenology were selected to measure the species-level traits in each pipe, and six leaves on each individual were selected to measure leaf traits once at full flowering, and the leaves in each pipe were collected on the same day. The traits were quantified using standard methods proposed by Pérez-Harguindeguy et al. [51]. Specific leaf area was calculated as the ratio of leaf area to its dry weight. Leaf area, length, width, and maximum width, spread leaves were scanned and analyzed by Li-Cor 310 (Li-Cor Inc., Lincoln, USA), and then leaves were oven-dried to a constant weight. The oven-dried leaf samples were ground to determine leaf C and N concentration with an elemental analyzer (PE 2400 II, PerkinElmer Ltd., CT, USA) and then to calculate the C/N ratio. To measure the biomass of *M. sativa*, the aboveground part of each pipe was clipped in early September (the peak of the growing season) in 2019. *M. sativa* clipped from each pipe were pooled together and then oven-dried at 65 °C to a constant weight.

Soil temperature and moisture at a depth of 10 cm were measured every week from April to October with a W. E. T sensor kit (Delta-T Devices Ltd, Cambridge, UK). Three soil cores were collected in each pipe in early September at a depth of 10 cm and then mixed into one sample. Available-soil N (Ammonium (NH_4^+) and nitrate (NO_3^-)) concentrations in the extracts were determined calorimetrically by automated segmented flow analysis (Bran + Luebbe AAIII, Bran + Luebbe Ltd, Hamburg, Germany).

4.4. Statistical Analyses

We analyzed experimental data with the following three steps. First, we scaled the species-level height to the community level by calculating the mean of abundance distributions (Equation (1) [52]):

$$Mean_c = \sum_i^n p_i T_i \qquad (1)$$

where p_i and T_i are the relative abundances and the plant height of the species i, respectively, and n is the number of plant species. Hence, the average height of *M. sativa* divided by $Mean_c$ is the relative height.

Second, we applied linear mixed-effects models using the *"lme"* function (package *"nlme"* [53]) to test the effects of N addition and plant species richness on soil temperature and moisture. We set N addition and species richness levels as fixed effects; the date, block, and plant combination were set as random effects in each model to account for variation among repeated measurements. In addition, linear mixed-effects models were also used to examine the effect of N addition and plant species richness on flowering phenology and functional traits. Nitrogen addition and species richness levels were treated as fixed effects, and the block and plant combinations were treated as random effects.

Third, variation partitioning analysis that partitioned the variance shared by all factors was then used to quantify the unique contribution of biotic and abiotic factors. Structural equation modeling was employed to evaluate which are the major factors that influence flowering phenology [54] by the package *'piecewise-SEM'* in R software [55]. The SEM requires establishing an a priori framework based on the hypothesized causal relationships among these variables. Second, the relationships between these variables were examined by bivariate correlations. Finally, models with lower *Fisher's C* and Akaike information criterion (*AIC*) and higher p values ($p \geq 0.05$) were selected in our analysis (Figure 5). All statistical analyses and graphs were prepared in R 3.2.2 [56]. Differences were considered to be statistically significant at $p \leq 0.05$.

5. Conclusions

The study highlighted the influence of functional traits on flowering phenology following nitrogen addition levels and plant species richness gradients in an assemblage grassland through a common garden experiment. It was observed that the first flowering day was delayed 0.31 days, the last flowering day advanced 0.64 days, and the flowering duration was shortened by 0.95 days with per-plant species increase, but the effects of plant species richness on flowering phenology did not interact with nitrogen addition, which indicates that nitrogen addition could change plant flowering phenology by changing biodiversity, but the effects would be independent with the effects of biodiversity. Moreover, flowering phenology changed following nitrogen addition levels, and plant species richness gradients were mainly driven by the intraspecific variation in functional traits, which suggests that variation in functional traits among communities may be a good predictor for the dynamic of plant phenology under global changes.

Supplementary Materials: The following supporting information can be downloaded at: https://www.mdpi.com/article/10.3390/plants12101994/s1. Figure S1: Design and photo of our experiments; Figure S2: Photos of experimental arrangement; Figure S3: The average biomass, relative biomass, relative abundance, and relative height of *M. Sativa* along the plant species richness gradients under different nitrogen (N) addition levels; Figure S4: Pearson's correlation between the biotic and abiotic factors; Figure S5: Relative contributions of light acquisition traits, nutrient acquisition traits, and abiotic factors to flowering events; Figure S6: Original structural equation modellings of N addition and plant species richness on the first flowering day, last flowering day, flowering duration, and flower number. Table S1: Assemblage types at different species richness levels.

Author Contributions: Conceptualization, C.W. and X.Z.; methodology, C.W., Y.M., X.L. and Y.H.; writing—original draft preparation, Y.M. and C.W.; writing—review and editing, Y.M., X.Z. and C.W.; funding acquisition, X.Z., C.W. and Y.H.; All authors have read and agreed to the published version of the manuscript.

Funding: This study was supported by the Beijing Natural Science Foundation (Grant No. 5232006), National Natural Science Foundation of China (Grant No. 31901173), China Key Technologies Research and Development Program (Grant No. 2020YFD1000201), and the Excellent Talents Innovation Project of Shanxi, China (Grant No. 201805D211018).

Data Availability Statement: The data that support the findings of this study are available from the corresponding author upon reasonable request.

Acknowledgments: We thank Wenjun Teng, Chao Han, and Ruibin Xue for their help with the fieldwork.

Conflicts of Interest: The authors declare that they have no known competing financial interests or personal relationships that could have appeared to influence the work reported in this paper.

References

1. Fowler, D.; Coyle, M.; Skiba, U.; Sutton, M.A.; Cape, J.N.; Reis, S.; Sheppard, L.J.; Jenkins, A.; Grizzetti, B.; Galloway, J.N.; et al. The global nitrogen cycle in the twenty-first century. *Philos. Trans. R. Soc. B Biol. Sci.* **2013**, *368*, 20130164. [CrossRef]
2. Ledford, H. Global plant extinctions mapped. *Nature* **2019**, *570*, 148–149. [CrossRef]
3. Blumenthal, D.M.; Carillo, Y.; Kray, J.A.; Parsons, M.C.; Morgan, J.A.; Pendall, E. Soil disturbance and invasion magnify CO_2 effects on grassland productivity, reducing diversity. *Glob. Chang. Biol.* **2022**, *28*, 6741–6751. [CrossRef] [PubMed]
4. Gubsch, M.; Buchmann, N.; Schmid, B.; Schulze, E.D.; Lipowsky, A.; Roscher, C. Differential effects of plant diversity on functional trait variation of grass species. *Ann. Bot.* **2011**, *107*, 157–169. [CrossRef] [PubMed]
5. Wolf, A.A.; Zavaleta, E.S.; Selmants, P.C. Flowering phenology shifts in response to biodiversity loss. *Proc. Natl. Acad. Sci. USA* **2017**, *114*, 3463–3468. [CrossRef] [PubMed]
6. Zhang, D.Y.; Peng, Y.F.; Li, F.; Yang, G.B.; Wang, J.; Yu, J.C.; Zhou, G.Y.; Yang, Y.H. Trait identity and functional diversity co-drive response of ecosystem productivity to nitrogen enrichment. *J. Ecol.* **2019**, *107*, 2402–2414. [CrossRef]
7. Piao, S.L.; Liu, Q.; Chen, A.P.; Janssens, I.A.; Fu, Y.S.; Dai, J.H.; Liu, L.L.; Lian, X.; Shen, M.G.; Zhu, X.L. Plant phenology and global climate change: Current progresses and challenges. *Glob. Chang. Biol.* **2019**, *25*, 1922–1940. [CrossRef]
8. Hulme, P.E. Contrasting impacts of climate-driven flowering phenology on changes in alien and native plant species distributions. *New Phytol.* **2011**, *189*, 272–281. [CrossRef] [PubMed]
9. Wang, S.; Meng, F.; Duan, J.; Wang, Y.; Cui, X.; Piao, S.; Niu, H.; Xu, G.; Luo, C.; Zhang, Z.; et al. Asymmetric sensitivity of first flowering date to warming and cooling in alpine plants. *Ecology* **2014**, *95*, 3387–3398. [CrossRef]
10. Dorji, T.; Totland, Ø.; Moe, S.R.; Hopping, K.A.; Pan, J.B.; Klein, J.A. Plant functional traits mediate reproductive phenology and success in response to experimental warming and snow addition in Tibet. *Glob. Chang. Biol.* **2013**, *19*, 459–472. [CrossRef]
11. Bucher, S.F.; Konig, P.; Menzel, A.; Migliavacca, M.; Ewald, J.; Romermann, C. Traits and climate are associated with first flowering day in herbaceous species along elevational gradients. *Ecol. Evol.* **2018**, *8*, 1147–1158. [CrossRef] [PubMed]
12. Pérez-Ramos, I.M.; Cambrollé, J.; Hidalgo-Galvez, M.D.; Matías, L.; Montero-Ramírez, A.; Santolaya, S.; Godoy, Ó. Phenological responses to climate change in communities of plants species with contrasting functional strategies. *Environ. Exp. Bot.* **2020**, *170*, 103852. [CrossRef]
13. Sun, S.; Frelich, L.E. Flowering phenology and height growth pattern are associated with maximum plant height, relative growth rate and stem tissue mass density in herbaceous grassland species. *J. Ecol.* **2011**, *99*, 991–1000. [CrossRef]
14. Sporbert, M.; Jakubka, D.; Bucher, S.F.; Hensen, I.; Freiberg, M.; Heubach, K.; Koenig, A.; Nordt, B.; Plos, C.; Blinova, I.; et al. Functional traits influence patterns in vegetative and reproductive plant phenology-a multi-botanical garden study. *New Phytol.* **2022**, *235*, 2199–2210. [CrossRef] [PubMed]
15. König, P.; Tautenhahn, S.; Cornelissen, J.H.C.; Kattge, J.; Bönisch, G.; Römermann, C. Advances in flowering phenology across the Northern Hemisphere are explained by functional traits. *Glob. Ecol. Biogeogr.* **2018**, *27*, 310–321. [CrossRef]
16. Larcher, W. *Ökophysiologie der Pflanzen*; VerlagEugen Ulmer: Stuttgart, Germany, 1994.
17. Ye, Y.; Kitayama, K.; Onoda, Y. A cost-benefit analysis of leaf carbon economy with consideration of seasonal changes in leaf traits for sympatric deciduous and evergreen congeners: Implications for their coexistence. *New Phytol.* **2022**, *234*, 1047–1058. [CrossRef]
18. Bolmgren, K.; Cowan, P.D. Time-size tradeoffs: A phylogenetic comparative study of flowering time, plant height and seed mass in a north-temperate flora. *Oikos* **2008**, *117*, 424–429. [CrossRef]
19. Cadotte, M.W. Functional traits explain ecosystem function through opposing mechanisms. *Ecol. Lett.* **2017**, *20*, 989–996. [CrossRef] [PubMed]
20. He, N.; Liu, C.; Piao, S.; Sack, L.; Xu, L.; Luo, Y.; He, J.; Han, X.; Zhou, G.; Zhou, X.; et al. Ecosystem traits linking functional traits to macroecology. *Trends Ecol. Evol.* **2019**, *34*, 200–210. [CrossRef]

21. Liu, H.; Li, Y.; Ren, F.; Lin, L.; Zhu, W.; He, J.; Niu, K. Trait-abundance relation in response to nutrient addition in a Tibetan alpine meadow: The importance of species trade-off in resource conservation and acquisition. *Ecol. Evol.* **2017**, *7*, 10575–10581. [CrossRef]
22. Wang, Y.; Niu, G.; Wang, R.; Rousk, K.; Li, A.; Hasi, M.; Wang, C.; Xue, J.; Yang, G.; Lu, X.; et al. Enhanced foliar ^{15}N enrichment with increasing nitrogen addition rates: Role of plant species and nitrogen compounds. *Glob. Chang. Biol.* **2023**, *29*, 1591–1605. [CrossRef] [PubMed]
23. Qiao, J.; Zuo, X.; Yue, P.; Wang, S.; Hu, Y.; Guo, X.; Li, X.; Lv, P.; Guo, A.; Sun, S. High nitrogen addition induces functional trait divergence of plant community in a temperate desert steppe. *Plant Soil* **2023**. [CrossRef]
24. Valladares, F.; Niinemets, Ü. Shade tolerance, a key plant feature of complex nature and consequences. *Annu. Rev. Ecol. Evol. Syst.* **2008**, *39*, 237–257. [CrossRef]
25. Anten, N.P.R.; Hirose, T. Shoot structure, leaf physiology, and daily carbon gain of plant species in a tallgrass meadow. *Ecology* **2003**, *84*, 955–968. [CrossRef]
26. Khan, A.; Yan, L.; Wang, W.; Xu, K.; Zou, G.; Liu, X.D.; Fang, X.W. Leaf traits and leaf nitrogen shift photosynthesis adaptive strategies among functional groups and diverse biomes. *Ecol. Indic.* **2022**, *141*, 109098. [CrossRef]
27. Smith, J.G.; Sconiers, W.; Spasojevic, M.J.; Ashton, I.W.; Suding, K.N. Phenological changes in alpine plants in response to increased snowpack, temperature, and nitrogen. *Arct. Antarct. Alp. Res.* **2012**, *44*, 135–142. [CrossRef]
28. Wang, C.; Tang, Y. Responses of plant phenology to nitrogen addition: A meta-analysis. *Oikos* **2019**, *128*, 1243–1253. [CrossRef]
29. Rheault, G.; Levesque, E.; Proulx, R. Diversity of plant assemblages dampens the variability of the growing season phenology in wetland landscapes. *BMC Ecol. Evol.* **2021**, *21*, 91. [CrossRef]
30. Rathcke, B.; Lacey, E.P. Phenological patterns of terrestrial plants. *Annu. Rev. Ecol. Syst.* **1985**, *16*, 179–214. [CrossRef]
31. Kudo, G.; Hirao, A.S. Habitat-specific responses in the flowering phenology and seed set of alpine plants to climate variation: Implications for global-change impacts. *Popul. Ecol.* **2006**, *48*, 49–58. [CrossRef]
32. Xi, Y.; Jiang, Y.; Zhang, G.; Zhang, T.; Zhang, Y.; Zhu, J. Nitrogen addition alters the phenology of a dominant alpine plant in northern Tibet. *Arct. Antarct. Alp. Res.* **2015**, *47*, 511–518. [CrossRef]
33. Hautier, Y.; Niklaus, P.A.; Hector, A. Competition for light causes plant biodiversity loss after eutrophication. *Science* **2009**, *324*, 636–638. [CrossRef] [PubMed]
34. Liu, Y.; Miao, R.; Chen, A.; Miao, Y.; Liu, Y.; Wu, X. Effects of nitrogen addition and mowing on reproductive phenology of three early-flowering forb species in a Tibetan alpine meadow. *Ecol. Eng.* **2017**, *99*, 119–125. [CrossRef]
35. Bordeleau, L.M.; Prévost, D. Nodulation and nitrogen fixation in extreme environments. *Plant Soil* **1994**, *161*, 115–125. [CrossRef]
36. Jones, H.G. *Plants and Microclimate: A Quantitative Approach to Environmental Plant Physiology*; Cambridge University Press: Cambridge, UK, 1992.
37. Spehn, E.M.; Hector, A.; Joshi, J.; Scherer-Lorenzen, M.; Schmid, B.; Bazeley-White, E.; Beierkuhnlein, C.; Caldeira, M.C.; Diemer, M.; Dimitrakopoulos, P.G.; et al. Ecosystem effects of biodiversity manipulations in European grasslands. *Ecol. Monogr.* **2005**, *75*, 37–63. [CrossRef]
38. Lorentzen, S.; Roscher, C.; Schumacher, J.; Schulze, E.D.; Schmid, B. Species richness and identity affect the use of aboveground space in experimental grasslands. *Perspect. Plant Ecol. Evol. Syst.* **2008**, *10*, 73–87. [CrossRef]
39. Cleland, E.E.; Allen, J.M.; Crimmins, T.M.; Dunne, J.A.; Pau, S.; Travers, S.E.; Zavaleta, E.S.; Wolkovich, E.M. Phenological tracking enables positive species responses to climate change. *Ecology* **2012**, *93*, 1765–1771. [CrossRef]
40. MacArthur, R.H.; Wilson, E.O. *The Theory of Island Biogeography*; Princeton University Press: Princeton, NJ, USA, 1967.
41. He, J.; Wang, X.; Flynn, D.F.B.; Wang, L.; Schmid, B.; Fang, J.Y. Taxonomic, phylogenetic, and environmental trade-offs between leaf productivity and persistence. *Ecology* **2009**, *90*, 2779–2791. [CrossRef]
42. Sun, S.C.; Jin, D.M.; Li, R.J. Leaf emergence in relation to leaf traits in temperate woody species in East-Chinese Quercus fabri forests. *Acta Oecol. Int. J. Ecol.* **2006**, *30*, 212–222. [CrossRef]
43. Hector, A.; Schmid, B.; Beierkuhnlein, C.; Caldeira, M.C.; Diemer, M.; Dimitrakopoulos, P.G.; Finn, J.A.; Freitas, H.; Giller, P.S.; Good, J.; et al. Plant diversity and productivity experiments in European grasslands. *Science* **1999**, *286*, 1123–1127. [CrossRef]
44. Tilman, D.; Reich, P.B.; Knops, J.; Wedin, D.; Mielke, T.; Lehman, C. Diversity and productivity in a long-term grassland experiment. *Science* **2001**, *297*, 843–845. [CrossRef]
45. Purschke, O.; Schmid, B.C.; Sykes, M.T.; Poschlod, P.; Michalski, S.G.; Durka, W.; Kuhn, I.; Winter, M.; Prentice, H.C. Contrasting changes in taxonomic, phylogenetic and functional diversity during a long-term succession: Insights into assembly processes. *J. Ecol.* **2013**, *101*, 857–866. [CrossRef]
46. Reich, P.B.; Cornelissen, H. The world-wide 'fast-slow' plant economics spectrum: A traits manifesto. *J. Ecol.* **2014**, *102*, 275–301. [CrossRef]
47. Török, P.; Matus, G.; Toth, E.; Papp, M.; Kelemen, A.; Sonkoly, J.; Tothmeresz, B. Both trait-neutrality and filtering effects are validated by the vegetation patterns detected in the functional recovery of sand grasslands. *Sci. Rep.* **2018**, *8*, 13703. [CrossRef] [PubMed]
48. Wang, C.; Li, X.; Hu, Y.; Zheng, R.; Hou, Y. Nitrogen addition weakens the biodiversity-multifunctionality relationships across soil profiles in a grassland assemblage. *Agric. Ecosyst. Environ.* **2023**, *342*, 108241. [CrossRef]
49. Lavorel, S.; Grigulis, K. How fundamental plant functional trait relationships scale-up to trade-offs and synergies in ecosystem services. *J. Ecol.* **2012**, *100*, 128–140. [CrossRef]

50. Grigulis, K.; Lavorel, S.; Krainer, U.; Legay, N.; Baxendale, C.; Dumont, M.; Kastl, E.; Arnoldi, C.; Bardgett, R.D.; Poly, F.; et al. Relative contributions of plant traits and soil microbial properties to mountain grassland ecosystem services. *J. Ecol.* **2013**, *101*, 47–57. [CrossRef]

51. Pérez-Harguindeguy, N.; Díaz, S.; Garnier, E.; Lavorel, S.; Poorter, H.; Jaureguiberry, P.; Bret-Harte, M.S.; Cornwell, W.K.; Craine, J.M.; Gurvich, D.E.; et al. New handbook for standardised measurement of plant functional traits worldwide. *Aust. J. Bot.* **2016**, *64*, 715–716. [CrossRef]

52. Gross, N.; Kunstler, G.; Liancourt, P.; De Bello, F.; Suding, K.N.; Lavorel, S. Linking individual response to biotic interactions with community structure: A trait-based framework. *Funct. Ecol.* **2009**, *23*, 1167–1178. [CrossRef]

53. Pinheiro, J.; Bates, D.; DebRoy, S.; Sarkar, D. Linear and nonlinear mixed effects models. *R Package Version* **2007**, *3*, 1–89.

54. Wang, C.; Tang, Y.J. A global meta-analyses of the response of multi-taxa diversity to grazing intensity in grasslands. *Environ. Res. Lett.* **2019**, *14*, 114003. [CrossRef]

55. Shipley, B. A new inferential test for path models based on directed acyclic Graphs. *Struct. Equ. Model.* **2000**, *7*, 206–218. [CrossRef]

56. R Development Core Team. *R: A Language and Environment for Statistical Computing*; R Foundation for Statistical Computing: Vienna, Austria, 2018.

Review

Research Progress of Grassland Ecosystem Structure and Stability and Inspiration for Improving Its Service Capacity in the Karst Desertification Control

Shuyu He [1,2], Kangning Xiong [1,2,*], Shuzhen Song [1,2], Yongkuan Chi [1,2], Jinzhong Fang [1,2] and Chen He [1,2]

[1] School of Karst Science, Guizhou Normal University, Guiyang 550001, China
[2] State Engineering Technology Institute for Karst Desertification Control of China, 116 Baoshan North Road, Guiyang 550001, China
* Correspondence: xiongkn@gznu.edu.cn; Tel.: +86-163-0854-7678

Abstract: The structure and stability of grassland ecosystems have a significant impact on biodiversity, material cycling and productivity for ecosystem services. However, the issue of the structure and stability of grassland ecosystems has not been systematically reviewed. Based on the Web of Science (WOS) and China National Knowledge Infrastructure (CNKI) databases, we used the systematic-review method and screened 133 papers to describe and analyze the frontiers of research into the structure and stability of grassland ecosystems. The research results showed that: (1) The number of articles about the structure and stability of grassland ecosystems is gradually increasing, and the research themes are becoming increasingly diverse. (2) There is a high degree of consistency between the study area and the spatial distribution of grassland. (3) Based on the changes in ecosystem patterns and their interrelationships with ecosystem processes, we reviewed the research progress and landmark results on the structure, stability, structure–stability relationship and their influencing factors of grassland ecosystems; among them, the study of structure is the main research focus (51.12%), followed by the study of the influencing factors of structure and stability (37.57%). (4) Key scientific questions on structural optimization, stability enhancement and harmonizing the relationship between structure and stability are explored. (5) Based on the background of karst desertification control (KDC) and its geographical characteristics, three insights are proposed to optimize the spatial allocation, enhance the stability of grassland for rocky desertification control and coordinate the regulation mechanism of grassland structure and stability. This study provided some references for grassland managers and relevant policy makers to optimize the structure and enhance the stability of grassland ecosystems. It also provided important insights to enhance the service capacity of grassland ecosystems in KDC.

Keywords: grassland ecosystem; structure; stability; ecosystem services; karst desertification control

Citation: He, S.; Xiong, K.; Song, S.; Chi, Y.; Fang, J.; He, C. Research Progress of Grassland Ecosystem Structure and Stability and Inspiration for Improving Its Service Capacity in the Karst Desertification Control. *Plants* **2023**, *12*, 770. https://doi.org/10.3390/plants12040770

Academic Editors: Bingcheng Xu and Zhongming Wen

Received: 11 January 2023
Revised: 4 February 2023
Accepted: 5 February 2023
Published: 8 February 2023

1. Introduction

Constituting almost 40% of the terrestrial biosphere, grasslands provide the habitat for a great number of diverse animals and plants and contribute to the livelihoods of more than 1 billion people worldwide [1]. Under the growing impact of global climate change and unreasonable human activities, grasslands are facing major problems that threaten the sustainable development of grassland ecosystems, such as a sharp decline in biodiversity, pasture degradation and reduced supply capacity [2–4]. Therefore, how to promote the healthy development of grassland ecosystems and enhance their service capacity has become an urgent issue [5–7]. Grassland ecosystem services are influenced by multiple factors such as their structure and stability. Furthermore, trade-offs and synergies between ecosystem services often depend on their structural and functional interactions [8]. A reasonable ecosystem structure can improve ecosystem productivity; promote material cycling, energy flow and information transfer; and increase the provision capacity of ecosystem service [9]. This plays a very important role in the healthy development of ecosystems and in human well-being. Rational allocation of grassland structure is one of the main measures to improve the stability and resilience of grassland ecosystems [10]. However, while the global concept of sustainability continues to spread, irrational structural configurations of grasslands (unreasonable cropping patterns and unscientific pasture management, such as planting density, species or grazing methods, etc.) continue to exist, which not only undermines the gains of ecological restoration and conservation, but also exacerbates the conflict between ecological conservation and economic development [11]. In particular, large-scale cultivation-based rangelands generally have a single planting structure, low biodiversity, high vulnerability and low stability, making it difficult to create a cascade of ecosystem service benefits [12]. As a result, grassland ecosystem structure and structural optimization are gradually becoming a research priority, with a focus on component structure, spatial and temporal structure, nutrient structure and their driving factors [13,14]. Optimizing the structure of grassland ecosystems is an important measure to improve and maintain the service capacity of grassland ecosystems [15,16], to balance grassland ecology and farmers' livelihoods, and to resolve the contradictory issues of grassland ecology and sustainable economic development [17].

Structural changes drive changes in stability in grassland ecosystems [18], which, in turn, change their ecological functions and the provisioning capacity of an ecosystem [19]. At present, there are many studies about revealing the mechanisms of species diversity on grassland stability through controlled experiments in grassland [20–23]. The application of basic principles and methods of biochemistry to quantitatively study nutrient limitation and nutrient balance in forage, as well as the regulation mechanism of water–fertilizer coupling on forage quality, productivity and stability in grassland, which is the focus of the current research [24–26]. At the same time, studies on assessing the stability of grassland ecosystems are gradually emerging [27], such as those using remote sensing data; indicators characterizing ecosystem vitality, resilience and organization; and landscape pattern indices to evaluate ecosystem stability [28,29]. However, there is a lack of research regarding the mechanisms by which the complementarity and diversity of functional traits regulate the stability of grasslands, which results in a structure–process–function–service cascade [30]. What is more, there are different methods for quantifying ecosystem stability indicators, and the evaluation models are uneven; therefore, whether they are universally applicable remains to be verified [31,32]. Biodiversity and species diversity have an important influence on the productivity, stability and nutrient cycling of grasslands and their resistance and resilience to disturbance. Resilience, resistance and restoration are the main elements in determining whether an ecosystem is stable [33]. The study of the role of biodiversity, species diversity and functional traits on stability has been the focus of ecosystem research [34–37], and they are influenced by multiple factors on multiple scales [38]. It has been shown that extreme weather and irrational human activities are the main drivers of structural changes in grassland ecosystems [39], which indirectly alter ecosystem stability, increase the overall vulnerability of grassland ecosystems and reduce their capacity to provide ecosystem services [40]. Therefore, a deep understanding of the relationship between the structure and stability of grassland ecosystems is a major part of maintaining the stability of grassland ecosystems and enhancing their service capacity [41–43].

Due to their fragile attributes, ecologically fragile areas are prone to high ecosystem sensitivity and structural vulnerability under climate change and irrational human activities, which, in turn, reduces the stability of their ecosystems and changes their ability to supply services [44]. Therefore, the mutual interaction between ecosystem structure and stability should be deeply understood to enhance their resistance to human disturbances and environmental changes [45,46]. However, in ecologically fragile areas, irrational land use reduces the diversity of grassland species and productivity stability and affects the sustainable development of the region [47]. Especially in the environmentally fragile areas of karst, unreasonable human activities have led to the degradation of vegetation, increased soil erosion, gradual exposure of rocks and degradation of land productivity, with the surface showing a visual evolution similar to that of a desert landscape [48,49]. For this reason, the first task in the comprehensive control of rocky desertification is to restore and re-establish vegetation. In order to solve the ecological problem of karst desertification, the Chinese government has carried out a lot of work on the issue of karst desertification in southwest China since 1989, such as grain for green and closing the land for reforestation (grass), etc. The area of rocky desertification generally exhibits a trend of "continuous net reduction" [50], which provides a Chinese solution to global greening [51–54]. However, the KDC ecosystem still suffers from its simple structure, incomplete system function, high ecosystem sensitivity, lagging ecosystem service capacity and difficulty in maintaining the results of rocky-desertification management [55]. The rugged and fragmented surface in the karst areas, coupled with the constraints of rocky desertification, has affected the biodiversity of grassland ecosystems, resulting in a homogeneous structure and low productivity for the grassland [56]. Therefore, how to maintain ecological stability and optimize the structure of the system in the grassland ecosystems of KDC has become a key issue to consolidate the achievements of rocky-desertification management, ensure the smooth flow of service supply and demand, and enhance the well-being of local people [57,58]. Therefore, the structure and stability of grassland ecosystems and their interactions are not only a global concern [59–61], but also an important element that needs attention in KDC [62,63]. In the natural vegetation succession, grasses are the pioneer plants for vegetation restoration and ecological improvement [64]. Meanwhile, artificial grass breeding can enrich grassland species diversity, improve grassland ecosystem stability, provide high-quality forage, reduce the risk of surface erosion, improve soil nutrient composition and provide multiple ecosystem services for humans [51,65,66]. Optimizing the spatial configuration of systems is an important means of improving the stability and resilience of grassland ecosystems [67]. Therefore, optimizing the structure of grassland ecosystems for KDC and enhancing their stability are of great significance in enhancing the service capacity of grassland ecosystems and promoting the sustainable development of the regional ecological environment.

Thus far, research on the structure and stability of grassland ecosystems is increasing and breakthroughs have been achieved. The structure and stability of grassland is an essential element to maintain the sustainable and healthy operation of grassland in KDC, which is an important part of its ecosystem. However, there is a lack of relevant studies on the structure and stability of grassland ecosystems in KDC. In view of this, based on the perspective of grassland-ecosystem pattern change and its relationship with ecosystem processes, and the systematic review method, this study systematically reviewed the main research progress and landmark results on the structure and stability of global grassland ecosystems, and summarized the key scientific issues on structure optimization, stability enhancement and the interaction between structure and stability, aiming to provide some insights into grassland-ecosystem structure optimization and stability enhancement in KDC. In this way, it can enhance the supply of grassland ecosystem services and promote the sustainable development of the regional ecological environment and social economy in KDC.

2. Results and Discussion

2.1. Research Characteristics

As shown in Figure 1, the number of studies on the structure and stability of grassland ecosystems showed an overall fluctuating upward trend. From 1995 to 2004, the average annual number of articles did not exceed two, and it was in the budding stage. The number of papers published from 2005–2014 was 26, and the number of papers published during this period increased rapidly with the UN Millennium Ecosystem Assessment (MEA) synthesis report [68], with large fluctuations; this is the development phase. The number of articles published from 2015 to present was 97, mainly due to the convening of the United Nations Sustainable Development Goals (SDGs) [69]; the number of studies on ecosystem structure and stability shows a rapid upward trend, and the degree of research on concepts and mechanisms is gradually expanding and entering a diversified development stage.

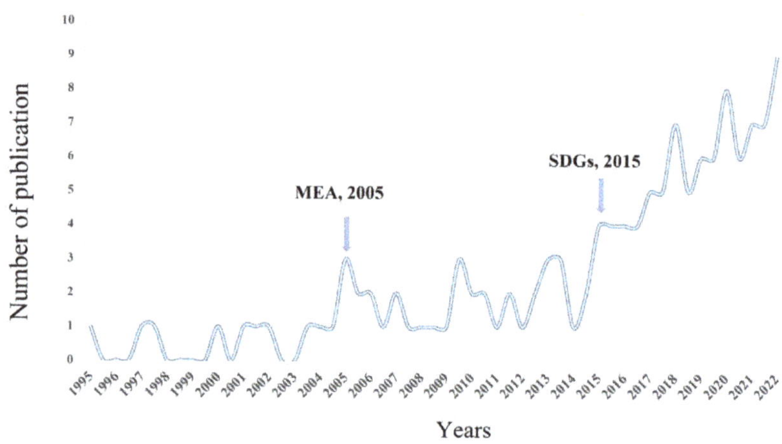

Figure 1. Trend in the annual distribution of literature related to the structure and stability of grassland ecosystems (MEA: Millennium Ecosystem Assessment; SDGs: Sustainable Development Goals).

2.1.1. Stage Characteristics

Budding Stage

Research during the budding stage was mainly focused on theoretical studies of ecosystem structure and stability, and mechanisms of impact. The aim is mainly to improve the yield and quality of grassland forage through these studies and to meet the needs of human food security. The relationship between biodiversity and ecosystem function and productivity and the mechanisms that influence them have long been debated and have been shown to be positively correlated in national and international studies [70]. For example, through field experiments controlling for plant species diversity, functional diversity and functional composition, Tilman et al. concluded that these three factors are the main determinants of plant productivity, total plant nitrogen and light infiltration capacity, which, in turn, have a significant impact on ecosystems [71]. Spehn et al. monitored range biomass production, resource use (space, light and nitrogen) and decomposition processes on eight European pastures with different plant species richness, showing that higher species and functional group diversity also resulted in higher yields and more efficient use of resources [72]. These scholars have mainly observed changes in grassland biodiversity and productivity from the field scale, which is only a single-sided influence process, and have not addressed the interaction mechanisms between grassland species diversity, and biodiversity, etc., and stability. Moreover, the scale of the study is limited to the field scale, which does not involve the regional or global scale. Therefore, it remained uncertain at this stage how the findings would be extended to the landscape and regional level, and how they would be extended across different ecosystem types and processes [73].

Development Stage

Facing the threat of increased environmental pollution and the endangerment of cherished flora and fauna, the United Nations hosted the MEA, which motivated the conservation and sustainable development of ecosystems and was a phase of development in the study of the structure and stability of grassland ecosystems.

The study of factors that influence the structure and stability of grassland ecosystems was becoming progressively more advanced. How to manage pastures scientifically so that they can maintain healthy grassland ecology while balancing pasture production were the main issue of research during this period. For example, Isbell et al. conducted a long-term N enrichment experiment on grasslands to analyze the effects of N enrichment on productivity, plant diversity and the species composition of natural grasslands. The results showed that in the early stages, nitrogen enrichment increased grassland productivity, but over time, nitrogen enrichment was negatively correlated with species diversity and productivity [74]. Thus, scientific anthropogenic fertilization, generalizable results derived from experimental fertilization, is an important factor driving grassland biodiversity and species composition. Facing the problem that the productivity of natural grasslands cannot meet the current demand for livestock feed, the improvement of natural grasslands and the optimization of the planting structure of artificial grasslands became the focus of research at this stage [75]. For example, Albayrak et al. conducted experiments with seed mixes of oats and vetch on pastures in the highlands of Madagascar. The experiments showed that the mixed planting not only enhanced forage yield and quality, but also improved resource utilization [76]. Meanwhile, methods for evaluating the stability of grassland ecosystems are gradually arising. Zheng et al. used the fuzzy comprehensive evaluation method to evaluate the stability of mixed-seeded grassland in terms of community components, function and resistance to invasion. The results showed that the mixed seeding species and the proportion of mixed seeding could influence the community stability; however, this did not play a decisive role [77]. Therefore, how to scientifically quantify the stability of legume–grass mixed grassland communities, considering the temporal scale, spatial scale and their corresponding sensitive indicators, etc., is one of the issues that needed to be urgently explored at this stage.

In summary, research on the structure and stability of grassland ecosystems has gone through a budding stage and a development stage, which have developed towards diversification. In the budding stage, qualitative theoretical studies and influence mechanisms are the focus; in the development stage, with the aim of maintaining the stability of grassland ecosystems and protecting and promoting their sustainable development, a series of studies on the structure, function, stability influence mechanisms and stability evaluation of grassland ecosystems were carried out. Since then, the research directions and themes of ecosystem structure and stability have gradually developed in a diversified manner.

Diversification Stage

In order to thoroughly solve the development problems in the social, economic and environmental dimensions and shift to a sustainable development path, the United Nations Sustainable Development Summit put forward the SDGs, and many countries and regions have responded to the goals of poverty eradication, hunger eradication and climate change, etc. How to optimize the structure and stability of grassland ecosystems to promote sustainable development of ecosystems has become an issue that needs to be solved at the stage of diversified development [78,79].

With the gradual progress of research, the research at the stage of diversified development is mainly focused on the control experiments of grassland, to reveal the influence mechanism of species diversity and stability [80], and to quantitatively study the nutrients and productivity of pasture, and the regulatory mechanism of stability [24]. In addition, with the development of technology, data disclosure and the rise of multiple research methods (GIS technology, etc.), different spatial and temporal comparative studies have been widely performed.

The exploration of the mechanisms influencing species configuration and productivity and stability is a prerequisite for exploring the scientific management of grasslands and improving their productivity and stability. For example, Prieto et al. studied the effects of grassland productivity and sustainable supply capacity through species and genetic diversity. The results show that the complementary effects of taxonomy and genetic diversity can increase productivity under conditions of multi-grass species configurations, which, in turn, increases the productivity and resilience of grasslands in the face of environmental hazards [81]. Quantitative studies of nutrient limitation and balance and water–fertilizer coupling in forages are important processes to improve forage quality, productivity, and resistance to invasion [23]. Niu et al. assessed the effect of nitrogen enrichment on the stability of semi-arid grassland ecosystems in northern China and its potential mechanisms by simulating atmospheric nitrogen enrichment. The results showed that community stability was non-linearly related to nitrogen enrichment and that this relationship was positively correlated with species asynchrony, species richness and species diversity, as well as the stability of dominant species and the stability of grassland functional groups [82]. Therefore, it is important to re-evaluate the mechanisms by which multiple levels of nitrogen deposition affect the stability of natural ecosystems to gain a deeper understanding of the multiple nutrient inputs to grasslands and their stability response mechanisms.

In summary, the optimization and stabilization of grassland ecosystem structure can provide important basic research to elucidate the mechanisms of the complementarity and functional diversity of functional traits of forage grasses, as well as the synergy and trade-off between productive and ecological functions of grassland, thus promoting the healthy and sustainable development of grassland. We believe that, based on the idea of cascading benefits of grassland ecosystem pattern, process, function, and services, we should explore the mechanisms of multi-species configurations or natural grassland improvement and stability maintenance, and clarify how the species configuration of grassland (leguminous–grass, annual–perennial, and deep-rooted–shallow-rooted, etc.) affects the nutrient flow of grassland, which, in turn, affects community stability, and ultimately promotes the synergistic development of grassland-ecosystem productivity and stability.

2.1.2. Stage Characteristics

Table 1 illustrates the global distribution of studies related to the structure and stability of grassland ecosystems. China and the USA have the highest number of published literature, with over 20 articles, which shows the concern for global issues such as the sustainable development of grasslands and food security in those countries [5,83]. Developed countries such as Europe and Oceania also published a relatively significant amount of literature. In addition, countries such as Brazil and South Africa account for a relatively small number of publications.

Table 1. Distribution of the number of publications issued by countries or region.

Country or Region	Number of Articles Issued
China	53
American	22
Australia	8
Canada, England	6
Germany, Switzerland	5
France, The Netherlands	4
South Africa, Brazil	3
Estonia, Poland, Austria, Japan, Mexico, Sweden, Senegal, New Zealand, Belgium, Argentina, Denmark, Romania, Spain, Italy	1

Due to regional differences in natural economic and social conditions, research on grassland ecosystems has developed unevenly and has prominent regional characteristics. In terms of the number of publications, Asia accounts for 47.8%, which is related to national policy support and the attention of research institutions [84]. In the Asian region, China is the main publication country, with most of the research focus on large-scale natural grasslands in the provinces of Tibet, Inner Mongolia and Xinjiang [85], where the restoration of these grasslands is of great importance to China in realizing the "two mountains theory" of "Lucid waters and lush mountains are invaluable assets." [86]. The emergence of global issues (global warming, land degradation and resource crises) has led to sustainable and green development being a crucial issue in the 21st century, as exemplified by the United

Nations SDGs. As a result, publications focused on the restoration of grassland ecosystems are gradually increasing, with particular attention being paid by countries in North America. In addition, with a large proportion of grasslands in North and South America, particularly in the Pampas, which has always been an export area for beef cattle, the sustainability of grassland ecosystems determines the economic lifeblood of the entire region. Similarly, although Europe does not have the same area of grassland as Asia or the Americas, food mainly comes from pastureland due to dietary habits, etc. As a result, the stable output of pastureland is vital to Europe's survival [87]. Oceania is the region with the highest exports of wool and dairy products, and the quality of forage is related to the value of trade in export commodities and is equally important to the economy of the region as a whole.

The low number of publications in Africa is mainly related to the economic development of Africa, where most of the countries are still developing countries and the primary concern is to solve the problem of food and maintain national security and stability. Thus, although the area of grassland in Africa is large, and the sustainable development of savannah, in particular, is of great importance for maintaining the global climate, it is mostly studied by developed countries, such as the USA and Denmark [88–90]. South Africa is one of the few African countries with a significant volume of publications due to its better economic conditions.

2.2. Research Progress and Major Landmark Results

Based on the titles, keywords, abstracts and previous studies of the literature [91], drawing on the studies of Wang et al. and Fu et al. and the changes in grassland ecosystem patterns and their relationship with ecological processes [92,93], the 133 papers were divided into five aspects: structural studies (component structure, trophic structure, spatiotemporal structure), structural optimization, stability studies, structure–stability relationships and influencing factors (Figure 2), according to structural composition, ecological processes (replaced by stability) [94,95], the relationship between structure and stability, and factors influenced by the environment. The largest portion of literature focused on the factors influencing ecosystem structure and stability, at 31.5%, followed by structure studies, which accounted for 30% of this type of literature. In addition, studies on ecosystem stability accounted for 17.2% of the literature, while studies on ecosystem structure optimization and structure–stability relationships accounted for 12.7% and 8%, respectively.

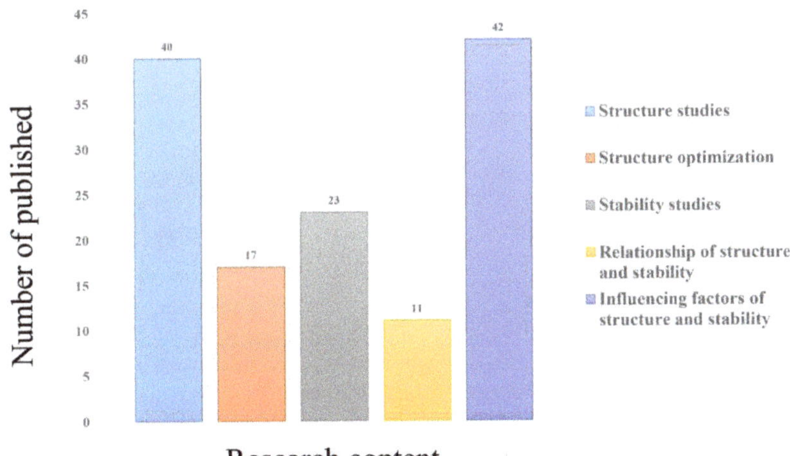

Figure 2. Grassland ecosystem structure and stability research content division. (From left to right, there is the legend of structure studies; structure optimization; stability studies; relationship between structure and stability; and influencing factors of structure and stability).

Grasslands include the degraded, natural, and artificial, which determine the type of ecosystem structure and species composition of their habitats. This alters the process of ecosystem material cycling and information transfer through the component structure, spatial and temporal structure and nutrient structure [96], influencing the degree of ecosystem stability and, thus, the supply of ecosystem service capacity [97]. It also combines with artificial capital to form multiple forms of animal husbandry, which enter the market and, further, form an industrial chain which provides ecosystem services and benefits to humans. Therefore, attention to the structural composition and stability of grassland ecosystems is a prerequisite for interpreting the processes that shape grassland-ecosystem services.

2.2.1. Structure Research

Grassland ecosystem structure refers to the relatively ordered and stable state of the various components of an ecosystem in space and time [98], including component structure, nutrient structure and spatial and temporal structure [99].

Component Structure

The structure of grassland ecosystems consists of biotic (plants, animals, microorganisms) and abiotic (water, air, soil) factors, and a certain hierarchy and structural pattern is maintained between the components of each system. The composition of grasslands is the basis of all grassland research, and the analysis of changes in species composition and their response to grassland has always been an important part of grassland research [100–102]. Silva Mota et al. evaluated changes in floristic composition, structure, diversity, and life-forms spectra along an altitudinal gradient in the rupestrian grasslands in the south-eastern Espinhaço Mountains Range, and the results showed that differences in vegetation structure, diversity, species composition, frequency and the richness of each life form in relation to soil attributes and elevation. Plant height, species richness, diversity and evenness, frequency and richness of phanerophytes and chamaephytes decreased with elevation. The related results indicated that soil is an important driver of community change [103]. At the same time, human activities affect the species composition of grassland, and, thus, changing the cascading benefits of research into ecosystem services has become an emerging research direction. However, it is difficult to solve the various problems of grassland composition alone at the present stage. Therefore, the above studies were conducted in combination with spatial and temporal changes [104].

Spatial and Temporal Structure

The characteristics of ecosystems depend on biodiversity, that is, the functional characteristics of organisms present in the ecosystem, and the distribution and abundance of these organisms in space and time [105]. Due to species evolution and turnover, and a lack of scientific grassland management measures, the species diversity of grassland is gradually homogenizing, the soils are gradually being degraded, and the productivity is gradually decreasing [106]. Therefore, study on the spatial and temporal structure of grassland ecosystem provides a good reference for improving the service capacity of grassland ecosystem in the KDC area.

Time Structure

Quantitative assessment of the interannual variability in grassland landscape patterns in response to their ecosystem service values is a current research hot spot [107]. Monitoring the spatial and temporal dynamics of grassland areas can help decision makers to plan the use of grassland in a rational manner and achieve the sustainable development of grassland [108]. Landscape changes directly affect changes in the demand and supply of ecosystem services. Compared to the spatial scale of ecosystem services, little attention has been paid to how interannual changes in land use affects ecosystem service trade-offs, but this is necessary to facilitate the rationalization of decisions, especially when those decisions aim to re-establish diverse services and restore biodiversity [109]. Therefore, it is important to analyze the spatial and temporal evolution characteristics of grasslands and elucidate the main drivers of grassland-ecosystem service capacity to develop reasonable ecological restoration measures for grasslands. Recent studies have shown that current research has focused too much on global, national, and provincial spatial and temporal dynamics at large scales, neglecting research on the mechanisms of trade-off synergy and value transfer of grassland-ecosystem service functions at small scales in counties and sample sites [107]. Therefore, it is important to conduct overall monitoring of grassland landscape patterns in different periods in a specific region, analyze the temporal variability and heterogeneity of grassland-ecosystem landscapes at spatial scales, and grasp the structural-change characteristics of grassland ecosystems, to promote the optimization of the overall ecosystem structure and stability enhancement.

Seasonal changes in grasslands are also a current research hotspot in ecological restoration. The physical characteristics of plant growth in grassland determine its seasonal variation, which can lead to changes in ecosystem structure in the short term [110]. Thus, the changes in the species composition of grasslands in the short term can affect not only the above- and below-ground productivity supply of grasslands, but also the quality of forage and, in turn, the supply of multiple ecosystem functions. Rodriguez Barrera et al. compared the effects of groundhog and grassland use types and their seasonal changes on grassland vegetation structure and diversity in the semi-arid grassland in northern Mexico. The results of this study showed that grassland use type and its seasonal variation were the main determinants of grassland vegetation structure and cover [111]. However, current research on ecosystem services is mainly focused on inter-annual variability, with less research on monthly and seasonal variability [112]. Therefore, the interannual and seasonal changes in grassland-ecosystem structure should be strengthened in future studies.

Spatial Structure

Some studies have specifically classified grassland structure into vertical canopy structure and horizontal planting structure [113,114]. Therefore, the spatial structure of grassland ecosystems is discussed in terms of vertical and horizontal structure.

In terms of vertical structure, the three-dimensional structure of forest grassland is globally recognized as a stable state of ecosystems [115,116]. Studies have shown that a reasonable three-dimensional structure can effectively use light, heat, water, air and other environmental factors to improve productivity and maintain the stability of ecosystems [117], and that a composite ecosystem structure has a better soil retention effect [118]. The ecosystems of economic fruit forests and agroforestry are typical composite ecosystems, of which grassland is an essential and important component. The shallow roots of grassland combined with the deep roots of trees provide effective use of natural elements such as light and hot water and air, and intercept water through the multi-layered canopy, thus preventing flooding and providing resistance to erosion and, thus, having a good soil and water conservation effect [119].

In terms of horizontal structure, the optimal combination of multiple species can improve the productivity of grassland ecosystems and promote the efficient recovery of degraded grasslands, thus enhancing the stability and service functions of grassland ecosystems [120]. It has been proven that the persistence and stability of forage production in grassland ecosystems is generally higher than mixed planting with perennial grass species and single-forage cultivation [121,122]. Mixing planting with leguminous forages and replanting natural grasslands with leguminous forages can increase the productivity of grasslands, enhance their resilience and biodiversity, maintain the stability of grassland ecosystems and improve their service capacity [123]. By monitoring light cut-offs in the canopy of grassland communities with high species diversity, researchers found a positive correlation between plant species richness and multiple ecosystem functions, at which time grassland resource utilization was high. The relationship between grassland biodiversity-productivity and management intensity was also positively correlated when supplemented by sound grassland management [124]. Thus, a rational vertical canopy structure and a scientific horizontal structure not only maintain the biodiversity of the grassland, but also ensure its productivity, thus providing the basis for multi-functional grassland management [125]. In a protected area in central Italy ("Laghi di Suviana e Brasimone" regional park), Cervasio et al. showed that mixed seeding had a significant contribution to improving grassland quality and had a positive impact on grassland ecosystem biodiversity by changing the vertical and horizontal characteristics of grassland ecosystem species [126].

In addition, ecological corridors with a predominantly grassland landscape structure are also an important part of current spatial-structure research [127,128]. Ecological corridors are important links between the structure and function of a system, and grasslands can be used as buffer zones to improve the connectivity of ecological functions, allowing species to use corridors for long-distance dispersal or to provide shelter for some animal movements [129], thus maintaining and protecting biodiversity and improving the quality of their habitats [130,131]. At the same time, the spatial pattern of the landscape influences the magnitude of ecosystem service provision at the landscape scale through grassland use patterns and connectivity [132–135]. By quantifying inter-annual and seasonal changes in grassland landscape patterns in northern China, Hao et al. showed that degraded vegetation can influence changes in ecosystem service functions by changing land use patterns, increasing or dividing patch sizes of grassland and agricultural land, and increasing the size of forest areas where appropriate [136]. This is a very good reference for the grassland ecological protection of KDC. At this stage, however, most studies are still at the stage of qualitative description, and these mainly explore the factors influencing ecological functions (such as biomass production and other ecosystem services) and plant community characteristics [137]. There is also an urgent need to use these insights to develop combinations of grass species for high-yielding, high-quality communities [138], as the research subjects and geographical environments change and whether regional research results can be adapted to global environments or specific geographical areas, such as karst desertification areas and alpine grassland regions.

The study of the spatial structure of grassland ecosystems should focus on the vertical and horizontal cropping structure of grasslands and the distribution of grassland landscape patterns. This not only helps researchers to monitor the dynamics of grasslands comprehensively from a micro to macro level, but also helps grasslands (as an ecological linkage corridor) to provide better connectivity to other ecosystems, guaranteeing the healthy development of the whole terrestrial ecosystem and providing more and better service functions for humans [139]. Studies have found that increasing the abundance, evenness and diversity of dominant species not only effectively improves the productivity of the system [140,141], but also ensures the harmonious development of the "production-living-ecological function" of grassland ecosystems [142,143], maintaining the stability of their productive functions and the diversity of biological species. In summary, the vertical and horizontal structural characteristics of grasslands change the species composition and biodiversity of grasslands, affecting the quality of grassland habitats and their ability to provide services.

Trophic Structure

The study of trophic structure, food webs and ecological networks in ecosystems is the theoretical basis for understanding the composition, structure and dynamics of ecosystems, and provides indispensable scientific support for biodiversity conservation, ecosystem management and restoration, as well as response to global change [99]. The trophic structure of an ecosystem refers to the food chains and food webs formed by producers, consumers and decomposers in a biome with food as the link, which constitutes the main pathway for material cycling and energy flow [144]. Food webs are based on the interactions between species at different trophic levels, forming upstream and downstream regulatory mechanisms to maintain their structural stability [145].

Species interactions or alterations are one of the most important areas in the study of food webs. Food webs, which depict networks of trophic relationships in ecosystems, provide complex yet tractable depictions of biodiversity, species interactions, and ecosystem structure and function [146]. Nearly all ecosystems have been altered by human activities; these changes may modify species interactions and food-web stability [147]. De Castro et al. started from the classic soil food-web model of Hunt to formulate a plausible topology of soil food-web models and then compare the effect of this topology with those of random topologies [148,149]. The results showed that the stronger the species interactions in the soil, the more stable the food web, and that disrupting the functional-group assemblages and strength of interactions in the soil food web inevitably has a significant impact on species and their relative abundance. Duchardt et al. applied structural equation models (SEM) to disentangle direct and indirect effects of prairie dogs on multiple trophic levels (vegetation, arthropods, and birds) in the Thunder Basin National Grassland, and the results indicated that prairie dogs directly or indirectly influence associated vegetation, arthropods, and avifauna [150].

In summary, changes in key traits that sustain interactions are one of the most important factors in determining the stability of food webs [151]. Therefore, the study of species interactions is important for clarifying inter-species relationships and food-web driving mechanisms, which is one of the priorities that should be focused on in the future.

2.2.2. Structure Optimization

Structural-optimization measures can change the species composition of grasslands, thus affecting their ecosystem biodiversity and altering their capacity to supply ecosystem functions. There are many ways to optimize structure, but this study will only discuss biological measures (optimal allocation of species structure) and engineering measures (pasture management systems).

Scientific and rational species allocation is the key to the high productivity and stability of grassland ecosystems [152]. Optimal species structure allocation includes the optimal vertical and horizontal allocation of grassland species. The three-dimensional structure of composite ecosystems has proven to be an important strategy for reconciling environmental protection and economic development in ecologically fragile areas [153], with grasslands being the most crucial aspect, and their strong renewal rate making them a large surface area within the composite ecosystem. The resulting three-dimensional ecosystem structure not only plays an important role in maintaining and providing ecosystem services, but its multiple ecosystem services also provide an effective way to promote the restoration of degraded ecosystems, which is an important biological measure to make full use of resources [117,154]. A diverse three-dimensional and horizontal structure is an effective measure to restore grassland ecosystems, that is, through a combination of different species of grasses from different families (in intercropping, crop rotation, and mixed sowing, etc.; see Figure 3) [155]. The most studied of these is the mixed cropping pattern of legumes and grasses. This mixed cropping pattern makes full use of sunlight, heat, water and air, creates complementarity between species, and promotes the uptake of soil nutrients by forage grasses, which, in turn, improves their annual growth and nitrogen use efficiency [156]. Multi-species planting with different habits is an effective means of restoring grassland ecosystems, and it can effectively improve the degradation of grasslands, protect and improve soil quality, enrich inter-species and intra-species species diversity, and effectively increase the resilience and recovery of degraded grassland ecosystems [157]. Scientific seeding is an effective measure to restore the dynamic balance of grass species [158]. Under the same circumstances, mixed-seeded grasslands have higher forage density and species richness, and it has more competitive and better grassland vegetation restoration compared to single-seeded grasslands. However, it has been demonstrated that species diversity and stability in grassland ecosystems with different rates of mixed seeding are positively correlated, and that seeding rates are negatively correlated with resistance to invasion [159]. Therefore, understanding forage seeding rates and their effective mix relationships is an important area of vegetation-restoration research [160]. Meanwhile, the mixing and intercropping of annual and perennial forages are also efficient management practices that take advantage of the temporal structure of grassland ecosystems, reducing inter-specific competition and increasing the persistence of forage production [161]. Mixing annual and perennial legume-grass is recognized as a highly productive and robust management practice [162], which can maintain a diverse supply of ecosystem services, and the leguminous forage grasses provide a more adequate supply of nitrogen to other grass species, thus maintaining high grass biomass and meeting the production requirements of farmers [163].

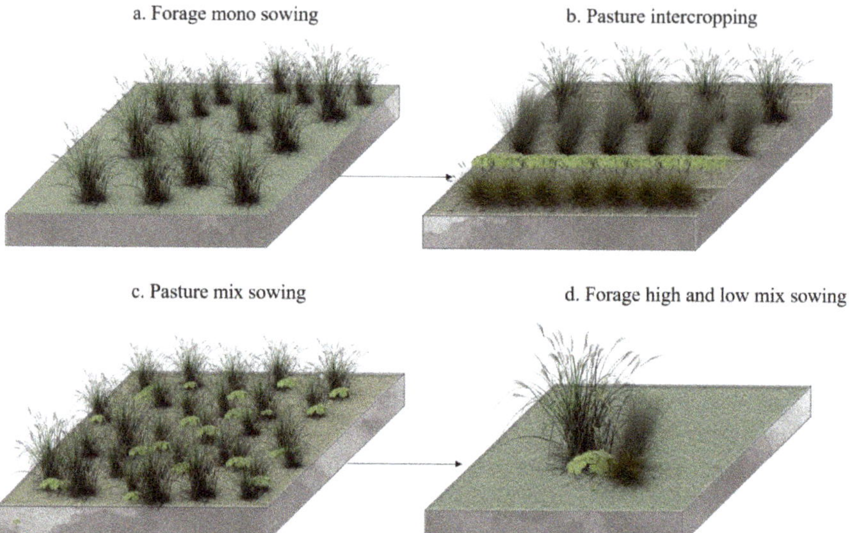

Figure 3. Artificial grassland planting structure. (**a**) Forage mono sowing: Grass forage, *Lolium perenne* L., (**b**) Pasture intercropping: Grasses-Leguminosae Forage, *Lolium perenne* L. + *Medicago sativa* L., *Trifolium repens* L.+ *Lolium perenne* L., (**c**) Pasture mix sowing: Grasses-Leguminosae Forage, *Lolium perenne* L. + *Medicago sativa* L., (**d**) Forage high and low mix sowing: *Lolium perenne* L. + *Medicago sativa* L. + *Trifolium repens* L. + *Lolium perenne* L. (The grass species of (**b**–**d**) are the same, but the arrangement of planting order is different. (**b**,**c**) are planted in a horizontal way, but the density and proportion of planting are different. (**b**) is good for personnel to manage, and (**c**) is good for the nutrient complementation of grass species. (**d**) is planted in a vertical way, which can fully utilize light, heat, water and air).

A scientific system of pasture management can effectively enhance the sustainable supply of grassland resources. Pasture management not only includes grass seed cultivation and fertilization, but also includes the full use of time and space, such as rotational grazing, rest grazing and ban grazing. In the early days, pasture management was mainly aimed at improving pasture production. In the new context of global climate change and increased environmental pollution, balancing ecological protection and green economic development has become a necessary path to improve the overall ecological quality of regions [164]. The general system of rotational and rest grazing is only an artificial strategic rest and tactical grazing of grasslands [165]. Therefore, some studies have proposed models such as the sustainable grazing system (SGS), DairyMod, GRAZPLAN, GrassGro, DairyNZ whole-farm model and EverGraze [166–172]. The above systems encourage the sustainable development of grassland ecosystems, and are key to preventing soil erosion and promoting soil improvement, as well as being important measures to conserve grassland plant biodiversity. Michalk et al. detailed sustainable and permanent grazing systems and answer the key questions currently facing Australia: (1) whether increasing the number of paddocks and implementing rotational grazing results in higher grazing rates, higher yields per hectare and better economic benefits; (2) which combination of grazing methods and grazing rates is most appropriate to create higher and more stable perennial grasslands to improve yields and environmental benefits in different parts of the landscape; and (3) can landscape variability be identified, mapped and effectively managed on native grasslands in high-rainfall areas [173]? These can provide some inspiration for grassland grazing and farmers' livelihoods in KDC.

Therefore, optimizing the structural configuration of grasslands in terms of both biological and engineering measures is currently an unavoidable and important issue for improving the sustainable development of grasslands and economic development. It will not only solve the shortage of supply capacity for ecosystem services and promote regional economic development, but also issues related to the global supply of food.

2.2.3. Stability Studies

Stability refers primarily to the ability of a community or ecosystem to maintain its original structural and functional state and resist disturbance after a disturbance, known as resistance stability, and the ability of a community or ecosystem to return to its original state after a disturbance, known as resilience stability.

Exploring the causes that affect the stability of grassland ecosystems and suggesting targeted restoration strategies for degraded outcomes is a major focus of current research [83]. Community stability increases species diversity, species heterogeneity and population size, so maintaining the long-term stability of communities is crucial to maintaining ecosystem function. Researchers have explored the potential drivers and mechanisms of ecosystem stability by studying changes in the relationship between biodiversity and ecosystem service functions [174–176]. Biodiversity is a major driving factor of ecosystem stability [177]. Biodiversity consists mainly of above- and below-ground components. In terms of above-ground biodiversity, an increasing number of studies have shown that the more above-ground biodiversity (plant diversity) there is, the more stable the ecosystem is [178]. The results of Garcia-Palacios et al. showed that the diversity of leaf traits may promote the stability of an ecosystem under low drought conditions, while the species richness may play a greater role in the stability of an ecosystem under high drought conditions [179]. Similarly, below-ground biodiversity (soil biodiversity) altered grassland-ecosystem resistance and resilience through direct and indirect effects on plant diversity, net ecosystem productivity and plant species interactions, and, thus, changed grassland ecosystem stability [180]. In addition, recent studies have shown that phenological variation can also reconcile plant diversity with the seasonal stability of ecosystems, and, thus, affect the stability of the whole ecosystem [181].

Evaluating grassland-ecosystem stability has also become an effective means of detecting changes in grassland landscapes or the effectiveness of grassland restoration [182]. When evaluating the stability of grassland ecosystems, timescale variation is a factor that must be considered. By comparing the changes (years, seasons, or months) in grassland landscape patterns in different periods combined with evaluation models and methods, researchers have selected indicators of ecosystem vitality, resistance, resilience and variability to evaluate ecosystem stability [32], revealing the processes of change in grassland ecosystems under extreme weather conditions and their driver factors, quantifying the correlation between grassland ecosystem stability and biodiversity and their spatial patterns, and providing important prerequisites for improving sustainable ecosystem services [183,184]. The traditional evaluation methods of ecosystem stability are mainly qualitative analysis methods, including empirical methods and expert consultation methods, which are, generally, using expert consultation methods to quantify measures. The other is mainly quantitative evaluation including the comprehensive-evaluation method, gray-information analysis, ecological models and other methods, among which the most common methods of comprehensive analysis are hierarchical analysis, the entropy weighting method, the weighted comprehensive average method, and the comprehensive index evaluation method, etc. New kinds of ecosystem stability evaluation are diverse, and an increasing number of methods are becoming more credible and scientific in their assessment of ecological stability. These methods for evaluating the stability of ecosystems can provide good theoretical references for the ecosystem evaluation of KDC, and they can provide insights into the establishment of stability evaluation index systems for karst grassland ecosystems.

2.2.4. The Relationship between Structure and Stability

A good system structure maintains its relative stability, and ecosystems with high stability generally have a more rational structure. Tilman et al. argued that diversity leads to higher community productivity, ecosystem stability and resistance to invasion [74]. Before the 1970s, ecologists believed that communities with higher diversity had more stable ecosystems [185]. Since then, ecologists have focused more on the relationship between species diversity and ecosystem stability [186,187].

Biodiversity affects ecosystem services by altering ecosystem function and stability [188]. Biodiversity–stability relationships showed that above-ground productivity and temporal stability increased significantly with increasing species richness, while biodiversity was largely influenced by ecosystem structure [85,189]. Therefore, understanding the relationship between grassland-ecosystem structure and stability and its influencing factors is essential for the sustainable development of grassland ecosystems [190]. However, how plant species diversity regulates the stability of ecosystems (such as biomass reproduction and nutrient cycling) has become one of the challenging questions in ecology [191,192]. Experiments have demonstrated that the higher the complexity of the ecological network of grasslands, the higher the ecological stability, by influencing plant physiological conditions, as well as species generation, diversity and variation [193,194]. Of course, grassland-ecosystem stability is also vulnerable to the influence of ecosystem components (ecosystem species diversity, composition), climate change and anthropogenic activities, mainly in the form of lowering the productivity of grassland, weakening biodiversity and declining service functions, which, in turn, can affect grassland-ecosystem services [195]. Therefore, studying the structure of grassland ecosystems and their material and energy flows and cycles among different components, and exploring the relationship between grassland-ecosystem structure, function, and stability is one of the research areas that should be focused on in the future [196]. Recent studies have shown that the positive relationship between biodiversity and stability is also influenced by spatial scale [197]. Therefore, the study of the relationship between structure and stability should also consider different spatial scales and timescales.

2.2.5. Factors Affecting Structure and Stability

The structure and stability of grassland ecosystems are influenced by multiple factors, such as the natural environment and human activities. Therefore, the heterogeneity of spatial environments inevitably leads to variability in the degree of stability. In the case of grassland ecosystems, their structure and stability are mainly influenced by climate change, nitrogen deposition or nutrient addition, grazing and grassland management, plant invasion and natural disasters (Figure 4).

Global climate change is affecting ecosystem function and stability, indirectly altering species diversity, species composition, and functional plant traits [198,199], thus reducing the capacity of ecosystem services and reducing the various benefits that humans derive from them [200]. The impact of climate change on ecosystem function and biodiversity is highly dependent on grazing history and natural conditions [190]. By comparing changes in the spatial pattern of grasslands and simulating different climatic conditions on grassland ecosystem resistance, recovery times and rates, White et al. concluded that future climate change will have a significant impact on the resistance and recovery of disturbed ecosystems; however, the spatial pattern of impacts varies widely [201].

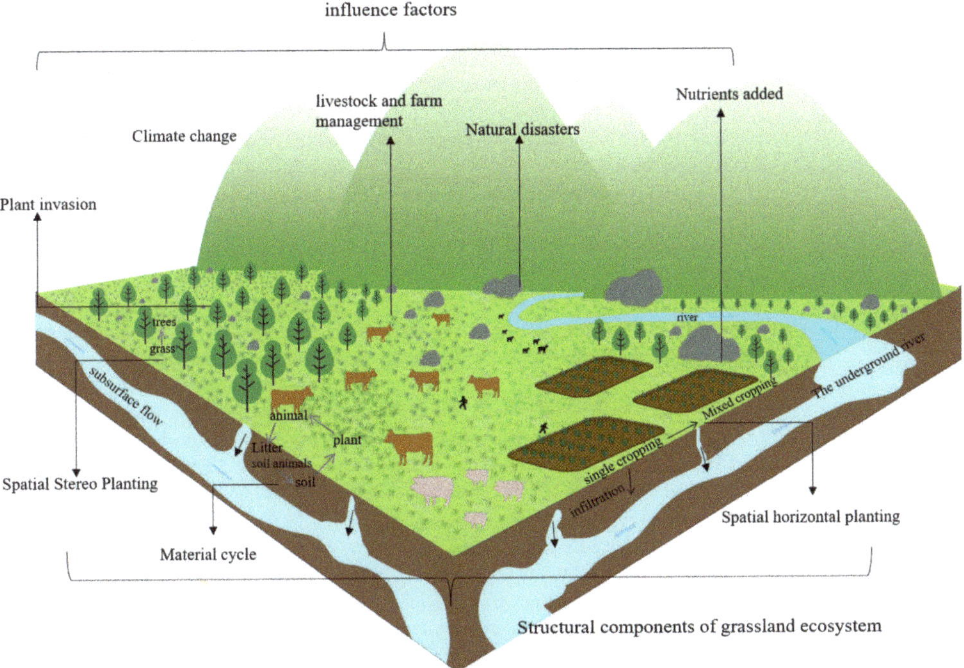

Figure 4. The structural components of grassland ecosystems and the factors that influence them combine to influence grassland ecosystem stability.

Key limiting elements can alter species interactions [202], change the spatial and temporal patterns of terrestrial plant communities [203], and influence ecosystem function [204]. Atmospheric nitrogen deposition and the application of nutrients (phosphorus, potassium, etc.) can affect species diversity in grassland communities [205]. The results of related studies show that the addition of nitrogen or phosphorus individually can increase the productivity of grasslands by 0–20%, while the addition of both nitrogen and phosphorus can increase the productivity of grasslands by 60% [206,207]. Nitrogen is an essential nutrient for plant growth, and an appropriate addition of nitrogen deposition will directly promote rapid plant growth and increase the net primary productivity of terrestrial ecosystems. However, excessive nitrogen deposition can also lead to soil acidification, altering the effectiveness of soil mineral nutrients and the structure and function of microbial communities, which, in turn, affects plant growth and changes plant diversity [208–210]. Plant diversity is closely related to ecosystem stability and productivity [211], and ultimately drives changes in ecosystem structure and function [212]. Therefore, paying attention to the differences in supply between different nutrient elements and the resulting differences in the resistance and resilience of plant diversity that result is extremely important for both productivity stability and the temporal stability of grasslands [213].

Plant invasion and vegetation succession are also among the factors that alter the species composition of grasslands and directly change the structure of grassland ecosystems [214,215]. Since the early 20th century, the invasion of woody plants into grasslands and their impact on the carrying capacity of grasslands has become a serious problem for savannas [216]. As reported in savanna ecosystems, the large-scale invasion of shrubs, trees, and other plants has led to significant changes in the functioning of global ecosystems, which will have profound impacts on the biodiversity, carbon storage capacity, and supply of these ecosystems [217]. Invasions of poisonous weeds has been considered one of the most important causes of economic losses by inhibiting forage production and killing livestock. However, a recent study concluded that toxic weeds can also have some potential positive ecological impacts on grasslands, such as promoting soil and water conservation, improving nutrient cycling and biodiversity, and protecting rangelands from excessive livestock damage [218]. Therefore, appropriate actions are needed by policy makers, managers and stakeholders to assess the ecological functions of invasive toxic weeds and to reconcile the long-term trade offs between livestock development and maintaining the ecological services provided by grasslands.

Animal husbandry plays an important role in eradicating hunger and improving malnutrition [219]. Protein and energy from livestock products are the main sources of human nutrition. Therefore, improving and sustaining livestock development is critical to advancing the United Nations SDGs, especially in addressing zero hunger and mitigating climate change [220]. Rational livestock management is a major initiative to promote healthy ecosystem development. Therefore, rangeland management, as a determinant of maintaining biodiversity, ecosystem services and landscape [221], is more important to focus on exploring its driving mechanisms on grassland species structure and ecosystem stability [222]. In addition, natural hazards are of great importance to our understanding of global biogenic burning and its emissions, carbon cycling, biodiversity, conservation and land management, especially as ecosystem succession and biodiversity changes associated with grassland fires are critical to the patterns and dynamics of ecosystem functions and services [223].

In summary, sustainable management of grasslands requires a deep understanding of the functional relationships among these factors due to the spatial heterogeneity of environmental conditions, production potential and flora composition [224]. It is more important to jointly explore the similarities and differences in the structural composition of grassland ecosystems in multiple factors and scales, extract the key factors affecting the service capacity of grassland ecosystems, and apply them to the restoration of the grassland vegetation of KDC.

3. Materials and Methods

Based on the research of Khan et al. and Chapman et al. [225,226], we performed a systematic literature search and review following the protocol from Preferred Reporting Items for Systematic Reviews and Meta-analyses (PRISMA) including quantitative statistics and qualitative content analysis. Systematic reviews have an advantage over traditional reviews and commentaries in that they cover studies by following an explicitly formulated procedure, which can help readers to understand the whole protocol followed for the literature review [227].

Literature Search and Selection

The first step was identifying records. We conducted a systematic search of peer-reviewed literature (articles, reviews) using the key words in WOS and CNKI databases (Table 2). A total of 99 repeated references were excluded; 3821 references were selected for review. The search timeframe for both databases was the maximum timeframe of the databases, and the search deadline was 1 November 2022.

The second step was screening. We screened the titles and abstracts of each article to select articles that assessed, depicted, and quantified or mapped grassland ecosystem structure and stability which were eligible for full-text reading ($n = 278$). The third step was reading the full text of each of the selected publications, and, finally, a total of 133 references were chosen as case studies. To summarize the information about main structural characteristics, stability, structure–stability relationship and influencing factors and suggest future directions of ecosystem service-capacity enhancement from case studies, we recorded the annual distribution of the literature, distribution of countries of publication, types of grassland structure, disturbance factors of grassland stability, and research methods (Figure 5).

Table 2. Literature search terms and results.

Literature Databases	Types	Search Terms (All Fields)	Number of Initially Acquired Publications
WOS and CNKI	Structure	"Grassland" OR "Meadow" OR "Pasture" OR "Rangeland" OR "Steppe" OR "Prairie" OR "Veld" OR "Savanna" AND "Ecosystem structure" "Food chains" OR "Food web" OR "Trophic levels" OR "Nutrient levels" OR "Community structure" OR "Species configuration" OR "Biodiversity" OR "Landscape pattern" OR "Forage mixes" OR "ecological corridors" OR "Time change" OR "Interannual variation" OR "Month change" AND "Ecosystem services"	1435
	Stability	"Grassland" OR "Meadow" OR "Pasture" OR "Rangeland" OR "Steppe" OR "Prairie" OR "Veld" OR "Savanna" AND "Ecosystem stability" OR "Stability assessment" OR "Flexibility" OR "Resistance" OR "Resilience" AND "Ecosystem services"	295
Total			1730

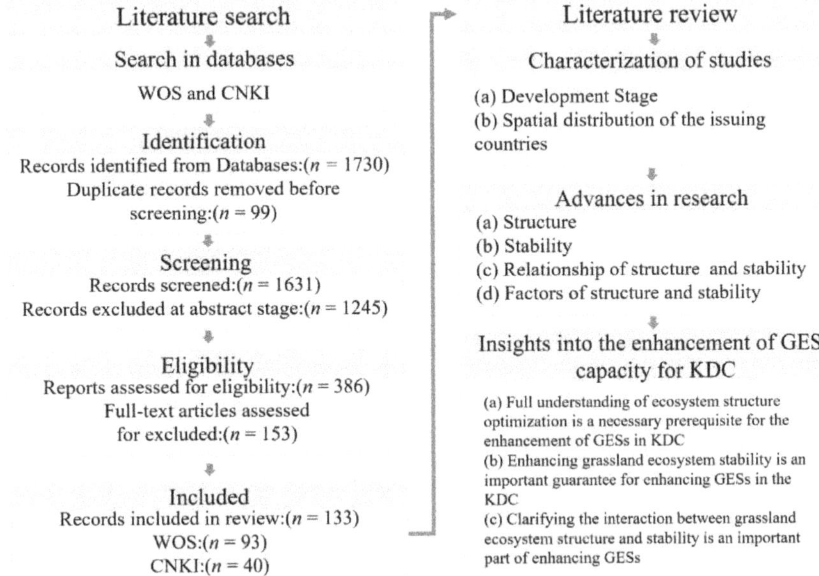

Figure 5. The flow diagram showing the steps of the methodology and selection process used for the systematic literature review on the left and the literature review process on the right.

4. Prospects

4.1. Key Scientific Issues That Need to Be Addressed

Research on the structure and stability of grassland ecosystems has achieved great success, but there are still many scientific questions that need to be addressed. Based on the previous systematic analysis, this paper categorizes the key scientific problems to be solved in the structure and stability of grassland ecosystem into three aspects: structure optimization and stability improvement and their relationship.

4.1.1. Structure Optimization

To address the problems of ecological imbalance, food-chain disruption, and multiple functions in grassland ecosystems, we investigate the interactions between population dynamics and community properties through synergistic intra- and inter-species differences and spatial landscape-scale differences, determine the mechanisms and effects of biodiversity on the stability of ecosystem functions, and clarify the overall operation of ecosystems [27].

To address the problem of the relatively homogeneous grassland-ecosystem structure, this study suggests exploring the maintenance mechanisms of grassland-species allocation and stability; combining the heterogeneity of grassland functions and spatial patterns; performing a comprehensive assessment of grasslands in terms of local stand conditions, climatic habitats, plant functional traits, and other conditions; and targeting the selection of grass species with high adaptability and resistance to stress for mixed seeding [228].

To address the problem of the optimal configuration of grassland ecological structure, different grassland management measures can be combined to form an efficient and integrated management system by organically combining biological, engineering and management aspects to promote the sustainable development of grassland ecological animal husbandry. For example, scientific methods such as rotational grazing, rest grazing and limited-term enclosure, and the use of artificially planted forage instead of grazing. Sustainable grazing management in adaptive multiple paddocks (AMP) can be employed to incorporate forage and ruminants into regeneration-management planting systems to strengthen the productivity, stability and resilience of agroecosystems and, thus, improve ecosystem service capacity.

4.1.2. Stability Improvement

The issue of the low stability of grassland ecosystems, can be explored through the mechanisms regulating the stability of grassland ecosystems, identify and quantify the factors affecting the stability of grassland ecosystems, select grass species suitable for local conditions, increase the diversity of grassland species, improve the resilience of grassland ecosystems, enhance ecological stability, and further promote the process of grassland ecological restoration, so as to reduce the ecological vulnerability of grassland and enhance the resilience and stability of grassland ecosystems [229].

To deal with the problem that the ecological stability index system and evaluation model have not been unified, it is necessary to establish and standardize the selection criteria of ecological stability evaluation indexes based on the analysis and summary of ecological stability research results, to select methods according to the basic attributes of grassland ecosystems, especially the key variables of ecosystem processes and functions, and to consider the validity, sensitivity and operability of alternative indexes. The evaluation indexes are selected using the expert consultation method, the evaluation index weights are determined using the hierarchical analysis method, and the mathematical model is constructed to specify them and quantify the factor percentages [230], so as to build a scientific grassland-ecosystem stability evaluation index system and promote the sustainable development of the system [231].

4.1.3. Interaction between Structure and Stability

To address the problem of unclear relationships between the structure and stability of grassland ecosystems, study is based on the idea of "structure-process-function-services", to strengthen the material cycle, energy flow and information transfer process of producer-consumer-decomposer in grassland ecosystems, to clarify the relationship between grassland species allocation and productivity and stability, and to clarify the response mode of structure and stability. It has been suggested that structure can directly alter the biodiversity of grassland ecosystems and further alter the stability of grassland ecosystems [232]. Therefore, the relationship between structure and stability can be explored through the medium of biodiversity.

To address the problem of single research methods (e.g., field surveys, field experiments) or models (statistics, modeling) for the structure and stability of grassland ecosystems, remote-sensing research into the process of interannual or monthly changes in grassland landscape patterns can be employed, combined with indoor experiments, field surveys, and field trials for comparative assessment [233,234].

4.2. The Current Status of Grassland Ecosystem for KDC and Inspiration for Improving the Ecosystem Service Capacity of Grasslands for KDC

4.2.1. The Current Status of Grassland Ecosystem for KDC

Karst areas are very fragile and prone to ecological degradation due to their unique binary three-dimensional geological structure, of which karst desertification is an extreme manifestation of ecological degradation [229]. In the past 30 years, the KDC in southern China has achieved remarkable results in ecological environment and ecological restoration, but there is an urgent need to improve the stability of the KDC ecosystem and ecosystem service function [52,235]. Grasslands of the KDC are dominated by artificial grasslands and supplemented by improved grasslands, which constitute an important ecosystem in karst areas [229]. The artificial grasses in the KDC areas mainly include monoculture *Dactylis glomerata* L., *Lolium perenne* L. and *Medicago sativa* L. Grasslands have a monoculture planting structure and weak disease resistance, and the resulting artificial grass ecosystem is extremely fragile. This is mainly due to farmers' pursuit of high forage yields. Thus, while the artificial grassland has brought high economic income to farmers for a short period of time, its ability to provide ecosystem services is gradually diminishing. In addition, the degraded grassland is dominated by replanting white clover by government aircrafts. However, due to the overgrazing behavior of farmers in recent years, the replanted degraded grassland is difficult to recover, and only some grass species that are resistant to gnawing and trampling are propagated, resulting in the difficult recovery of the grass species structure of degraded grassland and low stability of the ecosystem.

In the process of KDC and ecological restoration, grass is the pioneer plant for vegetation restoration and ecological environment improvement [64]. Therefore, optimizing the structure and enhancing the stability of grassland ecosystems for KDC is of great significance for the sustainable development of regional ecological environments and promoting the virtuous cycle of regional eco-industry.

4.2.2. Inspiration for Improving the Ecosystem Service Capacity of Grasslands for the KDC

Due to its inherent sensitivity and fragility, the ecosystem structure and function of the grassland ecosystem in the KDC area are improper and the stability of the ecosystem is poor [236,237]. In the context of promoting the world's desertification management and global sustainable greening, how to consolidate and stabilize the effectiveness of KDC and, at the same time, optimize the structure of desertification-control grassland ecosystems, guaranteeing the stability of the ecosystem and enhancing the ecosystem service capacity is a key issue that needs to be solved in the KDC area [238]. In view of this, based on the premise of sorting out the structure and stability of global grassland ecosystems, this paper drew on the above key scientific issues of structure optimization, stability enhancement and the interaction between structure and stability and, combined with the regional characteristics of karst desertification and the grassland status of KDC, proposes three insights for the enhancement of the ecosystem service capacity of grassland in the KDC area.

Adequate Understanding of Ecosystem Structure Optimization Is a Necessary Prerequisite for the Improvement of Ecosystem Services in Grasslands for the KDC

Although existing pasture cultivation has greatly increased forage production in the short term [239], it has made an outstanding contribution to KDC. However, the factors influencing the improvement of ecosystem services are not only determined by a single factor of production stability, and there is an urgent need to improve grassland ecosystem services through a variety of specific ways. Grassland ecosystem structure directly determines grassland ecosystem function and indirectly influences grassland ecosystem services. Research on multi-component mixes of legume grasses with different growth habits or morphological characteristics has concluded that the greater the number of components in the mix, the greater the likelihood of containing key species, and the greater the potential for community stability, thus achieving a seasonal balance in grassland output throughout the year and lasting stability in the number of years of use [240]. Similar experiments have been conducted in KDC areas, for example, the state replanted degraded grasslands with *Trifolium repens* L. and encouraged farmers to plant *Trifolium repens* L.+ *Lolium perenne* L. or *Dactylis glomerata* L. + *Medicago sativa* L. and other measures, which improved the ecosystem structure of grasslands of KDC to some extent. However, with the unique geographical environment of karst, seasonal drought and unreasonable human management, the ecological benefits of degraded and improved grasslands have gradually decreased [241]. Therefore, the following two points should be considered to enhance the service capacity of grassland ecosystems: (1) to optimize the planting structure of grassland ecosystems for rocky desertification management and enrich the spatial species composition of grassland, and to select stony, drought-tolerant and calcium-loving grasses for planting in response to the characteristics of karst desertification such as thin soil layers, easy soil erosion and calcareous lithology; and (2) strengthening research on the benefits of the structure–process–function–services cascade, so as to clarify the process of maintaining ecosystem productivity and stability [95].

Enhancing Grassland Ecosystem Stability Is an Important Guarantee for Enhancing Grassland Ecosystem Services in KDC

The more complex the structure of grassland ecosystem, the higher the degree of its ecological stability, which further affects the multifunctional functioning of grassland ecosystem and eventually changes the ecosystem service capacity. The existing grassland ecosystem structure of rocky desertification control is single and the grassland planting is mainly based on single high-yielding forage grasses, for example, *Pennisetum sinese Roxb* L. and *Lolium perenne* L., which have weak stability, low disease resistance and short yield supply time, which seriously hinders the sustainable development of local ecosystems and lacks effective measures to consolidate the achievement of rocky desertification control. Grassland ecosystem stability is an important factor that limits the sustainable development of grassland ecosystem productivity. Therefore, we should combine the needs of local farmers, and select suitable grass species for planting; conduct experiments on grass planting under different rocky desertification habitats; explore the best planting density of grass species and legume-grass planting ratio; and then improve the plant diversity of grassland ecosystems and enhance their stability and resilience. At the same time, we should combine various monitoring means (such as field sampling, drone aerial photography and remote sensing, etc.), and strengthen the research on the regulation mechanism of grassland ecosystem stability with the guidance of enhancing the sustainability of KDC, so as to improve the service capacity of the grassland ecosystems of KDC, further enhance its ecological recovery quality and maintain the sustainable development of the karst ecological environment.

Identifying the Interactions between Grassland Ecosystem Structure and Stability Is an Important Part of Enhancing Grassland Ecosystem Services

This paper showed that most of the literature focuses on multiple factors of the structure and stability of grassland ecosystems, but these studies have mainly focused on unidirectional effects, such as Gschwend et al. and Hu et al. showing that biodiversity can stabilize productivity through different mechanisms, such as the asynchronous responses of species to environmental change and stability [242,243]. What is more, exploring the mutual response mechanisms between grassland planting structures and their productive and ecological stability is an important element in clarifying the effects of current grass seed mixes in KDC and local sustainable development. Xu et al. explored the water–fertilizer coupling mechanism between soil and forage through intercropping in the southern karst region. The experiments proved that a reasonable intercropping system could coordinate the relationship between crops and environment and improve soil nutrients, which made a certain contribution to exploring soil–plant relationships in the grasslands of karst areas and improving the structure and stability of stone desertification grassland ecosystems [244].

However, from the perspective of the interaction mechanism of grassland ecosystem structure and stability, exploration into the changes in the supply of ecosystem service capacity of grassland managed by karst rock desertification is still lacking. Therefore, through field control experiments in the context of KDC, the interaction mechanisms of grassland-ecosystem structure and stability should be clarified in the context of KDC, and the exploration of above-ground-subsurface material flow and nutrient circulation fluency should provide important insights to elucidate the trade offs and synergistic relationships among service flows (Figure 6).

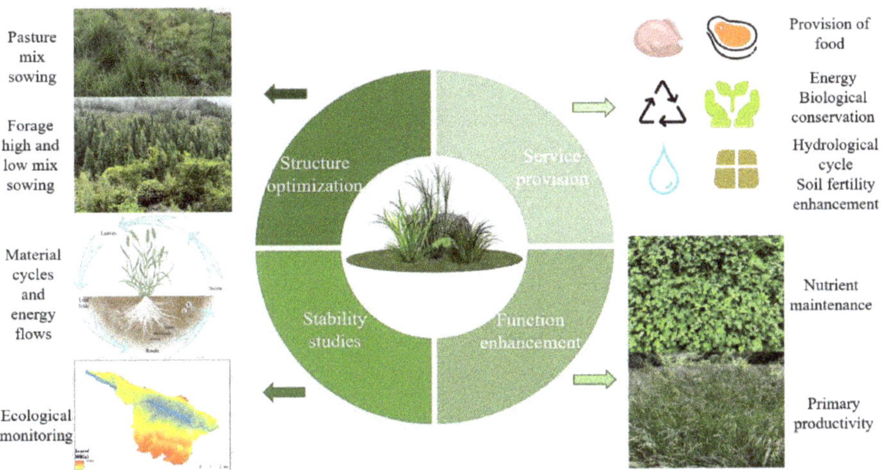

Figure 6. Processes of energy and service flows in grassland ecosystems (structure affects stability and thus changes the level of function and ultimately service provision).

5. Conclusions

Based on the WOS and CNKI databases, this study systematically reviewed the research progress of grassland-ecosystem structure and stability, and concluded the following: (1) There is a significant increase in the number of annual publications on the structure and stability of grassland ecosystems, and the research directions and themes are becoming increasingly diverse. The research mainly focuses on revealing the mechanisms influencing species diversity and stability, quantitatively studying the nutrient and productivity of forage grasses, and the regulatory mechanisms of stability. (2) The spatial pattern of the study countries is highly consistent with the spatial distribution characteristics of grassland ecosystems, mainly dominated by countries with wider grassland areas such as China and the United States. However, the number of national publications in Africa contradicts the distribution pattern of grasslands. (3) The research on the structure and stability of grassland ecosystems focuses on the structural characteristics of grassland ecosystems, structural optimization, ecosystem stability, and the structure–stability relationship and its influencing factors, in which the grassland-ecosystem structure and structure and stability influencing factors are the main research directions (4) The key scientific issues that need to be addressed nowadays were grouped according to whether they addressed structure optimization, stability enhancement, and structure–stability relationship. We should strengthen the multi-perspective exploration of grasslands under different spatial and temporal conditions, to enhance the functional output of grassland ecosystems from optimizing the structure of grassland ecosystems, to enhance the stability of grassland ecosystems, and to form a structure–process–function–services cascade benefit study. Based on the above conclusions, three insights can be provided for the enhancement of grassland ecosystem services in the KDC: (i) fully understanding ecosystem structure optimization is a necessary prerequisite for karst grassland stability; (ii) enhancing grassland ecosystem stability is an important guarantee for enhancing grassland ecosystem services in KDC; and (iii) clarifying the interaction between grassland ecosystem structure and stability is an important link to enhance grassland ecosystem services.

Author Contributions: Conceptualization, S.H.; methodology, S.H.; software, S.H.; validation, S.H.; formal analysis, S.H.; investigation, S.H.; resources, S.H.; data curation, S.H.; writing—original draft preparation, S.H.; writing—review and editing, S.H., K.X., Y.C., S.S., J.F. and C.H.; visualization, S.H.; supervision, K.X.; project administration, K.X. and Y.C.; funding acquisition, K.X. and Y.C. All authors have read and agreed to the published version of the manuscript.

Funding: This research was funded by the Key Project of Science and Technology Program of Guizhou Province (No. 5411 2017 Qiankehe Pingtai Rencai), the China Overseas Expertise Introduction Program for Discipline Innovation (No. D17016) and Natural Science Research Project of Education Department of Guizhou Province [Qianjiaohe KY Zi (2022) 157].

Data Availability Statement: Not applicable.

Acknowledgments: We would like to thank all the editors for their contributions to this paper and the anonymous reviewers for their thoughtful comments, which enriched the paper.

Conflicts of Interest: The authors declare no conflict of interest.

References

1. Buisson, E.; Archibald, S.; Fidelis, A.; Suding, K.N. Ancient grasslands guide ambitious goals in grassland restoration. *Science* **2022**, *377*, 594–598. [CrossRef] [PubMed]
2. Tilman, D.; Wedin, D.; Knops, J. Productivity and sustainability influenced by biodiversity in grassland ecosystems. *Nature* **1996**, *379*, 718–720. [CrossRef]
3. O'Mara, F.P. The role of grasslands in food security and climate change. *Ann. Bot.* **2012**, *110*, 1263–1270. [CrossRef] [PubMed]
4. Millard, J.; Outhwaite, C.L.; Kinnersley, R.; Freeman, R.; Gregory, R.D.; Adedoja, O.; Gavini, S.; Kioko, E.; Kuhlmann, M.; Ollerton, J.; et al. Global effects of land-use intensity on local pollinator biodiversity. *Nat. Commun.* **2021**, *12*, 2902. [CrossRef] [PubMed]
5. Michalk, D.L.; Kemp, D.R.; Badgery, W.B.; Wu, J.; Zhang, Y.; Thomassin, P.J. Sustainability and future food security—A global perspective for livestock production. *Land Degrad. Dev.* **2019**, *30*, 561–573. [CrossRef]
6. Bardgett, R.D.; Bullock, J.M.; Lavorel, S.; Manning, P.; Schaffner, U.; Ostle, N.; Chomel, M.; Durigan, G.; Fry, E.L.; Johnson, D.; et al. Combatting global grassland degradation. *Nat. Rev. Earth Environ.* **2021**, *2*, 720–735. [CrossRef]
7. Bai, Y.; Cotrufo, M.F. Grassland soil carbon sequestration: Current understanding, challenges, and solutions. *Science* **2022**, *377*, 603–608. [CrossRef]
8. Hönigová, I.; Vačkář, D.; Lorencová, E.K.; Melichar, J.K.; Götzl, M.; Sonderegger, G.; Oušková, V.; Hošek, M.; Chobot, K. Survey on Grassland Ecosystem Services Report of the European Topic Centre on Biological Diversity. Nature Conservation Agency of the Czech Republic: Prague, Czech Republic, 2012.
9. Denning, K.R.; Foster, B.L. Taxon-specific associations of tallgrass prairie flower visitors with site-scale forb communities and landscape composition and configuration. *Biol. Conserv.* **2018**, *227*, 74–81. [CrossRef]
10. Evans, I.S.; Robinson, D.T.; Rooney, R.C. A methodology for relating wetland configuration to human disturbance in Alberta. *Landsc. Ecol.* **2017**, *32*, 2059–2076. [CrossRef]
11. Chomel, M.; Lavallee, J.M.; Alvarez-Segura, N.; Baggs, E.M.; Caruso, T.; de Castro, F.; Emmerson, M.C.; Magilton, M.; Rhymes, J.M.; de Vries, F.T.; et al. Intensive grassland management disrupts below-ground multi-trophic resource transfer in response to drought. *Nat. Commun.* **2022**, *13*, 6991. [CrossRef]
12. Vides-Borrell, E.; Porter-Bolland, L.; Ferguson, B.G.; Gasselin, P.; Vaca, R.; Valle-Mora, J.; Vandame, R. Polycultures, pastures and monocultures: Effects of land use intensity on wild bee diversity in tropical landscapes of southeastern Mexico. *Biol. Conserv.* **2019**, *236*, 269–280. [CrossRef]
13. Smith, W.K.; Dannenberg, M.P.; Yan, D.; Herrmann, S.; Barnes, M.L.; Barron-Gafford, G.A.; Biederman, J.A.; Ferrenberg, S.; Fox, A.M.; Hudson, A.; et al. Remote sensing of dryland ecosystem structure and function: Progress, challenges, and opportunities. *Remote Sens. Environ.* **2019**, *233*, 111401. [CrossRef]
14. Berdugo, M.; Delgado-Baquerizo, M.; Soliveres, S.; Hernández-Clemente, R.; Zhao, Y.; Gaitán, J.J.; Gross, N.; Saiz, H.; Maire, V.; Lehmann, A.; et al. Global ecosystem thresholds driven by aridity. *Science* **2020**, *367*, 787–790. [CrossRef]
15. Carlier, J.; Moran, J. Landscape typology and ecological connectivity assessment to inform Greenway design. *Sci. Total Environ.* **2019**, *651*, 3241–3252. [CrossRef]
16. Zhang, L.; Zhang, H.; Xu, E. Information entropy and elasticity analysis of the land use structure change influencing eco-environmental quality in Qinghai-Tibet Plateau from 1990 to 2015. *Environ. Sci. Pollut. Res.* **2022**, *29*, 18348–18364. [CrossRef]

17. Meuret, M.; Provenza, F. How French shepherds create meal sequences to stimulate intake and optimise use of forage diversity on rangeland. *Anim. Prod. Sci.* **2015**, *55*, 309–318. [CrossRef]

18. Xu, Z.; Ren, H.; Li, M.-H.; van Ruijven, J.; Han, X.; Wan, S.; Li, H.; Yu, Q.; Jiang, Y.; Jiang, L. Environmental changes drive the temporal stability of semi-arid natural grasslands through altering species asynchrony. *J. Ecol.* **2015**, *103*, 1308–1316. [CrossRef]

19. Shi, Z.; Xu, X.; Souza, L.; Wilcox, K.; Jiang, L.; Liang, J.; Xia, J.; García-Palacios, P.; Luo, Y. Dual mechanisms regulate ecosystem stability under decade-long warming and hay harvest. *Nat. Commun.* **2016**, *7*, 11973. [CrossRef]

20. Isbell, F.I.; Polley, H.W.; Wilsey, B.J. Biodiversity, productivity and the temporal stability of productivity: Patterns and processes. *Ecol. Lett.* **2009**, *12*, 443–451. [CrossRef]

21. Borer, E.T.; Grace, J.B.; Harpole, W.S.; MacDougall, A.S.; Seabloom, E.W. A decade of insights into grassland ecosystem responses to global environmental change. *Nat. Ecol. Evol.* **2017**, *1*, 0118. [CrossRef]

22. Sasaki, T.; Lu, X.; Hirota, M.; Bai, Y. Species asynchrony and response diversity determine multifunctional stability of natural grasslands. *J. Ecol.* **2019**, *107*, 1862–1875. [CrossRef]

23. Hautier, Y.; Zhang, P.; Loreau, M.; Wilcox, K.R.; Seabloom, E.W.; Borer, E.T.; Byrnes, J.E.K.; Koerner, S.E.; Komatsu, K.J.; Lefcheck, J.S.; et al. General destabilizing effects of eutrophication on grassland productivity at multiple spatial scales. *Nat. Commun.* **2020**, *11*, 5375. [CrossRef] [PubMed]

24. Bai, Y.F.; Yu, Z.; Yang, Q.C. Mechanisms regulating the productivity and stability of artificial grasslands in China: Issues, progress, and prospects (in Chinese). *Chin. Sci. Bull.* **2018**, *63*, 511–520. [CrossRef]

25. Philp, J.N.M.; Cornish, P.S.; Te, K.S.H.; Bell, R.W.; Vance, W.; Lim, V.; Li, X.; Kamphayae, S.; Denton, M.D. Insufficient potassium and sulfur supply threaten the productivity of perennial forage grasses in smallholder farms on tropical sandy soils. *Plant Soil* **2021**, *461*, 617–630. [CrossRef]

26. Vázquez, E.; Schleuss, P.-M.; Borer, E.T.; Bugalho, M.N.; Caldeira, M.C.; Eisenhauer, N.; Eskelinen, A.; Fay, P.A.; Haider, S.; Jentsch, A.; et al. Nitrogen but not phosphorus addition affects symbiotic N2 fixation by legumes in natural and semi-natural grasslands located on four continents. *Plant Soil* **2022**, *478*, 689–707. [CrossRef]

27. Pretty, J.; Benton, T.G.; Bharucha, Z.P.; Dicks, L.V.; Flora, C.B.; Godfray, H.C.J.; Goulson, D.; Hartley, S.; Lampkin, N.; Morris, C.; et al. Global assessment of agricultural system redesign for sustainable intensification. *Nat. Sustain.* **2018**, *1*, 441–446. [CrossRef]

28. White, H.J.; Gaul, W.; Sadykova, D.; León-Sánchez, L.; Caplat, P.; Emmerson, M.C.; Yearsley, J.M. Quantifying large-scale ecosystem stability with remote sensing data. *Remote Sens. Ecol. Conserv.* **2020**, *6*, 354–365. [CrossRef]

29. Wu, Z.; Zhu, D.; Xiong, K.; Wang, X. Dynamics of landscape ecological quality based on benefit evaluation coupled with the rocky desertification control in South China Karst. *Ecol. Indic.* **2022**, *138*, 108870. [CrossRef]

30. De Bello, F.; Lavorel, S.; Hallett, L.M.; Valencia, E.; Garnier, E.; Roscher, C.; Conti, L.; Galland, T.; Goberna, M.; Májeková, M.; et al. Functional trait effects on ecosystem stability: Assembling the jigsaw puzzle. *Trends Ecol. Evol.* **2021**, *36*, 822–836. [CrossRef]

31. Kang, W.; Liu, S.; Chen, X.; Feng, K.; Guo, Z.; Wang, T. Evaluation of ecosystem stability against climate changes via satellite data in the eastern sandy area of northern China. *J. Environ. Manag.* **2022**, *308*, 114596. [CrossRef]

32. Feng, H.B. Study on the Stability and Ecogeology Mechanism of Grassland Ecosystem under the Influence of Coal Mining Activities—A Case Study of Chenqi Basin. Master's Thesis, China University of Geosciences, Wuhan, China, 2021. [CrossRef]

33. Donohue, I.; Hillebrand, H.; Montoya, J.M.; Petchey, O.L.; Pimm, S.L.; Fowler, M.S.; Healy, K.; Jackson, A.L.; Lurgi, M.; McClean, D.; et al. Navigating the complexity of ecological stability. *Ecol. Lett.* **2016**, *19*, 1172–1185. [CrossRef]

34. McNaughton, S.J. Relationships among Functional Properties of Californian Grassland. *Nature* **1967**, *216*, 168–169. [CrossRef]

35. Polley, H.W.; Yang, C.; Wilsey, B.J.; Fay, P.A. Temporal stability of grassland metacommunities is regulated more by community functional traits than species diversity. *Ecosphere* **2020**, *11*, e03178. [CrossRef]

36. Bai, Y.; Han, X.; Wu, J.; Chen, Z.; Li, L. Ecosystem stability and compensatory effects in the Inner Mongolia grassland. *Nature* **2004**, *431*, 181–184. [CrossRef]

37. Hautier, Y.; Tilman, D.; Isbell, F.; Seabloom, E.W.; Borer, E.T.; Reich, P.B. Anthropogenic environmental changes affect ecosystem stability via biodiversity. *Science* **2015**, *348*, 336–340. [CrossRef]

38. Melts, I.; Lanno, K.; Sammul, M.; Uchida, K.; Heinsoo, K.; Kull, T.; Laanisto, L. Fertilising semi-natural grasslands may cause long-term negative effects on both biodiversity and ecosystem stability. *J. Appl. Ecol.* **2018**, *55*, 1951–1955. [CrossRef]

39. Zhang, H.; Fan, J.; Wang, J.; Cao, W.; Harris, W. Spatial and temporal variability of grassland yield and its response to climate change and anthropogenic activities on the Tibetan Plateau from 1988 to 2013. *Ecol. Indic.* **2018**, *95*, 141–151. [CrossRef]

40. Li, X.; Lyu, X.; Dou, H.; Dang, D.; Li, S.; Li, X.; Li, M.; Xuan, X. Strengthening grazing pressure management to improve grassland ecosystem services. *Glob. Ecol. Conserv.* **2021**, *31*, e01782. [CrossRef]

41. Ives, A.R.; Carpenter, S.R. Stability and Diversity of Ecosystems. *Science* **2007**, *317*, 58–62. [CrossRef]

42. Pennekamp, F.; Pontarp, M.; Tabi, A.; Altermatt, F.; Alther, R.; Choffat, Y.; Fronhofer, E.A.; Ganesanandamoorthy, P.; Garnier, A.; Griffiths, J.I.; et al. Biodiversity increases and decreases ecosystem stability. *Nature* **2018**, *563*, 109–112. [CrossRef]

43. Ratzke, C.; Barrere, J.; Gore, J. Strength of species interactions determines biodiversity and stability in microbial communities. *Nat. Ecol. Evol.* **2020**, *4*, 376–383. [CrossRef] [PubMed]

44. Yang, M.D. On the Fragility of Karst Environment. *Yunnan Geogr. Environ. Res.* **1990**, *2*, 21–29.

45. Geekiyanage, N.; Goodale, U.M.; Cao, K.; Kitajima, K. Plant ecology of tropical and subtropical karst ecosystems. *Biotropica* **2019**, *51*, 626–640. [CrossRef]

46. Yuan, J.; Li, R.; Huang, K. Driving factors of the variation of ecosystem service and the trade-off and synergistic relationships in typical karst basin. *Ecol. Indic.* **2022**, *142*, 109253. [CrossRef]

47. Shackelford, N.; Hobbs, R.J.; Burgar, J.M.; Erickson, T.E.; Fontaine, J.B.; Laliberté, E.; Ramalho, C.E.; Perring, M.P.; Standish, R.J. Primed for Change: Developing Ecological Restoration for the 21st Century. *Restor. Ecol.* **2013**, *21*, 297–304. [CrossRef]

48. Xiong, K.; Li, P.; Zhou, Z.; An, Y.; Lv, T.; Lan, A. *Remote Sensing of Karst Rocky Desertification—A Typical Research of GIS: Taking Guizhou Province as an Example*; Geology Publishing House: Beijing, China, 2002; p. 18.

49. Xiong, K.N.; Li, J.; Long, M.Z. Features of Soil and Water Loss and Key Issues in Demonstration Areas for Combating Karst Rocky Desertification. *Acta Geographica Sinica* **2012**, *67*, 878–888. [CrossRef]

50. Xiong, K.; Zhu, D.; Peng, T.; Yu, L.; Xue, J.; Li, P. Study on Ecological industry technology and demonstration for Karst rocky desertification control of the Karst Plateau-Gorge. *Acta Ecol. Sin.* **2016**, *36*, 7109–7113. [CrossRef]

51. Li, X.; Xiong, K.; Gong, J.; Chen, Y.B. Advances in research on the function of artificial grassland in karst rock desertification control. *Acta Prataculture Sin.* **2011**, *20*, 279–286. Available online: http://cyxb.magtech.com.cn/EN/abstract/abstract4206.shtml (accessed on 1 January 2023).

52. Tong, X.; Brandt, M.; Yue, Y.; Horion, S.; Wang, K.; Keersmaecker, W.D.; Tian, F.; Schurgers, G.; Xiao, X.; Luo, Y.; et al. Increased vegetation growth and carbon stock in China karst via ecological engineering. *Nat. Sustain.* **2018**, *1*, 44–50. [CrossRef]

53. Chen, C.; Park, T.; Wang, X.; Piao, S.; Xu, B.; Chaturvedi, R.K.; Fuchs, R.; Brovkin, V.; Ciais, P.; Fensholt, R.; et al. China and India lead in greening of the world through land-use management. *Nat. Sustain.* **2019**, *2*, 122–129. [CrossRef]

54. Tang, X.; Xiao, J.; Ma, M.; Yang, H.; Li, X.; Ding, Z.; Yu, P.; Zhang, Y.; Wu, C.; Huang, J.; et al. Satellite evidence for China's leading role in restoring vegetation productivity over global karst ecosystems. *For. Ecol. Manag.* **2022**, *507*, 120000. [CrossRef]

55. Wang, K.L.; Yue, Y.M.; Chen, H.S.; Wu, X.B.; Xiao, J.; Qi, X.K.; Zhang, W.; Du, H. The comprehensive treatment of karst rocky desertification and its regional restoration effects. *Acta Ecol. Sin.* **2019**, *39*, 7432–7440.

56. Xu, L.; Xiong, K.; Zhang, J.; Chi, Y.; Chen, Y.; Liu, C. The problems and resolutions of grassland ecosystem in karst of southwest China. *Pratacultural Sci.* **2015**, *32*, 828–836.

57. Xiong, K.; Xu, L.; Liu, K.; Guo, W.; Yang, S.; Liu, C. Coupling relationship between the development of ecological animal husbandry and the Control of Rocky Desertification in Karst Mountain Area. *Acta Ecol. Anim. Domastici.* **2016**, *32*, 8. [CrossRef]

58. Xiong, K.N.; Xiao, J.; Zhu, D.Y. Research progress of agroforestry ecosystem services and its implications for industrial revitalization in karst regions. *Acta Ecol. Sin.* **2022**, *42*, 851–861. [CrossRef]

59. Andrej, K. Dinaric Karst—An Example of Deforestation and Desertification of Limestone Terrain. In *Deforestation around the World*; Paulo, M., Ed.; IntechOpen: Rijeka, Croatia, 2012; p. 5. [CrossRef]

60. Guo, B.; Zang, W.; Luo, W. Spatial-temporal shifts of ecological vulnerability of Karst Mountain ecosystem-impacts of global change and anthropogenic interference. *Sci. Total Environ.* **2020**, *741*, 140256. [CrossRef]

61. Zhou, Y.; Fu, D.; Lu, C.; Xu, X.; Tang, Q. Positive effects of ecological restoration policies on the vegetation dynamics in a typical ecologically vulnerable area of China. *Ecol. Eng.* **2021**, *159*, 106087. [CrossRef]

62. Zhang, J.Y.; Dai, M.H.; Wang, L.C.; Zeng, C.F.; Su, W.C. The challenge and future of rocky desertification control in karst areas in southwest China. *Solid Earth* **2016**, *7*, 83–91. [CrossRef]

63. Gao, J.; Du, F.; Zuo, L.; Jiang, Y. Integrating ecosystem services and rocky desertification into identification of karst ecological security pattern. *Landsc. Ecol.* **2021**, *36*, 2113–2133. [CrossRef]

64. Qi, X.; Wang, K.; Zhang, C. Effectiveness of ecological restoration projects in a karst region of southwest China assessed using vegetation succession mapping. *Ecol. Eng.* **2013**, *54*, 245–253. [CrossRef]

65. Chi, Y.; Xiong, K.; Zhang, Y.; Dong, Y.; Liu, C.; Xu, L. The beneficial results, problems and suggestions of grass-planting and livestock-raising to bring rocky desertification under control in the Karst areas of southwest China. *Eilongjiang Anim.* **2015**, *11*, 143–147.

66. Song, S.; Xiong, K.; Chi, Y.; He, C.; Fang, J.; He, S. Effect of Cultivated Pastures on Soil Bacterial Communities in the Karst Rocky Desertification Area. *Front. Microbiol.* **2022**, *13*, 922989. [CrossRef] [PubMed]

67. Staal, A.; van Nes, E.H.; Hantson, S.; Holmgren, M.; Dekker, S.C.; Pueyo, S.; Xu, C.; Scheffer, M. Resilience of tropical tree cover: The roles of climate, fire, and herbivory. *Glob. Change Biol.* **2018**, *24*, 5096–5109. [CrossRef] [PubMed]

68. Mengist, W.; Soromessa, T.; Legese, G. Ecosystem services research in mountainous regions: A systematic literature review on current knowledge and research gaps. *Sci. Total Environ.* **2020**, *702*, 134581. [CrossRef]

69. World Health Organization. *Health in 2015: From MDGs, Millennium Development Goals to SDGs, Sustainable Development Goals*; World Health Organization: Geneva, Switzerland, 2015.

70. Li, K.; Hu, Y.; Maidi, A.; Yu, J.; Wu, Q. Research Advances of Diversity in Grassland Communities. *Arid Zone Res.* **2005**, *22*, 165–169. [CrossRef]

71. Tilman, D.; Knops, J.; Wedin, D.; Reich, P.; Ritchie, M.; Siemann, E. The Influence of Functional Diversity and Composition on Ecosystem Processes. *Science* **1997**, *277*, 1300–1302. [CrossRef]

72. Spehn, E.M.; Hector, A.; Joshi, J.; Scherer-Lorenzen, M.; Schmid, B.; Bazeley-White, E.; Beierkuhnlein, C.; Caldeira, M.C.; Diemer, M.; Dimitrakopoulos, P.G.; et al. Ecosystem effects of biodiversity manipulations in european grasslands. *Ecol. Monogr.* **2005**, *75*, 37–63. [CrossRef]

73. Loreau, M.; Naeem, S.; Inchausti, P.; Bengtsson, J.; Grime, J.P.; Hector, A.; Hooper, D.U.; Huston, M.A.; Raffaelli, D.; Schmid, B.; et al. Biodiversity and Ecosystem Functioning: Current Knowledge and Future Challenges. *Science* **2001**, *294*, 804–808. [CrossRef]

74. Isbell, F.; Reich, P.B.; Tilman, D.; Hobbie, S.E.; Polasky, S.; Binder, S. Nutrient enrichment, biodiversity loss, and consequent declines in ecosystem productivity. *Proc. Natl. Acad. Sci. USA* **2013**, *110*, 11911–11916. [CrossRef]

75. Gabruck, D. Legume-Grass Forage Mixes for Maximizing Yield and Competitiveness against Weeds in Early Establishment. Master's Thesis, University of Alberta, Edmonton, AB, Canada, 2010. [CrossRef]

76. Albayrak, S.; Türk, M.; Yüksel, O.; Yilmaz, M. Forage Yield and the Quality of Perennial Legume-Grass Mixtures under Rainfed Conditions. *Not. Bot. Horti Agrobot.* **2011**, *39*, 114–118. [CrossRef]

77. Zheng, W.; Jianaerguli; Tang, G.R.; Zhu, J.Z. Determination and comparison of community stability in different legume-grass mixes. *Acta Prataculturae Sin.* **2015**, *24*, 155–167. [CrossRef]

78. Wang, L.; Delgado-Baquerizo, M.; Wang, D.; Isbell, F.; Liu, J.; Feng, C.; Liu, J.; Zhong, Z.; Zhu, H.; Yuan, X.; et al. Diversifying livestock promotes multidiversity and multifunctionality in managed grasslands. *Proc. Natl. Acad. Sci. USA* **2019**, *116*, 6187–6192. [CrossRef]

79. Song, M.H.; Zong, N.; Jiang, J.; Shi, P.L.; Zhang, X.Z.; Gao, J.Q.; Zhou, H.K.; Li, Y.K.; Loreau, M. Nutrient-induced shifts of dominant species reduce ecosystem stability via increases in species synchrony and population variability. *Sci. Total Environ.* **2019**, *692*, 441–449. [CrossRef]

80. Wang, Y.; Cadotte, M.W.; Chen, Y.; Fraser, L.H.; Zhang, Y.; Huang, F.; Luo, S.; Shi, N.; Loreau, M. Global evidence of positive biodiversity effects on spatial ecosystem stability in natural grasslands. *Nat. Commun.* **2019**, *10*, 3207. [CrossRef]

81. Prieto, I.; Violle, C.; Barre, P.; Durand, J.-L.; Ghesquiere, M.; Litrico, I. Complementary effects of species and genetic diversity on productivity and stability of sown grasslands. *Nat. Plants* **2015**, *1*, 15033. [CrossRef]

82. Niu, D.; Yuan, X.; Cease, A.J.; Wen, H.; Zhang, C.; Fu, H.; Elser, J.J. The impact of nitrogen enrichment on grassland ecosystem stability depends on nitrogen addition level. *Sci. Total Environ.* **2018**, *618*, 1529–1538. [CrossRef]

83. Li, M.; Liu, S.; Sun, Y.; Liu, Y. Agriculture and animal husbandry increased carbon footprint on the Qinghai-Tibet Plateau during past three decades. *J. Clean Prod.* **2021**, *278*, 123963. [CrossRef]

84. Li, S.; Xu, J.; Tang, S.; Zhan, Q.; Gao, Q.; Ren, L.; Shao, Q.; Chen, L.; Du, J.; Hao, B. A meta-analysis of carbon, nitrogen and phosphorus change in response to conversion of grassland to agricultural land. *Geoderma* **2020**, *363*, 114149. [CrossRef]

85. Liu, B.; Zhao, W.Z.; Meng, Y.Y.; Liu, C. Biodiversity, productivity, and temporal stability in a natural grassland ecosystem of China. *Sci. Cold Arid Reg.* **2018**, *10*, 293–304. [CrossRef]

86. Xiang-chao, P. The Theoretical Innovation and Practical Significance of the Theory about "Two Mountains" by Xi Jinping. *IOP Conf. Ser. Earth Environ. Sci.* **2018**, *199*, 022047. [CrossRef]

87. McLaughlin, M.J.; Parker, D.R.; Clarke, J.M. Metals and micronutrients—food safety issues. *Field Crop. Res.* **1999**, *60*, 143–163. [CrossRef]

88. Sankaran, M.; Hanan, N.P.; Scholes, R.J.; Ratnam, J.; Augustine, D.J.; Cade, B.S.; Gignoux, J.; Higgins, S.I.; Le Roux, X.; Ludwig, F.; et al. Determinants of woody cover in African savannas. *Nature* **2005**, *438*, 846–849. [CrossRef] [PubMed]

89. Abreu, R.C.R.; Hoffmann, W.A.; Vasconcelos, H.L.; Pilon, N.A.; Rossatto, D.R.; Durigan, G. The biodiversity cost of carbon sequestration in tropical savanna. *Sci. Adv.* **2017**, *3*, e1701284. [CrossRef] [PubMed]

90. Zhang, W.; Brandt, M.; Penuelas, J.; Guichard, F.; Tong, X.; Tian, F.; Fensholt, R. Ecosystem structural changes controlled by altered rainfall climatology in tropical savannas. *Nat. Commun.* **2019**, *10*, 671. [CrossRef] [PubMed]

91. Huang, Y.H. *Land Ecology*; China Agricultural Press: Beijing, China, 2013; p. 83.

92. Wang, K.; Zhang, C.; Chen, H.; Yue, Y.; Zhang, W.; Zhang, M.; Qi, X.; Fu, Z. Karst landscapes of China: Patterns, ecosystem processes and services. *Landsc. Ecol.* **2019**, *34*, 2743–2763. [CrossRef]

93. Fu, B.; Wang, S.; Su, C.; Forsius, M. Linking ecosystem processes and ecosystem services. *Curr. Opin. Environ. Sustain.* **2013**, *5*, 4–10. [CrossRef]

94. Tilman, D. Biodiversity: Population Versus Ecosystem Stability. *Ecology* **1996**, *77*, 350–363. [CrossRef]

95. Bardgett, R.D.; Mommer, L.; De Vries, F.T. Going underground: Root traits as drivers of ecosystem processes. *Trends Ecol. Evol.* **2014**, *29*, 692–699. [CrossRef]

96. Laws, A.N.; Joern, A. Predator–prey interactions in a grassland food chain vary with temperature and food quality. *Oikos* **2013**, *122*, 977–986. [CrossRef]

97. Lemaire, G.; Hodgson, J.; Chabbi, A. *Grassland Productivity and Ecosystem Services*; CABI: Oxfordshire, UK, 2011.

98. Guo, Y.J. *Ecological Grassland Establishment and Management in the Three Gorges Reservoir Area*; Guizhou People's Press: Guiyang, China, 2006; p. 6.

99. Wang, Z.H.; Liu, L.L. Ecosystem structure and functioning: Current knowledge and perspectives. *Chin. J. Plant Ecol.* **2021**, *45*, 1033–1035. [CrossRef]

100. Kahmen, S.; Poschlod, P.; Schreiber, K.-F. Conservation management of calcareous grasslands. Changes in plant species composition and response of functional traits during 25 years. *Biol. Conserv.* **2002**, *104*, 319–328. [CrossRef]

101. Xu, X.; Polley, H.W.; Hofmockel, K.; Wilsey, B.J. Species composition but not diversity explains recovery from the 2011 drought in Texas grasslands. *Ecosphere* **2017**, *8*, e01704. [CrossRef]

102. Ramos, C.S.; Isabel Bellocq, M.; Paris, C.I.; Filloy, J. Environmental drivers of ant species richness and composition across the Argentine Pampas grassland. *Austral Ecol.* **2018**, *43*, 424–434. [CrossRef]

103. Silva Mota, G.; Luz, G.R.; Mota, N.M.; Silva Coutinho, E.; das Dores Magalhães Veloso, M.; Fernandes, G.W.; Nunes, Y.R.F. Changes in species composition, vegetation structure, and life forms along an altitudinal gradient of rupestrian grasslands in south-eastern Brazil. *Flora* **2018**, *238*, 32–42. [CrossRef]
104. Silva, V.; Catry, F.X.; Fernandes, P.M.; Rego, F.C.; Paes, P.; Nunes, L.; Caperta, A.D.; Sérgio, C.; Bugalho, M.N. Effects of grazing on plant composition, conservation status and ecosystem services of Natura 2000 shrub-grassland habitat types. *Biodivers. Conserv.* **2019**, *28*, 1205–1224. [CrossRef]
105. Hooper, D.U.; Chapin Iii, F.S.; Ewel, J.J.; Hector, A.; Inchausti, P.; Lavorel, S.; Lawton, J.H.; Lodge, D.M.; Loreau, M.; Naeem, S.; et al. Effects of biodiversity on ecosystem functioning: A consensus of current knowledge. *Ecol. Monogr.* **2005**, *75*, 3–35. [CrossRef]
106. Song, M.; Liu, L.; Chen, J.; Zhang, X. Biology, multi-function and optimized management in grassland ecosystem. *Ecol. Environ. Sci.* **2018**, *27*, 1179–1188. [CrossRef]
107. Zheng, D.; Wang, Y.; Hao, S.; Xu, W.; Lv, L.; Yu, S. Spatial-temporal variation and tradeoffs/synergies analysis on multiple ecosystem services: A case study in the Three-River Headwaters region of China. *Ecol. Indic.* **2020**, *116*, 106494. [CrossRef]
108. Chen, P.; Wang, S.; Liu, Y.; Wang, Y.; Li, Z.; Wang, Y.; Zhang, H.; Zhang, Y. Spatio-temporal patterns of oasis dynamics in China's drylands between 1987 and 2017. *Environ. Res. Lett.* **2022**, *17*, 064044. [CrossRef]
109. Zhang, G.; Yan, J.; Zhu, X.; Ling, H.; Xu, H. Spatio-temporal variation in grassland degradation and its main drivers, based on biomass: Case study in the Altay Prefecture, China. *Glob. Ecol. Conserv.* **2019**, *20*, e00723. [CrossRef]
110. Locatelli, B.; Lavorel, S.; Sloan, S.; Tappeiner, U.; Geneletti, D. Characteristic trajectories of ecosystem services in mountains. *Front. Ecol. Environ.* **2017**, *15*, 150–159. [CrossRef]
111. Rodriguez-Barrera, M.G.; Kühn, I.; Estrada-Castillón, E.; Cord, A.F. Grassland type and seasonal effects have a bigger influence on plant functional and taxonomical diversity than prairie dog disturbances in semiarid grasslands. *Ecol. Evol.* **2022**, *12*, e9040. [CrossRef]
112. Hou, Y.; Lü, Y.; Chen, W.; Fu, B. Temporal variation and spatial scale dependency of ecosystem service interactions: A case study on the central Loess Plateau of China. *Landsc. Ecol.* **2017**, *32*, 1201–1217. [CrossRef]
113. Takaragawa, H.; Dinh, H.T.; Horie, M.; Kawamitsu, Y. Effects of mixed planting of horizontal-and erect-leafed varieties on canopy light use and growth in sugarcane. *Sugar Tech.* **2019**, *21*, 596–604. [CrossRef]
114. Zhang, X.; Zhang, C.; Dong, Q.; Yang, X.; Liu, W.; Yu, S.; Yang, Z.; Wei, L.; Zhang, Y.; Xu, H.; et al. Study on Plant Community Structure Characteristics of Alpine Mixed-seeding Grassland in Three Rivers Source Regions. *Acta Agrestia Sin.* **2020**, *28*, 1090–1099. [CrossRef]
115. Beckage, B.; Gross, L.J.; Platt, W.J. Modelling responses of pine savannas to climate change and large-scale disturbance. *Appl. Veg. Sci.* **2006**, *9*, 75–82. [CrossRef]
116. Fair, K.R.; Anand, M.; Bauch, C.T. Spatial structure in protected forest-grassland mosaics: Exploring futures under climate change. *Glob. Change Biol.* **2020**, *26*, 6097–6115. [CrossRef]
117. Jiang, S.; Xiong, K.; Xiao, J. Structure and Stability of Agroforestry Ecosystems: Insights into the Improvement of Service Supply Capacity of Agroforestry Ecosystems under the Karst Desertification Control. *Forests* **2022**, *13*, 878. [CrossRef]
118. Zhen, Y.; Zhuo, M.; Li, D.; Li, S.; Zhen, X. Application of mixture-planting of shrub and grass to side-slope greening and protection. *Ecol. Environ.* **2007**, *16*, 149–151.
119. Ludwig, J.A.; Wilcox, B.P.; Breshears, D.D.; Tongway, D.J.; Imeson, A.C. Vegetation patches and runoff–erosion as interacting ecohydrological processes in semiarid landscapes. *Ecology* **2005**, *86*, 288–297. [CrossRef]
120. Zhang, H.; Li, X.; Li, L.; Zhang, J. Effects of species combination on community diversity and productivity of alpine artificial grassland. *Acta Agrestia Sin.* **2020**, *28*, 1436–1443. [CrossRef]
121. Pyle, L.; Hall, L.M.; Bork, E.W. Linking management practices with range health in northern temperate pastures. *Can. J. Plant Sci.* **2017**, *98*, 657–671. [CrossRef]
122. Aryal, P.; Islam, M.A. Establishment of forage kochia in seeding mixtures with perennial grasses. *Grassl. Sci.* **2019**, *65*, 147–154. [CrossRef]
123. Szymura, T.H.; Szymura, M. Spatial structure of grassland patches in Poland: Implications for nature conservation. *Acta Soc. Bot. Pol.* **2019**, *88*, 3615. [CrossRef]
124. Guimarães-Steinicke, C.; Weigelt, A.; Proulx, R.; Lanners, T.; Eisenhauer, N.; Duque-Lazo, J.; Reu, B.; Roscher, C.; Wagg, C.; Buchmann, N.; et al. Biodiversity facets affect community surface temperature via 3D canopy structure in grassland communities. *J. Ecol.* **2021**, *109*, 1969–1985. [CrossRef]
125. Weigelt, A.; Weisser, W.W.; Buchmann, N.; Scherer-Lorenzen, M. Biodiversity for multifunctional grasslands: Equal productivity in high-diversity low-input and low-diversity high-input systems. *Biogeosciences* **2009**, *6*, 1695–1706. [CrossRef]
126. Cervasio, F.; Argenti, G.; Genghini, M.; Ponzetta, M.P. Agronomic methods for mountain grassland habitat restoration for faunistic purposes in a protected area of the northern Apennines (Italy). *iForest-Biogeosciences For.* **2016**, *9*, 490–496. [CrossRef]
127. Gao, Y.; Zhao, M.; Wang, H.; Xiong, M. Zhao, T. Analysis of spatial differentiation of grassland ecological quality and its affecting factors based on landscape ecology perspective: A case study of Siziwang Banner. *Acta Ecol. Sin.* **2019**, *39*, 5288–5300. [CrossRef]
128. Yang, J. Study on Grassland Ecosystem Service and Its Trade-Off and Synergy in The Yellow River Basin. Master's Thesis, Gansu Agriculture University, Lanzhou, China, 2021. [CrossRef]
129. Habel, J.C.; Ulrich, W.; Schmitt, T. Butterflies in corridors: Quality matters for specialists. *Insect Conserv. Divers.* **2020**, *13*, 91–98. [CrossRef]

130. Travers, E.; Härdtle, W.; Matthies, D. Corridors as a tool for linking habitats—Shortcomings and perspectives for plant conservation. *J. Nat. Conserv.* **2021**, *60*, 125974. [CrossRef]
131. Theron, K.J.; Pryke, J.S.; Samways, M.J. Maintaining functional connectivity in grassland corridors between plantation forests promotes high-quality habitat and conserves range restricted grasshoppers. *Landsc. Ecol.* **2022**, *37*, 2081–2097. [CrossRef]
132. Grêt-Regamey, A.; Sven-Erik, R.; Crespo, R.; Lautenbach, S.; Ryffel, A.; Schlup, B. On the importance of non-linear relationships between landscape patterns and the sustainable provision of ecosystem services. *Landsc. Ecol.* **2014**, *29*, 201–212. [CrossRef]
133. Yushanjiang, A.; Zhang, F.; Yu, H.; Kung, H.-t. Quantifying the spatial correlations between landscape pattern and ecosystem service value: A case study in Ebinur Lake Basin, Xinjiang, China. *Ecol. Eng.* **2018**, *113*, 94–104. [CrossRef]
134. Díaz-Reviriego, I.; Turnhout, E.; Beck, S. Participation and inclusiveness in the Intergovernmental Science–Policy Platform on Biodiversity and Ecosystem Services. *Nat. Sustain.* **2019**, *2*, 457–464. [CrossRef]
135. Sun, S.; Shi, Q. Global Spatio-Temporal Assessment of Changes in Multiple Ecosystem Services Under Four IPCC SRES Land-use Scenarios. *Earth's Future* **2020**, *8*, e2020EF001668. [CrossRef]
136. Hao, R.; Yu, D.; Liu, Y.; Liu, Y.; Qiao, J.; Wang, X.; Du, J. Impacts of changes in climate and landscape pattern on ecosystem services. *Sci Total. Environ.* **2017**, *579*, 718–728. [CrossRef]
137. Reiss, J.; Bridle, J.R.; Montoya, J.M.; Woodward, G. Emerging horizons in biodiversity and ecosystem functioning research. *Trends Ecol. Evol.* **2009**, *24*, 505–514. [CrossRef]
138. Mischkolz, J.M.; Schellenberg, M.P.; Lamb, E.G. Assembling productive communities of native grass and legume species: Finding the right mix. *Appl. Veg. Sci.* **2016**, *19*, 111–121. [CrossRef]
139. Brown, J.; Barton, P.; Cunningham, S.A. How bioregional history could shape the future of agriculture. In *Advances in Ecological Research*; Academic Press: Cambridge, MA, USA, 2021; Volume 64, pp. 149–189. [CrossRef]
140. Lamb, E.G.; Kennedy, N.; Siciliano, S.D. Effects of plant species richness and evenness on soil microbial community diversity and function. *Plant Soil* **2011**, *338*, 483–495. [CrossRef]
141. Balvanera, P.; Pfisterer, A.B.; Buchmann, N.; He, J.S.; Nakashizuka, T.; Raffaelli, D.; Schmid, B. Quantifying the evidence for biodiversity effects on ecosystem functioning and services. *Ecol. Lett.* **2006**, *9*, 1146–1156. [CrossRef]
142. Feng, X.; Lei, G.; Ma, Q.; Yang, C.; Zhang, Y. Spatial-Temporal of Productional-Linving-Ecological Function in Huanghuaihai Plain of He'nan Province during 1990-2020. *Bull. Soil Water Conserv.* **2022**, *42*, 357–364. [CrossRef]
143. Pang, X.; Lu, R.; Zhang, L.; Li, S.; Tan, L.; Wei, S.; Qin, Q. Study on Coordination and Zoning Optimization of Land Use Based on Production-living-ecological Function in Border Areas of Guangxi. *Res. Soil Water Conserv.* **2022**, *30*, 1–9. [CrossRef]
144. Elton, C.S. *Animal Ecology*; Sidgwick & Jackson: London, UK, 1927.
145. Wang, Q.Q.; Gao, Y.; Wang, R. Review on impacts of global change on food web structure. *Chin. J. Plant Ecol.* **2021**, *45*, 1064–1074. [CrossRef]
146. Dunne, J.A.; Williams, R.J.; Martinez, N.D. Food-web structure and network theory: The role of connectance and size. *Proc. Natl. Acad. Sci. USA* **2002**, *99*, 12917–12922. [CrossRef] [PubMed]
147. Cross, W.F.; Baxter, C.V.; Rosi-Marshall, E.J.; Hall, R.O., Jr.; Kennedy, T.A.; Donner, K.C.; Wellard Kelly, H.A.; Seegert, S.E.Z.; Behn, K.E.; Yard, M.D. Food-web dynamics in a large river discontinuum. *Ecol. Monogr.* **2013**, *83*, 311–337. [CrossRef]
148. Hunt, H.W.; Coleman, D.C.; Ingham, E.R.; Ingham, R.E.; Elliott, E.T.; Moore, J.C.; Rose, S.L.; Reid, C.P.P.; Morley, C.R. The detrital food web in a shortgrass prairie. *Biol. Fertil. Soils* **1987**, *3*, 57–68. [CrossRef]
149. De Castro, F.; Adl, S.M.; Allesina, S.; Bardgett, R.D.; Bolger, T.; Dalzell, J.J.; Emmerson, M.; Fleming, T.; Garlaschelli, D.; Grilli, J.; et al. Local stability properties of complex, species-rich soil food webs with functional block structure. *Ecol. Evol.* **2021**, *11*, 16070–16081. [CrossRef]
150. Duchardt, C.J.; Porensky, L.M.; Pearse, I.S. Direct and indirect effects of a keystone engineer on a shrubland-prairie food web. *Ecology* **2021**, *102*, e03195. [CrossRef]
151. Traveset, A.; Richardson, D.M. Mutualistic interactions and biological invasions. *Annu. Rev. Ecol. Evol. Syst.* **2014**, *45*, 89–113. Available online: http://hdl.handle.net/10019.1/116887 (accessed on 1 January 2023). [CrossRef]
152. Fan, Y.; Li, B.; Dai, X.; Ma, L.; Tai, X.; Bi, X.; Yang, Z.; Zhang, X. Optimizing Cropping Systems of Cultivated Pastures in the Mountain–Basin Systems in Northwest China. *Appl. Sci.* **2020**, *10*, 6949. [CrossRef]
153. Luo, X. Coupling Mechanism and Regulation of Karst World Heritage Value Protection and Agroforestry Development in Buffer Zone. Master's Thesis, Guizhou Normal University, Guiyang, China, 2022. [CrossRef]
154. Xiao, J.; Xiong, K. A review of agroforestry ecosystem services and its enlightenment on the ecosystem improvement of rocky desertification control. *Sci. Total Environ.* **2022**, *852*, 158538. [CrossRef]
155. Dominschek, R.; Barroso, A.A.M.; Lang, C.R.; de Moraes, A.; Sulc, R.M.; Schuster, M.Z. Crop rotations with temporary grassland shifts weed patterns and allows herbicide-free management without crop yield loss. *J. Clean. Prod.* **2021**, *306*, 127140. [CrossRef]
156. Vibart, R.E.; Vogeler, I.; Dodd, M.; Koolaard, J. Simple versus Diverse Temperate Pastures: Aspects of Soil–Plant–Animal Interrelationships Central to Nitrogen Leaching Losses. *Agron. J.* **2016**, *108*, 2174–2188. [CrossRef]
157. Lüscher, A.; Barkaoui, K.; Finn, J.A.; Suter, D.; Suter, M.; Volaire, F. Using plant diversity to reduce vulnerability and increase drought resilience of permanent and sown productive grasslands. *Grass Forage Sci.* **2022**, *77*, 235–246. [CrossRef]
158. Wagner, M.; Walker, K.J.; Pywell, R.F. Seed bank dynamics in restored grassland following the sowing of high- and low-diversity seed mixtures. *Restor. Ecol.* **2018**, *26*, S189–S199. [CrossRef]

159. Nemec, K.T.; Allen, C.R.; Helzer, C.J.; Wedin, D.A. Influence of Richness and Seeding Density on Invasion Resistance in Experimental Tallgrass Prairie Restorations. *Ecol. Restor.* **2013**, *31*, 168–185. [CrossRef]

160. Barr, S.; Jonas, J.L.; Paschke, M.W. Optimizing seed mixture diversity and seeding rates for grassland restoration. *Restor. Ecol.* **2017**, *25*, 396–404. [CrossRef]

161. Amiro, B.D.; Tenuta, M.; Gervais, M.; Glenn, A.J.; Gao, X. A decade of carbon flux measurements with annual and perennial crop rotations on the Canadian Prairies. *Agric. For. Meteorol.* **2017**, *247*, 491–502. [CrossRef]

162. Hayes, R.C.; Li, G.D.; Norton, M.R.; Culvenor, R.A. Effects of contrasting seasonal growth patterns on composition and persistence of mixed grass-legume pastures over 5 years in a semi-arid Australian cropping environment. *J. Agron. Crop Sci.* **2018**, *204*, 228–242. [CrossRef]

163. Hayes, R.C.; Newell, M.T.; Crews, T.E.; Peoples, M.B. Perennial cereal crops: An initial evaluation of wheat derivatives grown in mixtures with a regenerating annual legume. *Renew. Agric. Food Syst.* **2017**, *32*, 276–290. [CrossRef]

164. Clay, N.; Garnett, T.; Lorimer, J. Dairy intensification: Drivers, impacts and alternatives. *Ambio* **2020**, *49*, 35–48. [CrossRef] [PubMed]

165. Meyer, R.S.; Cullen, B.R.; Whetton, P.H.; Robertson, F.A.; Eckard, R.J. Potential impacts of climate change on soil organic carbon and productivity in pastures of south eastern Australia. *Agric. Syst.* **2018**, *167*, 34–46. [CrossRef]

166. Mason, W.K.; Lamb, K.; Russell, B. The Sustainable Grazing Systems Program: New solutions for livestock producers. *Aust. J. Exp. Agric.* **2003**, *43*, 663–672. [CrossRef]

167. Johnson, I.R.; Lodge, G.M.; White, R.E. The Sustainable Grazing Systems Pasture Model: Description, philosophy and application to the SGS National Experiment. *Aust. J. Exp. Agric.* **2003**, *43*, 711–728. [CrossRef]

168. Johnson, I.R.; Chapman, D.F.; Snow, V.O.; Eckard, R.J.; Parsons, A.J.; Lambert, M.G.; Cullen, B.R. DairyMod and EcoMod: Biophysical pasture-simulation models for Australia and New Zealand. *Aust. J. Exp. Agric.* **2008**, *48*, 621–631. [CrossRef]

169. Donnelly, J.R.; Moore, A.D.; Freer, M. GRAZPLAN: Decision support systems for Australian grazing enterprises—I. Overview of the GRAZPLAN project, and a description of the MetAccess and LambAlive DSS. *Agric. Syst.* **1997**, *54*, 57–76. [CrossRef]

170. Moore, A.D.; Donnelly, J.R.; Freer, M. GRAZPLAN: Decision support systems for Australian grazing enterprises—III. Pasture growth and soil moisture submodels, and the GrassGro DSS. *Agric. Syst.* **1997**, *55*, 535–582. [CrossRef]

171. Beukes, P.C.; Palliser, C.C.; Macdonald, K.A.; Lancaster, J.A.S.; Levy, G.; Thorrold, B.S.; Wastney, M.E. Evaluation of a whole-farm model for pasture-based dairy systems. *J. Dairy Sci.* **2008**, *91*, 2353–2360. [CrossRef]

172. Friend, M.; Robertson, S.; Masters, D.; Avery, A. EverGraze—A project to achieve profit and environmental outcomes in the Australian grazing industries. *J. Anim. Feed Sci.* **2007**, *16*, 70–75. [CrossRef]

173. Michalk, D.L.; Badgery, W.B.; Kemp, D.R. Balancing animal, pasture and environmental outcomes in grazing management experiments. *Anim. Prod. Sci.* **2017**, *57*, 1775–1784. [CrossRef]

174. Shi, Y.; Wang, Y.; Ma, Y.; Ma, W.; Liang, C.; Flynn, D.F.B.; Schmid, B.; Fang, J.; He, J.S. Field-based observations of regional-scale, temporal variation in net primary production in Tibetan alpine grasslands. *Biogeosciences* **2014**, *11*, 2003–2016. [CrossRef]

175. Ma, Z.; Liu, H.; Mi, Z.; Zhang, Z.; Wang, Y.; Xu, W.; Jiang, L.; He, J.-S. Climate warming reduces the temporal stability of plant community biomass production. *Nat. Commun.* **2017**, *8*, 15378. [CrossRef]

176. Li, L.; Zhang, Y.; Wu, J.; Li, S.; Zhang, B.; Zu, J.; Zhang, H.; Ding, M.; Paudel, B. Increasing sensitivity of alpine grasslands to climate variability along an elevational gradient on the Qinghai-Tibet Plateau. *Sci. Total Environ.* **2019**, *678*, 21–29. [CrossRef]

177. Craven, D.; Eisenhauer, N.; Pearse, W.D.; Hautier, Y.; Isbell, F.; Roscher, C.; Bahn, M.; Beierkuhnlein, C.; Bönisch, G.; Buchmann, N.; et al. Multiple facets of biodiversity drive the diversity–stability relationship. *Nat. Ecol. Evol.* **2018**, *2*, 1579–1587. [CrossRef]

178. Petermann, J.S.; Buzhdygan, O.Y. Grassland biodiversity. *Curr. Biol.* **2021**, *31*, R1195–R1201. [CrossRef]

179. García-Palacios, P.; Gross, N.; Gaitán, J.; Maestre, F.T. Climate mediates the biodiversity–ecosystem stability relationship globally. *Proc. Natl. Acad. Sci. USA* **2018**, *115*, 8400–8405. [CrossRef]

180. Yang, G.; Wagg, C.; Veresoglou, S.D.; Hempel, S.; Rillig, M.C. How Soil Biota Drive Ecosystem Stability. *Trends Plant Sci.* **2018**, *23*, 1057–1067. [CrossRef]

181. Dronova, I.; Taddeo, S.; Harris, K. Plant diversity reduces satellite-observed phenological variability in wetlands at a national scale. *Sci. Adv.* **2022**, *8*, eabl8214. [CrossRef]

182. Li, M.; Wu, J.; He, Y.; Wu, L.; Niu, B.; Song, M.; Zhang, X. Dimensionality of grassland stability shifts along with altitudes on the Tibetan Plateau. *Agric. For. Meteorol.* **2020**, *291*, 108080. [CrossRef]

183. Van Oijen, M.; Bellocchi, G.; Höglind, M. Effects of Climate Change on Grassland Biodiversity and Productivity: The Need for a Diversity of Models. *Agronomy* **2018**, *8*, 14. [CrossRef]

184. Wilsey, B. Restoration in the face of changing climate: Importance of persistence, priority effects, and species diversity. *Restor. Ecol.* **2021**, *29*, e13132. [CrossRef]

185. MacArthur, R. Fluctuations of Animal Populations and a Measure of Community Stability. *Ecology* **1955**, *36*, 533–536. [CrossRef]

186. May, R.M. Thresholds and breakpoints in ecosystems with a multiplicity of stable states. *Nature* **1977**, *269*, 471–477. [CrossRef]

187. Yodzis, P. The stability of real ecosystems. *Nature* **1981**, *289*, 674–676. [CrossRef]

188. Isbell, F.; Gonzalez, A.; Loreau, M.; Cowles, J.; Díaz, S.; Hector, A.; Mace, G.M.; Wardle, D.A.; O'Connor, M.I.; Duffy, J.E.; et al. Linking the influence and dependence of people on biodiversity across scales. *Nature* **2017**, *546*, 65–72. [CrossRef]

189. Bai, X.; Zhao, W.; Wang, J.; Ferreira, C.S.S. Reducing plant community variability and improving resilience for sustainable restoration of temperate grassland. *Environ. Res.* **2022**, *207*, 112149. [CrossRef]

190. Li, W.; Li, X.; Zhao, Y.; Zheng, S.; Bai, Y. Ecosystem structure, functioning and stability under climate change and grazing in grasslands: Current status and future prospects. *Curr. Opin. Environ. Sustain.* **2018**, *33*, 124–135. [CrossRef]
191. Van Rooijen, N.M.; de Keersmaecker, W.; Ozinga, W.A.; Coppin, P.; Hennekens, S.M.; Schaminée, J.H.J.; Somers, B.; Honnay, O. Plant Species Diversity Mediates Ecosystem Stability of Natural Dune Grasslands in Response to Drought. *Ecosystems* **2015**, *18*, 1383–1394. [CrossRef]
192. Li, T.; Kamran, M.; Chang, S.; Peng, Z.; Wang, Z.; Ran, L.; Jiang, W.Q.; Jin, Y.; Zhang, X.; You, Y.; et al. Climate-soil interactions improve the stability of grassland ecosystem by driving alpine plant diversity. *Ecol. Indic.* **2022**, *141*, 109002. [CrossRef]
193. Yin, J.; Bauerle, T.L. A global analysis of plant recovery performance from water stress. *Oikos* **2017**, *126*, 1377–1388. [CrossRef]
194. Gilbert, B.; MacDougall, A.S.; Kadoya, T.; Akasaka, M.; Bennett, J.R.; Lind, E.M.; Flores-Moreno, H.; Firn, J.; Hautier, Y.; Borer, E.T.; et al. Climate and local environment structure asynchrony and the stability of primary production in grasslands. *Glob. Ecol. Biogeogr.* **2020**, *29*, 1177–1188. [CrossRef]
195. Chen, L.; Jiang, L.; Jing, X.; Wang, J.; Shi, Y.; Chu, H.; He, J.-S. Above- and belowground biodiversity jointly drive ecosystem stability in natural alpine grasslands on the Tibetan Plateau. *Glob. Ecol. Biogeogr.* **2021**, *30*, 1418–1429. [CrossRef]
196. Li, D.H.; Wu, X.W.; Xiao, Z.S. Assembly, ecosystem functions, and stability in species interaction networks. *Chin. J. Plant Ecol.* **2021**, *45*, 1049–1063. [CrossRef]
197. Zhang, Y.; He, N.; Loreau, M.; Pan, Q.; Han, X. Scale dependence of the diversity–stability relationship in a temperate grassland. *J. Ecol.* **2018**, *106*, 1277–1285. [CrossRef]
198. Rozema, J.; Notten, M.J.M.; Aerts, R.; van Gestel, C.A.M.; Hobbelen, P.H.F.; Hamers, T.H.M. Do high levels of diffuse and chronic metal pollution in sediments of Rhine and Meuse floodplains affect structure and functioning of terrestrial ecosystems? *Sci. Total Environ.* **2008**, *406*, 443–448. [CrossRef]
199. Holmgren, M.; Stapp, P.; Dickman, C.R.; Gracia, C.; Graham, S.; Gutiérrez, J.R.; Hice, C.; Jaksic, F.; Kelt, D.A.; Letnic, M.; et al. Extreme climatic events shape arid and semiarid ecosystems. *Front. Ecol. Environ.* **2006**, *4*, 87–95. [CrossRef]
200. Dong, S.; Shang, Z.; Gao, J.; Boone, R.B. Enhancing sustainability of grassland ecosystems through ecological restoration and grazing management in an era of climate change on Qinghai-Tibetan Plateau. *Agric. Ecosyst. Environ.* **2020**, *287*, 106684. [CrossRef]
201. White, H.J.; Caplat, P.; Emmerson, M.C.; Yearsley, J.M. Predicting future stability of ecosystem functioning under climate change. *Agric. Ecosyst. Environ.* **2021**, *320*, 107600. [CrossRef]
202. LeBauer, D.S.; Treseder, K.K. Nitrogen limitation of net primary productivity in terrestrial ecosystems is globally distributed. *Ecology* **2008**, *89*, 371–379. [CrossRef]
203. Avolio, M.L.; Pierre, K.J.L.; Houseman, G.R.; Koerner, S.E.; Grman, E.; Isbell, F.; Johnson, D.S.; Wilcox, K.R. A framework for quantifying the magnitude and variability of community responses to global change drivers. *Ecosphere* **2015**, *6*, art280. [CrossRef]
204. Koerner, S.E.; Avolio, M.L.; La Pierre, K.J.; Wilcox, K.R.; Smith, M.D.; Collins, S.L. Nutrient additions cause divergence of tallgrass prairie plant communities resulting in loss of ecosystem stability. *J. Ecol.* **2016**, *104*, 1478–1487. [CrossRef]
205. Xiankai, L.; Jiangming, M.; Shaofeng, D. Effects of nitrogen deposition on forest biodiversity. *Acta Ecol. Sin.* **2008**, *28*, 5532–5548. [CrossRef]
206. Elser, J.J.; Bracken, M.E.S.; Cleland, E.E.; Gruner, D.S.; Harpole, W.S.; Hillebrand, H.; Ngai, J.T.; Seabloom, E.W.; Shurin, J.B.; Smith, J.E. Global analysis of nitrogen and phosphorus limitation of primary producers in freshwater, marine and terrestrial ecosystems. *Ecol. Lett.* **2007**, *10*, 1135–1142. [CrossRef] [PubMed]
207. Cleland, E.E.; Harpole, W.S. Nitrogen enrichment and plant communities. *Ann. N. Y. Acad. Sci.* **2010**, *1195*, 46–61. [CrossRef]
208. Sala, O.E.; Stuart Chapin, F.; Iii, N.; Armesto, J.J.; Berlow, E.; Bloomfield, J.; Dirzo, R.; Huber-Sanwald, E.; Huenneke, L.F.; Jackson, R.B.; et al. Global Biodiversity Scenarios for the Year 2100. *Science* **2000**, *287*, 1770–1774. [CrossRef]
209. Bobbink, R.; Hicks, K.; Galloway, J.; Spranger, T.; Alkemade, R.; Ashmore, M.; Bustamante, M.; Cinderby, S.; Davidson, E.; Dentener, F.; et al. Global assessment of nitrogen deposition effects on terrestrial plant diversity: A synthesis. *Ecol. Appl.* **2010**, *20*, 30–59. [CrossRef]
210. Porter, E.M.; Bowman, W.D.; Clark, C.M.; Compton, J.E.; Pardo, L.H.; Soong, J.L. Interactive effects of anthropogenic nitrogen enrichment and climate change on terrestrial and aquatic biodiversity. *Biogeochemistry* **2013**, *114*, 93–120. [CrossRef]
211. Tilman, D.; Reich, P.B.; Knops, J.M.H. Biodiversity and ecosystem stability in a decade-long grassland experiment. *Nature* **2006**, *441*, 629–632. [CrossRef]
212. Simkin, S.M.; Allen, E.B.; Bowman, W.D.; Clark, C.M.; Belnap, J.; Brooks, M.L.; Cade, B.S.; Collins, S.L.; Geiser, L.H.; Gilliam, F.S.; et al. Conditional vulnerability of plant diversity to atmospheric nitrogen deposition across the United States. *Proc. Natl. Acad. Sci. USA* **2016**, *113*, 4086–4091. [CrossRef]
213. Isbell, F.; Craven, D.; Connolly, J.; Loreau, M.; Schmid, B.; Beierkuhnlein, C.; Bezemer, T.M.; Bonin, C.; Bruelheide, H.; de Luca, E.; et al. Biodiversity increases the resistance of ecosystem productivity to climate extremes. *Nature* **2015**, *526*, 574–577. [CrossRef]
214. Kulmatiski, A.; Beard, K.H. Woody plant encroachment facilitated by increased precipitation intensity. *Nat. Clim. Chang.* **2013**, *3*, 833–837. [CrossRef]
215. Stevens, N.; Lehmann, C.E.R.; Murphy, B.P.; Durigan, G. Savanna woody encroachment is widespread across three continents. *Glob. Change Biol.* **2017**, *23*, 235–244. [CrossRef]
216. Briske, D. *Rangeland Systems: Processes, Management and Challenges*; Springer Nature: Berlin/Heidelberg, Germany, 2017; p. 661. [CrossRef]

217. Venter, Z.S.; Cramer, M.D.; Hawkins, H.J. Drivers of woody plant encroachment over Africa. *Nat. Commun.* **2018**, *9*, 2272. [CrossRef]

218. Zhang, Z.; Sun, J.; Liu, M.; Xu, M.; Wang, Y.; Wu, G.-l.; Zhou, H.; Ye, C.; Tsechoe, D.; Wei, T. Don't judge toxic weeds on whether they are native but on their ecological effects. *Ecol. Evol.* **2020**, *10*, 9014–9025. [CrossRef]

219. Yitbarek, M.B. Livestock and livestock product trends by 2050. *IJAR* **2019**, *4*, 30. Available online: https://escipub.com/international-journal-of-animal-research/ (accessed on 1 January 2023).

220. Lal, R. Integrating Animal Husbandry With Crops and Trees. *Front. Sustain. Food Syst.* **2020**, *4*, 113. [CrossRef]

221. García-Ruiz, J.M.; Tomás-Faci, G.; Diarte-Blasco, P.; Montes, L.; Domingo, R.; Sebastián, M.; Lasanta, T.; González-Sampériz, P.; López-Moreno, J.I.; Arnáez, J.; et al. Transhumance and long-term deforestation in the subalpine belt of the central Spanish Pyrenees: An interdisciplinary approach. *Catena* **2020**, *195*, 104744. [CrossRef]

222. Kuhn, T.; Domokos, P.; Kiss, R.; Ruprecht, E. Grassland management and land use history shape species composition and diversity in Transylvanian semi-natural grasslands. *Appl. Veg. Sci.* **2021**, *24*, e12585. [CrossRef]

223. Yan, H.; Liu, G. Fire's Effects on Grassland Restoration and Biodiversity Conservation. *Sustainability* **2021**, *13*, 12016. [CrossRef]

224. Da Silveira Pontes, L.; Maire, V.; Schellberg, J.; Louault, F. Grass strategies and grassland community responses to environmental drivers: A review. *Agron. Sustain. Dev.* **2015**, *35*, 1297–1318. [CrossRef]

225. Khan, K.S.; Kunz, R.; Kleijnen, J.; Antes, G.J.J. Five steps to conducting a systematic review. *J. R. Soc. Med.* **2003**, *96*, 118–121. Available online: https://xs.studiodahu.com/ (accessed on 1 January 2023). [CrossRef] [PubMed]

226. Chapman, S.; Watson, J.E.M.; Salazar, A.; Thatcher, M.; McAlpine, C.A. The impact of urbanization and climate change on urban temperatures: A systematic review. *Landsc. Ecol.* **2017**, *32*, 1921–1935. [CrossRef]

227. Page, M.J.; McKenzie, J.E.; Bossuyt, P.M.; Boutron, I.; Hoffmann, T.C.; Mulrow, C.D.; Shamseer, L.; Tetzlaff, J.M.; Akl, E.A.; Brennan, S.E.; et al. The PRISMA 2020 statement: An updated guideline for reporting systematic reviews. *Rev. Esp. Cardiol.* **2021**, *74*, 790–799. [CrossRef] [PubMed]

228. Deli, W.; Ling, W.; Jushan, L.; Hui, Z.; Zhiwei, Z. Grassland ecology in China: Perspectives and challenges. *Front. Agr. Sci. Eng.* **2018**, *5*, 24–43. [CrossRef]

229. Fang, J.; Xiong, K.; Chi, Y.; Song, S.; He, C.; He, S. Research Advancement in Grassland Ecosystem Vulnerability and Ecological Resilience and Its Inspiration for Improving Grassland Ecosystem Services in the Karst Desertification Control. *Plants* **2022**, *11*, 1290. [CrossRef]

230. Liao, Y.; Song, C.; Guo, Y.; Wang, L.; Wang, L. Index system and methodology for wetland ecosystem stability in Sanjiang Plain. *J. Arid. Land Resour. Environ.* **2009**, *23*, 89–94.

231. Pu, G.; Lv, Y.; Xu, G.; Zeng, D.; Huang, Y. Research Progress on Karst Tiankeng Ecosystems. *Bot. Rev.* **2017**, *83*, 5–37. [CrossRef]

232. De Boeck, H.J.; Bloor, J.M.G.; Kreyling, J.; Ransijn, J.C.G.; Nijs, I.; Jentsch, A.; Zeiter, M. Patterns and drivers of biodiversity–stability relationships under climate extremes. *J. Ecol.* **2018**, *106*, 890–902. [CrossRef]

233. Xu, D.; Guo, X. Some Insights on Grassland Health Assessment Based on Remote Sensing. *Sensors* **2015**, *15*, 3070–3089. [CrossRef]

234. Zhong, F.; Xu, X.; Li, Z.; Zeng, X.; Yi, R.; Luo, W.; Zhang, Y.; Xu, C. Relationships between lithology, topography, soil, and vegetation, and their implications for karst vegetation restoration. *Catena* **2022**, *209*, 105831. [CrossRef]

235. Li, S.; Zhao, X.; Pu, J.; Miao, P.; Wang, Q.; Tan, K. Optimize and control territorial spatial functional areas to improve the ecological stability and total environment in karst areas of Southwest China. *Land Use Policy* **2021**, *100*, 104940. [CrossRef]

236. Zhang, D.; Wang, S.; Li, R. Study on the Eco-Environmental Vulnerability in Guizhou Karst Mountains. *Geogr. Territ. Res.* **2002**, *18*, 77–79.

237. Wan, J.; Cai, Y. Land degradation and eco-reconstruction in fragile karst ecosystem: The case of guanling county, guizhou province. *China Popul. Resour. Environ.* **2003**, *13*, 5. [CrossRef]

238. Zhong, G.; Hai, Y.; Zhen, H.; Xu, W.; Ouyang, Z. Current Situation and Measures of Karst Rocky Desertification Control in Southwest China. *J. Yangtze River Sci. Res. Inst.* **2021**, *38*, 38–43. [CrossRef]

239. Wa, Q. Accelerating grassland restoration in rocky desertification area, promoting eco-environmental reconstruction in karst region. *Pratacultural Sci.* **2008**, *25*, 4.

240. Wang, Y.; Li, L.; Wang, K. Dynamics of Community Composition of Mixed Clover-grass Pasture during20 Years in Karst Region, Southwest China. *Acta Agrestia Sin.* **2014**, *22*, 475–480. [CrossRef]

241. Song, S.Z.; Xiong, K.N.; Chi, Y.K.; Guo, W.; Liao, J.J. Study on Improvement of Degraded Grassland in Rocky Desertification control in the Karst Areas of Southern China. *Acta Ecol. Anim. Domastici* **2019**, *40*, 82–96.

242. Hu, L.; Robert, C.A.M.; Cadot, S.; Zhang, X.; Ye, M.; Li, B.; Manzo, D.; Chervet, N.; Steinger, T.; van der Heijden, M.G.A.; et al. Root exudate metabolites drive plant-soil feedbacks on growth and defense by shaping the rhizosphere microbiota. *Nat. Commun.* **2018**, *9*, 2738. [CrossRef]

243. Gschwend, F.; Hartmann, M.; Hug, A.-S.; Enkerli, J.; Gubler, A.; Frey, B.; Meuli, R.G.; Widmer, F. Long-term stability of soil bacterial and fungal community structures revealed in their abundant and rare fractions. *Mol. Ecol.* **2021**, *30*, 4305–4320. [CrossRef]
244. Xu, Q.; Xiong, K.; Chi, Y. Effects of Intercropping on Fractal Dimension and Physicochemical Properties of Soil in Karst Areas. *Forests* **2021**, *12*, 1422. [CrossRef]

MDPI

St. Alban-Anlage 66

4052 Basel

Switzerland

Tel. +41 61 683 77 34

Fax +41 61 302 89 18

www.mdpi.com

Plants Editorial Office

E-mail: plants@mdpi.com

www.mdpi.com/journal/plants